# PHYSIOLOGICAL PROCESSES IN PLANT

*ADVANCES IN PLANT PHYSIOLOGY SERIES—7*

# Physiological Processes in Plant

**D. K. Arora**
**Seema Gupta**

**Anmol Publications Pvt Ltd**
**NEW DELHI-110002 (INDIA)**

**Anmol Publications Pvt. Ltd.**
4374/4B, Ansari Road, Daryanganj,
New Delhi-110002

*Physiological Processes in Plant*



**ISBN 81-7488-241-3**



**PRINTED IN INDIA**

Published by J.L. Kumar for Anmol Publications Pvt. Ltd., New Delhi-110 002 and Printed at Mehra Offset Press, Delhi.

# Preface

'Physiological Processes in Plant' offers a comprehensive account of the fundamental principles of plant physiology. It incorporates and organizes the information concerning Physiological Processes in Plant from relevant authentic sources. It is an humble attempt to present a correct and precise account of the subject matter of this important branch of Botany which forms an integral part of the studies undertaken by the undergraduate and post-graduate students of this subject. The principles have been outlined in considerable detail alongwith this fundamental facts and theories that explain the life processes of plants. The main object of the present book is to provide a comprehensive and well illustrated account of the prescribed subject matter in a very simple and lucid language and concise manner yet--still maintaining its desired academic standard.

Most of illustrations have been taken from professional papers, standard books and journals. Subject matter has also been obtained from a large number of standard books, reviews and research papers by internationally renowned workers and established authors. That is why there can be no claim of originality except in the manner of treatment.

Though the authors have taken special care to present a current account, yet they are fully aware of their limitations, and the readers may come across the mistakes of various types. For all types of mistakes they extend their due apology.

The authors express their thanks to their friends and colleagues whose continuous inspirations have invited them to bring out this text.

Contents include : Stomatal Apparatus, Stomatal Control, Biological Nitrogen Fixation, Nitrate Reduction in higher plants, Net-Gas Exchange, Photoperiodic Effect, Seed and Bud Dormancy, Role of Potassium in Plants and so on.

**Authors**

# Contents

# 1

# The Stomatal Apparatus

The most important physiological fact about the stomates is that they are sometimes open and sometimes closed. When open, they serve as the principal pathways through which gaseous exchanges take place between the intercellular spaces of the leaf and the surrounding atmosphere. When closed, gaseous exchanges between a leaf and its environment are greatly retarded. The gases of greatest physiological importance which enter or depart from a leaf principally through the stomates are oxygen, carbon dioxide, and water-vapor. The movement of gases through the stomates in either direction is primarily a diffusion phenomenon, although under certain conditions, mass movement of gases may occur through the stomates. Although the stomates are the principal portals through which entry and escape of gases take place, the fact should not be overlooked that at least small quantities of gases pass directly through the epidermis and cuticular layers of all leaves. This appears to be true, in particular, of carbon dioxide. In submerged vascular aquatics all gaseous exchanges between the plant and its environment occur through the epidermis.

*Structure of the Stomates.* The *stomata* (singular *stoma*) are minute pores which occur in the epidermis of plants. They are surrounded by two distinctive epidermal cells known as the *guard cells*. Stomates may occur on any part of a plant except the roots, but in most species are most abundant upon the leaves. The size of the stomatal pore varies in most plants depending upon the turgidity of the guard cells and often, especially at night, it is entirely closed. In Fig. 1.1 are depicted surface and cross-sectional views of several of the commoner types of stomates. The structure of stomatal apparatus shows marked variations in detail in different species of plants, but the essential feature of a pore between two guard cells is common to all species of vascular plants. Guard cells which are roughly kidney or bean-shaped as seen in surface view are typical of more species of plants than any other kind (Fig. 1.1A). In some species the epidermal cells bordering on the guard cells are different in configuration from other cells in the same tissue; these are called *subsidiary cells* or *accessory cells* (Fig. 1.1B). In many species of the grass and

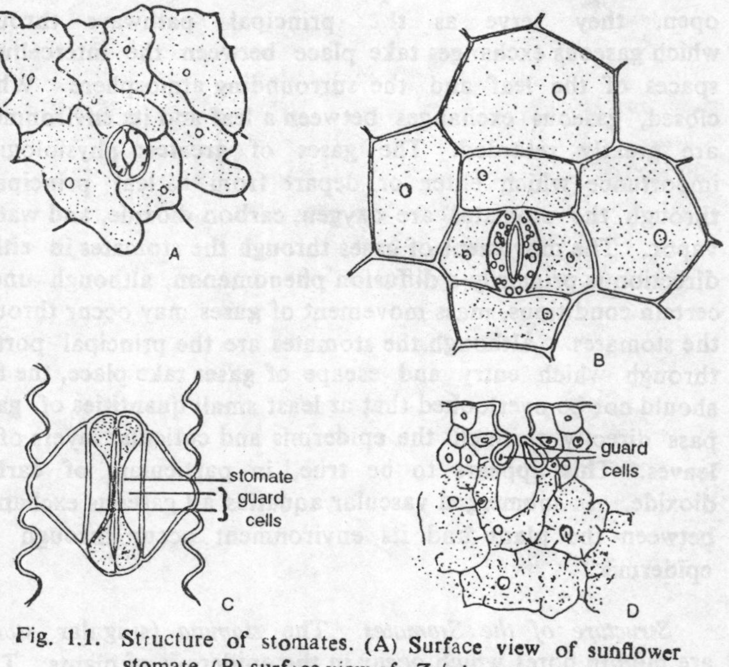


Fig. 1.1. Structure of stomates. (A) Surface view of sunflower stomate, (B) surface view of *Zebrina* stomate, showing four subsidiary cells around the guard cells, (C) surface view of maize stomate, (D) cross sectional view of Austrian pine stomate.

sedge families the guard cells are distinctly elongate (Fig. 1.1C). In most species of conifers and in certain other species, stomates are of the "sunken" type (Fig. 1.1D). A perspective view of a stomate and the surrounding guard cells is shown in Fig. 1.2. Unlike other epidermal cells, the guard cells contain chloroplasts They also appear to contain a larger proportion of cytoplasm than the epidermal cells.

*Size and Distribution of Stomates.* The size of the stomatal pore varies greatly according to the species of plant, and somewhat among the individual stomates on any one plant. The pores are always very minute, however, their dimensions being expressed in terms of microns (Table 1.1).

Minute as these openings appear to be from a human scale of values, they are enormous when compared with the size of the gas molecules which diffuse through them. The calculated diameter of a water molecule is 0.000454 $\mu$. More than 2000 water molecules would have to be placed side by side to measure a distance of 1 $\mu$. The molecules of both carbon dioxide and oxygen are larger than water molecules. Since the stomatal diameters usually are considerably in excess

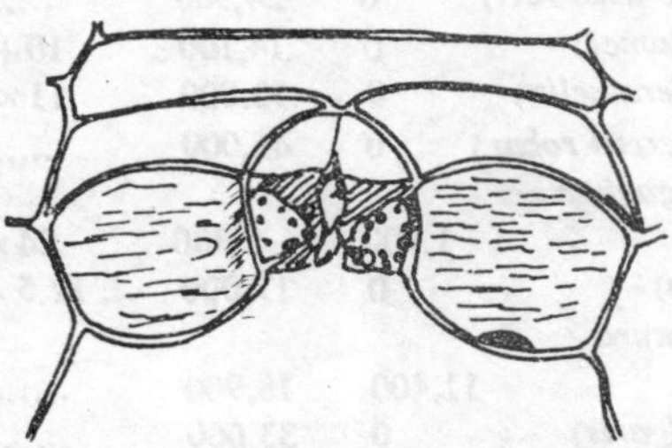

Fig. 1.2. Perspective view of a stomate and adjacent cells (semidiagrammatic).

of 1 $\mu$, it is evident that the stomates afford relatively enormous portals to the gas molecules which diffuse through them.

In general, the number of stomates present in the epidermis of leaves may range from a few thousand to over a hundred thousand per square centimeter, the exact number depending upon the species and upon the environmental conditions under

**Table 1.1.** Size and distribution of stomates on the leaves of various species of plants (data of Eckerson, 1908; Salisbury, 1927, Kisser, 1929; Miller, 1931; and Yocum, 1635).

| *Species* | *Ave. no. of stomates per cm²* | | *Size (length × breadth) of pore when fully open (lower epidermis)* | *Reference* |
|---|---|---|---|---|
| | *Upper epidermis* | *Lower epidermis* | | |
| Alfalfa (*Medicago sativa*) | 16,900 | 13,800 | .......... | M |
| Apple (*Pyrus malus* var,) | 0 | 29,400 | .......... | M |
| Barberry (*Berberis vulgaris*) | 0 | 22,900 | .......... | K |
| Bean (*Phaseolus vulgaris*) | 4,000 | 28,100 | 7 × 3μ | E |
| Begonia (*Begonia coccinea*) | 0 | 4,000 | 21 × 8μ | E |
| Black Oak (*Quercus velutina*) | 0 | 58,000 | .......... | Y |
| Black Poplar (*Populus nigra*) | 2,000 | 11,500 | .......... | S |
| Black Walnut (*Juglans nigra*) | 0 | 46,100 | .......... | K |
| Cabbage (*Brassica oleracea*) | 14,100 | 22,600 | .......... | M |
| Castor Bean (*Ricinus communis*) | 6,400 | 17,600 | 10 × 4μ | E |
| Cherry (*Prunus cerasus* var.) | 0 | 24,900 | .......... | M |
| Coleus (*Coleus blumei*) | 0 | 14,100 | 10+5μ | E |
| English Ivy (*Hedera helix*) | 0 | 15.800 | 11 × 4μ | E |
| English Oak (*Quercus robur*) | 0 | 45,000 | .......... | S |
| Geranium (*Pelargonium domesticum*) | 1,900 | 5,900 | 24 × 9μ | E |
| Holly (*Ilex opaca*) | 0 | 17,000 | 12.5 × 6.5μ | K |
| Jimson Weed (*Datura stramoniun*) | 11,400 | 18,900 | .......... | K |
| Lilac (*Syringa vulgaris*) | 0 | 33,000 | .......... | K |
| Linden (*Tilia vulgaris*) | 0 | 13.000 | .......... | S |
| Maize (*Zea mais*) | 5,200 | 6,800 | 19 × 5μ | E |
| Mulberry (*Morus alba*) | 0 | 48,000 | .......... | K |
| Nasturtium (*Tropacolum majus*) | 0 | 13,000 | 12 × 6μ | E |
| Nightshade (*Solanum dulcamara*) | 6,000 | 26,600 | .......... | K |
| Oat (*Avena sativa*) | 2,500 | 2,300 | 38 × 8μ | E |
| Pea (*Pisum sativum*) | 10,100 | 21,000 | .......... | K |
| Peach (*Prunus persica* var.) | 0 | 22,500 | .......... | M |

| | | | | |
|---|---|---|---|---|
| Potato (*Solanum tuberosun.*) | 5,100 | 16,100 | ........... | M |
| Red Oak (*Quercus rubra*) | 0 | 68,000 | ........... | Y |
| Scarlet Oak (*Quercus coccinea*) | 0 | 103,800 | ........ | Y |
| Scilla (*Scilla nutans*) | 5,500 | 5,100 | ........... | S |
| Sunflower (*Helianthus annuus*) | 8,500 | 15,600 | 22 × 8μ | E |
| Sycamore (*Platanus occidentalis*) | 0 | 27,800 | ........... | K |
| Tomato (*Lycopersicon esculentun.*) | 1,200 | 13,000 | 13 × 6μ | E |
| Tree of Heaven (*Ailanthus glandulosa*) | 0 | 38,600 | ........... | K |
| Wandering Jew (*Zebrina pendula*) | 0 | 1,400 | 31 × 12μ | E |
| Wheat (*Triticum sativum*) | 3,300 | 1,400 | 38 × 7μ | E |
| Willow Oak (*Quercus phellos*) | 0 | 72,300 | .......... | Y |
| Wood Sorrel (*Oxalis acetosella*) | 0 | 4,500 | ........... | S |
| Yew (*Taxus baccata*) | 0 | 11,500 | ........... | S |

which the leaf has developed. A single maize plant has been estimated to bear from 140 to 240 million stomates, whereas the number on a large tree could be expressed only by a figure of astronomical dimensions.

The average numbers of stomata which have been found per square centimeter on the leaves of a number of representative species are listed in Table 1.1. However, marked deviations from such average values are possible for any species, depending upon the environmental conditions under which the leaves have developed. The number of stomates per unit area of leaf surface may be quite different on leaves of two plants of the same species, if one grew in a greenhouse, and the other grew in the open, or upon the leaves of plants of the same species which have developed during different seasons.

The following types of stomata are found based on the morphology of guard cells and subsidiary cells :

(*i*) *Ranunculaceous or Anomocytic type.* The guard cells are surrounded by a number of similar cells which cannot be distinguished from other epidermal cells (Fig. 1.3A).

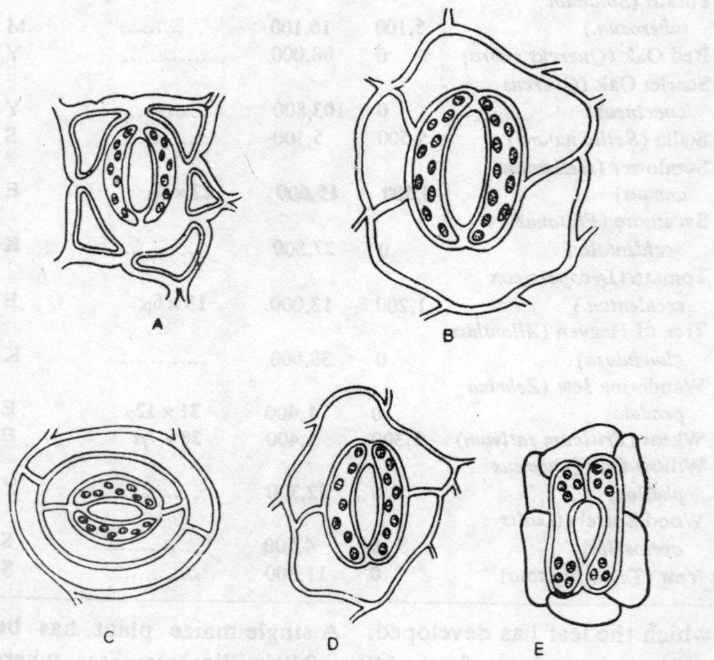


Fig. 1.3. Different types of stomata.

(*ii*) *Cruciferous or Anisocytic type*. The kidney-shaped guard cells are surrounded by three subsidiary cells, two of them being larger and one smaller (Fig. 1.3B).

(*iii*) *Rubiaceous or Paracytic type*. In this type the guard cells are surrounded by one or more subsidiary cells that are parallel to the stomata (Fig. 1.3C).

(*iv*) *Caryophyllaceous or Diacytic type*. The guard cells are surrounded by two cells which lie at right angles to each other (Fig. 1.3D).

(*v*) *Graminaceous type*. In this type the gaurd cells are dumbell shaped. They are surrounded by a number of subsidiary cells which lie parallel to the guard cells (Fig. 1.3E).

The guard cells control the opening and closing of stomata. A fully opened stoma may measure from 3—12$\mu$ across and 10—40$\mu$ in length.

Depending upon the species, the surface of a leaf may have from 1,000 to 60,000 stomata per square centimeter. Some workers have reported that 1—2% area of the total leaf surface may be under the open area of the leaf. Stomata are more common on the under surface than on the dorsal surface. Depending upon the relative distribution of stomata on the two sides of the leaf, the plants can be grouped into five categories :

(*i*) *Apple or Mulberry type*. Stomata are present only on the ventral surface of the leaf, *e.g.*, Apple, *Quercus* sp., Walnut, Peach, Oxalis, Pipal, Mulberry, etc.

(*ii*) *Potato type*. Stomata are more numerous on the ventral, than on the dorsal surface of the leaf. Pea, Bean, Potato, Tomato belong to this group.

(*iii*) *Oat type*. The number of stomata on the dorsal and ventral surface of the leaf is almost equal. Oat, Maize, *Verbena*, etc., belong to this category.

(*iv*) *Water Lily type*. Stomata are distributed only on the dorsal surface. *Nelumbium, Nymphaéa* and other water plants with floating leaves belong to this category.

(*v*) *Potamogeton type*. Stomata are completely absent or if present are vestigeal. *Vallisneria, Potamogeton* and most of the other submerged aquatic plants belong to this group.

**Diffusive Capacity of the Stomata**

Stomatal openings may be considered as pores of exchange between the outer atmosphere and interior of the leaf. Minute openings of stomata are very efficient for diffusion of water vapours out of the leaf, though the total open area under the stomatal pores is only 1—2% of the total leaf area. The diffusion of water through stomata exceeds 50—60% of that evaporating from a free water surface of the same area.

Brown and Escombe (1900) for the first time found that diffusion through a pore is more proportional to the perimeter of the pore than to its area. The reason for more rapid diffusion through small pores per unit area than from a large free water surface is that in both cases the concentration of water vapour increases over the evaporating surface; this lowers the vapour pressure gradient and in turn the rate of diffusion is decreased. Water vapours form several overlapping shells or caps (Fig. 1.4) over a large free water surface.

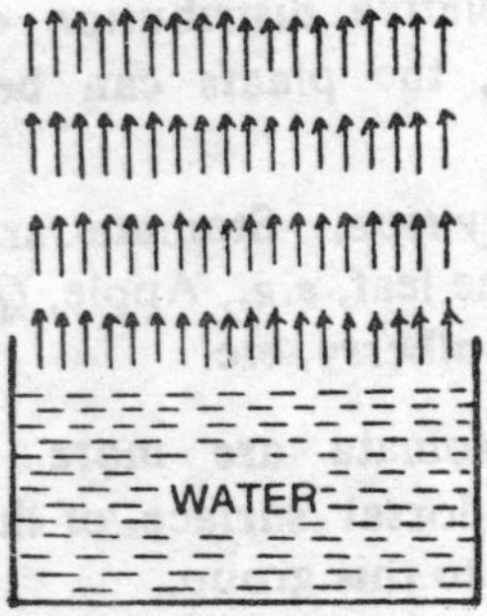


Fig. 1.4. Overlapping shells of water vapours.

This heavy blanket of water vapours lowers the vapour pressure gradient over the whole area. The area around the perimeter is still less saturated and offers lesser resistance to diffusing water vapours. However, the amount of this perimeter diffusion is relatively insignificant compared to the total free surface. If we study the loss of water through a small isolated pore, the perimeter diffusion in such a case becomes significant since the perimeter is more than the area of the pore (Fig. 1.5A).

Now suppose there is a water surface covered over by a membrane containing numerous small pores. If these pores

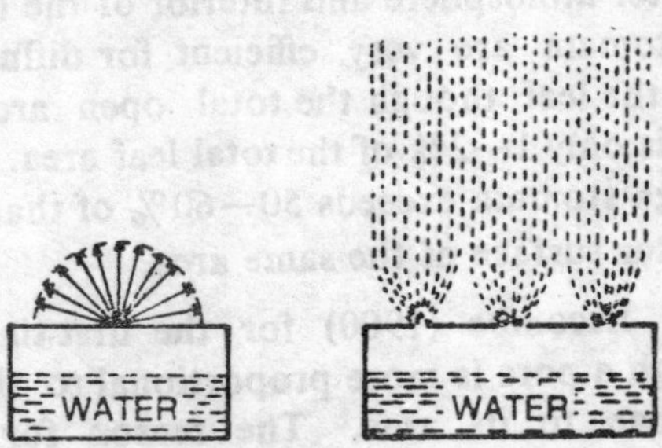


A. Perimeter diffusion of water vapours. B. Independent diffusion shells of water vapours.

Fig. 1.5.

are spaced far enough so that their diffusion shells do not interfere with one another (Fig. 1.5B). It can be imagined that perimeter diffusion in such a case will be appreciably high. This situation is analogous to that of water diffusing out through stomata of a leaf.

As shown in Table 1.1, stomates occur in both the upper and lower epidermis of many species of plants. In numerous others, especially woody species, they are confined to the lower epidermis. Even in those species in which stomates occur on both surfaces of the leaf they are commonly, but not always, more abundant in the lower epidermis. In floating leaves, such as those of the water lily, stomates occur only in the upper epidermis. Species in which the stomates are relatively small usually have more per unit area than species in which the stomates are relatively large.

In general, no correlation has been found between transpiration rates and either the size or distribution of the stomates, other factors being much more important in determining the rate of loss of water-vapor from the intercellular spaces.

*Principles Govering Diffusion of Gases through the Stomates.* Since gaseous exchanges between the intercellular spaces and the atmosphere take place principally through the stomates, the problem of the *diffusive capacity* of the stomates is an important one. Although the aggregate area of the fully open stomates of a leaf is probably never more than 3 per cent of the stomate-bearing surface and is often less than 1 per cent, the rate of water-vapor loss from leaves per unit of area may be 50 per cent or even more of the evaporation from an exposed water surface of the same area. Stomatal transpiration from a leaf of birch (*Betula pubescens*), for example, under the most favourable conditions, may be as much as 65 per cent of the evaporation from the same area of an evaporating surface (Stalfelt, 1932). Leaves of this species bear stomates only in the lower epidermis. Assuming the aggregate area of the open stomates to be 1 per cent of the leaf area, it is evident that water-vapor often diffuses through the stomates at rates ranging up to at least fifty times greater than it diffuses away from an equal area of exposed evaporating surface (epiderms).

**Table 1.2.** Diffusion of water-vapor through small openings under uniform conditions (data of Sayre, 1926).

| Septum | *Diameter of pores, man.* | *Loss of water-vapor, grams* | *Relative areas of pores.* | *Relative perimeters (circumferences) of pores* | *Relative amounts of water-vapor lost* |
|---|---|---|---|---|---|
| 1 | 2.64 | 2.655 | 1.00 | 1.00 | 1.00 |
| 2 | 1 60 | 1.583 | 0.37 | 0.61 | 0.59 |
| 3 | 0.95 | 0.928 | 0.13 | 0.36 | 0.35 |
| 4 | 0.81 | 0.762 | 0.09 | 0.31 | 0.29 |
| 5 | 0.72 | 0.672 | 0.07 | 0.27 | 0.25 |
| 6 | 0.65 | 0.590 | 0.06 | 0.25 | 0.22 |
| 7 | 0.56 | 0.492 | 0.05 | 0.21 | 0.18 |
| 8 | 0.48 | 0.455 | 0.03 | 0.18 | 0.17 |
| 9 | 0.41 | 0.393 | 0.02 | 0.15 | 0.15 |
| 10 | 0.35 | 0.364 | 0.01 | 0.13 | 0.14 |

This unexpectedly high diffusive capacity of the stomates is intelligible in terms of the results of studies upon the principles of diffusion through small apertures. The classical investigation of this problem was made by Brown and Escombe, (1900) who studied the rate of diffusion of carbon dioxide through tubes of known dimensions. They first made the important discovery that, if a septum which had been pierced with a small circular aperture was interposed across a column of diffusing gas, the rate at which carbon dioxide diffused through this aperture was much greater than the rate at which it passed through an equal area of the unobstructed tube.

Since the problem at present under consideration is the diffusion of water-vapor rather than carbon dioxide through small apertures, the data presented in Table 1.2 will be used to illustrate more fully the principles regarding the diffusion of gases through small openings. These data were obtained by sealing thin septa, through the center of which were cut round openings of known dimensions, across the circular months of small bottles which had previously been filled to a certain level with water, and then determining the loss in weight of each bottle after all of them had been kept under uniform conditions for the same period of time.

The result of this experiment illustrate two important general principles : (1) The quantities of water-vapor diffusing through small openings in a given period of time are proportional (essentially) to the perimeters (circumferences) and not to the areas of the pores. This is shown by the close correspondence of the figures in the last two columns of Table 1.2. (2) The smaller the pore, the greater the water loss per unit area. The pore in septum ., for example, has an exposed area only slightly more than one-third as great as the pore in septum 1, but diffusion through it is nearly two-thirds as great as through septum 1. Similarly, the pore in septum 7, with only 5 per cent of the area of the pore in septum 1, nevertheless permits 18 per cent as much diffusion as the pore in the latter septum.

It has also been found that the diffusion of water-vapor through small pores of elliptical cross section in more nearly proportional to their perimeter than to their area. A stomate attains almost its maximum diffusive capacity considerably before it is fully open Fig. 1.6, because the perimeter of a stomatal pore does not increase greatly after the aperture between the guard cells has once widened appreciable.

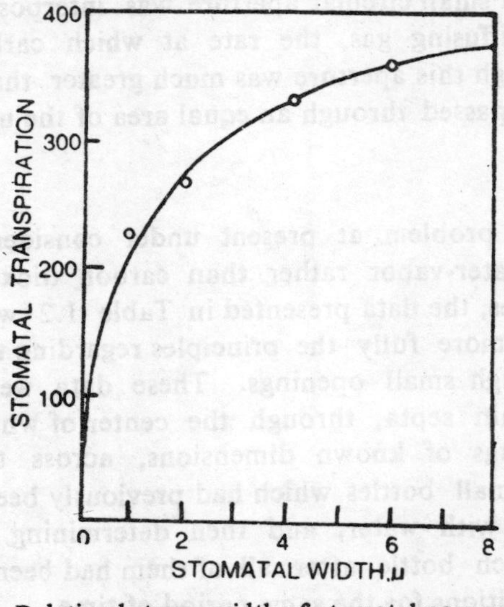


Fig. 1.6. **Relation between width of stomatal pore and rate of stomatal transpiration in leaf of birch in mg. per 25 cm.² leaf area per hr.**

Diffusion of gases through small apertures is much more nearly proportional to their perimeters than to their areas because of the more rapid diffusion of the molecules through an opening at its periphery than through its center. The concentration of diffusing molecules is much less in outward directions from the rim than it is just above the pore. Hence the number of molecules escaping through the hole near the rim per unit time interval greatly exceeds the number escaping near the center of the opening. Reduction in the area of a circular or elliptical aperture results in increasing its perimeter relative to its area. Hence in small pores such a large proportion of the diffusion is "rim diffusion" and such a small proportion is diffusion through the pore centers that measurements show the rate of diffusion to be essentially proportional to the perimeters of the openings.

Diffusion of gases from or into a leaf through the stomates involves a much more complex system than is represented by a septum pierced by a single aperture. Diffusion is occurring, not through a single opening, but simultaneously through thousands of minute apertures which are relatively close together. Experiments have been performed in analogous physical systems in which multiperforate instead of uniperforate diaphragms have been used. The results of experiments on different spacings of the pores in a septum upon the rate of diffusion of water-vapor per pore and pre septum are depicted graphically in Fig. 1.7 and Fig. 1.8 respectively.

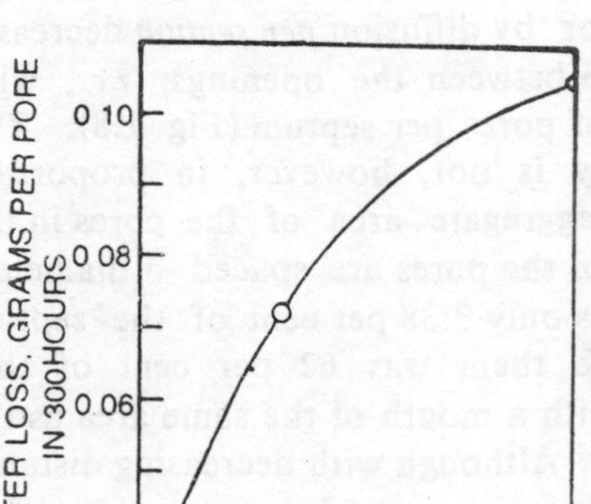


**Fig. 1.7. Relation between water loss per pore and distance in diameters between pores.**

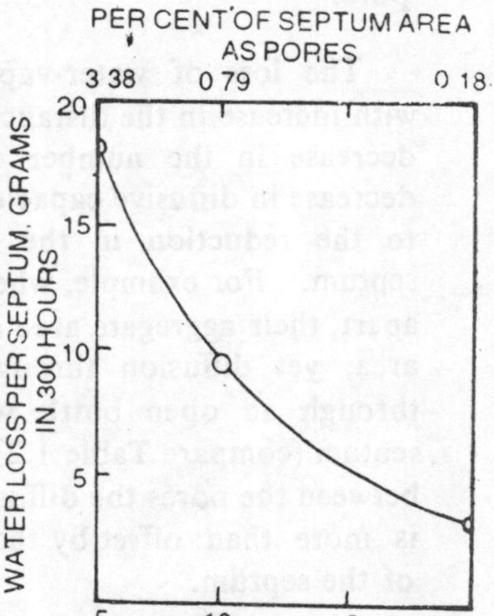


**Fig. 1.8. Relation between water loss per septum and distance in diameters between pores.**

As shown in Fig. 1.7, the diffusion *per pore* increases with increase in the distance apart of the apertures, although not proportionately. With pores of this diameter (0.3 mm.) nearly the maximum diffusive capacity per pore is attained when they are spaced at intervals of 20 diameters. In multiperforate septa the molecules diffusing through each aperture invade in part the zones into which the molecules passing through neighboring pores would diffuse were each opening the only one in the septum. Also, the number of molecules diffusing through each pore is not as great as if it were the only pore in the septum because some of those which would diffuse through the pore by chance if it were solitary are now diffusing by chance through neighboring pores. Hence the diffusion pressure gradient is not as steep through each pore as it would be through a single pore, and the rate of diffusion through each pore is correspondingly reduced. The closer together the pores, the less the rate of diffusion through each pore.

The loss of water-vapor by diffusion *per septum* decreases with increase in the distance between the openings, *i.e.*, with decrease in the number of pores per septum (Fig. 1.8). The decrease in diffusive capacity is not, however, in proportion to the reduction in the aggregate area of the pores in the septum. For example, when the pores are spaced 5 diameters apart, their aggregate area is only 3.38 per cent of the septum area, yet diffusion through them was 62 per cent of that through an open bottle with a mouth of the same area as the septum (compare Table 1.2). Although with decreasing distance between the pores the diffusion per pore decreases, this effect is more than offset by the increased number of pores per unit of the septum.

The dimensions of the pores (0.3 mm. diameter, 0.071 mm.$^2$ area) used in the experiments just discussed are much greater than those of the average stomate. Pore systems in which the individual pores approach or are within the stomatal range of sizes have still higher diffusive capacities. Huber (1930) found the diffusion through a septum in which the pores were 0.05 mm. (50 $\mu$) in diameter and occupied 3.2 per cent of the septum area to be 72 per cent as great as that from an open evaporating surface of the same area. Similarly

Sierp and Seybold (1929) found diffusional water loss through multiperforate diaphragms in which the pores were 10 $\mu$, 20 $\mu$, or 50 $\mu$ in diameter and occupied 0.8 per cent of the septum area to be about 70-75 per cent of the water loss from an open evaporating surface of the same area as the septum. Diffusion through a multiperforate septum may thus approach, although it never equals, that from a freely exposed surface.

The relatively high rates of water-vapor loss which occur from leaves in proportion to the aggregate area of the stomatal pores are thus explicable in terms of the principles of diffusion through multiperforate septa. Known rates of transpiration rarely if ever attain the values which calculations based on the principles just discussed indicate to be theoretically possible, even when the stomates are fully open. Evidently some factor other than the diffusive capacity of the stomates is often the limiting factor in stomatal transpiration.

The foregoing discussion of the diffusion of gases through small pores has been based on data obtained when diffusion was allowed to occur into "quiet" air. Even in so-called "quiet" air, it should be realized that considerable convectional movement of air is often taking place. If an air current is blowing across the surface of a multiperforate septum, the usual effect is an increase in the rate of diffusion through the pores. In pore systems within the size and frequency ranges of the stomates of most species, however, the effect of such air movement on the diffusion rate of water-vapor does not appear to be very great.

In applying the principles of diffusion through small pores to stomates, the further complication that the distance from the rim of one stomate to the next is variable must be considered (Verduin, 1949). The effective diameter of the stomates on tomato leaves, for example, is about 10 m$\mu$, and they are spaced about 10 diameters apart when fully open. At this spacing there is considerable interference with the diffusion of the molecules through one pore by the diffusion through other pores in its vicinity. If these stomates close to half their maximum effective diameter, they will be spaced 20 diameters apart, at which distance there is much less interference with diffusion through one pore by that through another. Decrease in interference during stomatal closure permits the maintenance of a higher diffusive capacity when the stomates are partly closed than would otherwise be possible.

Not all movement of gases through the stomates occurs by diffusion. In a wind, for example, back and forth bending of leaves causes alternate compression and expansion of the intercellular spaces with corresponding outward or inward mass flow of gases. Mass flows of gases also undoubtedly occur when leaf temperatures are rapidly fluctuating. This is commonly the situation when wind velocities are variable on a bright day, or when leaves are alternately exposed to direct sunlight and shade, as on a day with scattered clouds in the sky. With each increase in leaf temperature an outflow of gases through the stomates undoubtedly occurs ; with each decrease in temperature, at inflow.

*Mechanism of the Opening and Closing of the Stomates.* The degree of stomatal opening is influenced both by changes in the turgor of the guard cells and by changes in the turgor of the epidermal cells, although the former usually play a predominant role. In general, an increase in the turgor of the guard cells relative to that of the epidermal cells leads to a widening of the stomatal aperture, and *vice versa.* The greater this turgor difference, the wider the stomatal aperture.

The mechanism of the effect of changes in the turgor of the guard cells upon the size of the stomatal aperture varies with the structure, form, and position of the stomates. In one type of guard cell, found in many different species of plants, the cell wall is thicker on the side bordering the stomatal pore than on the side bordering the epidermal cells (Fig. 1.2). With an increase in turgor the thinner walls of the guard cells are stretched more than the thicker ; this causes the thicker-walled sides to assume a concave shape and results in the appearance of a gap—the stomatal pore—between the two guard cells. Opening of the stomates typical of the grass and sedge families (Fig. 1.1C), appears to result from swelling of the ends of the guard cells thus separating the abutting walls of the middle portion of the two adjacent guard cells. In the sunken stomates typical of conifers (Fig. 1.1D), opening of the stomates seems to result largely from a change in the shape of the guard cells as a result of an increased turgor which is unaccompanied by any appreciable stretching of the walls. These various types of stomatal mechanisms are discussed by Copeland (1902).

The three principal factors which influence the opening and closing of the stomates are : (1) light, (2) the internal water relations of the leaf, and (3) temperature.

1. *Influence of the Light Factor in Stomatan Opening and Closing*. Unless other conditions, to be discussed later, are limiting, the stomates of most species open upon exposure to light and close in its absence. Most commonly, therefore, the stomates are open in the daytime and closed at night, although there are many exceptions to this statement. The reactivity of the guard cells to light undoubtedly varies according to species. It probably differs considerably in shade species, for example, from sun species. Within limits the guard cells appear to react quantitatively to the amount of light they absorb. Stomatal opening apparently will occur in all wave lengths of the visible spectrum, although the influence of radiations in the red region appears to be weaker than the influence of other wave lengths (Sierp, 1933).

Stomatal closure upon the cessation of illumination is generally a gradual process and, according to Stalfelt (1929), the greater the quantity of light which has been absorbed by the guard cells in the course of the day, the longer it takes, at least under some conditions for the completion of stomatal closure.

Although several theories have been proposed in explanation of the mechanism of stomatal action, as conditioned by light and other factors ; the only one for which there is any substantial evidence is the osmotic theory. This does not, of course, eliminate the possibility that other mechanisms may also be operative in this process. The guard cells usually contain starch, but the quantity present is not constant. Sayre (1926) showed that the starch content of the guard cells of dock (*Rumex patientia*) is at its maximum during the night, decreases rapidly during the daylight hours, and increases again toward evening. When the starch content of the guard cells was high the sugar content was low, and *vice versa*. Similarly Alvim (1949) found that opening of stomates of several species as induced by various factors was accompanied by a diminution in the starch content of the guard cells.

The major reaction influencing the proportion of soluble and insoluble carbohydrates present which occurs in the guard cells appears to be the reversible transformation of starch to glucose-1-phosphate in the presence of inorganic phosphates.

$$\text{Starch} + \text{Phosphate} \overset{\text{phosphorylase}}{\rightleftharpoons} \text{glucose-1-phosphate}$$

The enzyme phosphorylase, which catalyzes this reaction, is known to be present in the chloroplasts of the guard cells (Yin and Tung, 1948). The equilibrium position of this reaction is affected by the *p*H of the medium, being much further to the right at *p*H values about 7 than at *p*H values of about 5 (Hanes, 1940). The concentration of soluble glucose-1-phosphate is thus greater at the higher of these two *p*H values than at the lower.

Illumination of the guard cells of a number of species has been found to result in an increase in their *p*H ; failure of illumination, in a decrease (Scarth, 1932 ; Small *et. al.*, 1942). Scarth, for example, found the *p*H of the guard cells of zebrina to range from 4.0 or less in the dark to between 6.0 and 7.4 in the light. There is some evidence that the higher *p*H of the guard cells in the light as compared with the dark may result from a reduction in the carbon dioxide concentration within the leaf as a result of photosynthesis (Scarth and Shaw, 1951). Whether or not photosynthesis occurs in guard cells is not known with certainty, although chlorophyll appears to be present (Freeland, 1951). Reduction of carbon dioxide concentration throughout the leaf could result, however, from occurrence of photosynthesis in the mesophyll cells. In the light, therefore, increase in the *p*H of the guard cells appears to favor conversion of starch into glucose-1-phosphate under the influence of phosphorylase ; in the dark decrease in *p*H appears to favor the reverse reaction.

Increase in the soluble carbohydrate concentration of the guard cells results in an increase in their osmotic pressure while a decrease in their soluble carbohydrate concentration has the opposite effect. That such changes in the osmotic pressure of the guard cells actually occur has been shown by many investigators. The dicurnal changes in the osmotic pressures at incipient plasmolysis of the guard cells and epidermal cells of the English Ivy (*Hedera helix*) are shown in Fig. 1.9. In general, the osmotic pressure of the guard cells is usually relatively high during the daylight hours and relatively low at night. The osmotic pressure of the epidermal cells does not change appreciably during the course of the day and approximates that of the guard cells at night.

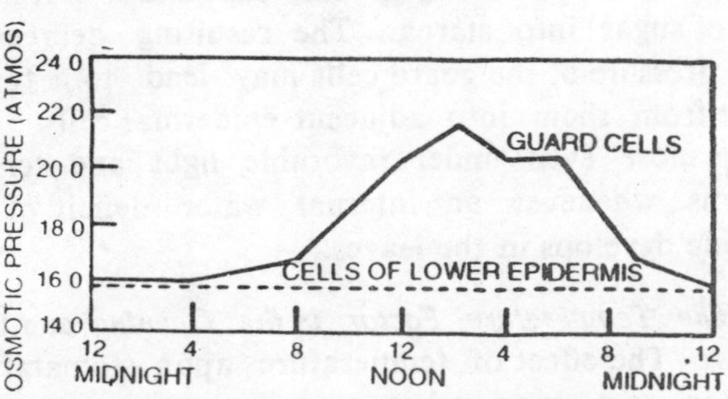


Fig. 1.9. Daily variations in osmotic pressure of the guard cells and epidermal cells of English ivy (*Hedera helix*). Data of Beek (1931).

Increase in the osmotic pressure of the guard cells in the morning results in an increase in their diffusion pressure deficit relative to that of the contiguous cells. Water therefore moves into the guard cells, increasing their turgor, which in turn leads to a widening of the stomatal aperture. Movement of water into the guard cells from neighboring cells results in a loss of turgor by the latter which probably facilitates opening of the stomate. Decrease in the osmotic pressure of the guard cells leads to the reverse series of processes and results in a narrowing of the stomatal aperture.

2. *Influence of the Water Factor in Stomatal Opening and Closing*. As discussion shows, development of an internal water deficit in plants is of frequent occurrence, especially on clear, warm days. A shrinkage in the total volume of water in a plant results in general in a dimioution in the volume of water in each individual cell, although all cells will not necessarily be affected equally. Such a decrease in the water content of the leaf cells, not sufficient to induce visible wilting, is called *incipient wilting*. Under such conditions the guard cells usually decrease in turgor as a result of osmotic movement of water into contiguous cells. Reduction in the turgor pressure of the guard cells as a result of the diminution of the volume of water in them will bring about a partial to complete closure of the stomates. There is also some evidence that diminution in the water content of the guard cells induces a decrease in the *p*H of their cell sap and a correlated conversion of sugar into starch. The resulting decrease in the osmotic pressure of the guard cells may lead to a further loss of water from them into adjacent epidermal cells. Stomates may thus close even under favorable light and temperature conditions whenever an internal water deficit of sufficient magnitude develops in the leaves.

3. *The Temperature Factor in the Opening and Closing of Stomates.* The effect of temperature upon stomatal opening depends in part upon other prevailing environmental conditions. Under constant and favorable light and other environmental conditions, stomatal opening in cotton (Fig. 1.10) and tobacco increases with rise in temperature up to 25°-30°C and decreases at still higher temperatures (Wilson 1948). In most species stomatal opening fails to occur at temperatures approaching 0°C, or lower. Relatively high temperatures (about 40°C.) induce opening of the stomates of some species in the dark.

*Daily Periodicity of Stomatal Opening and Closing.* No kind of living organism is more inescapably at the mercy of its

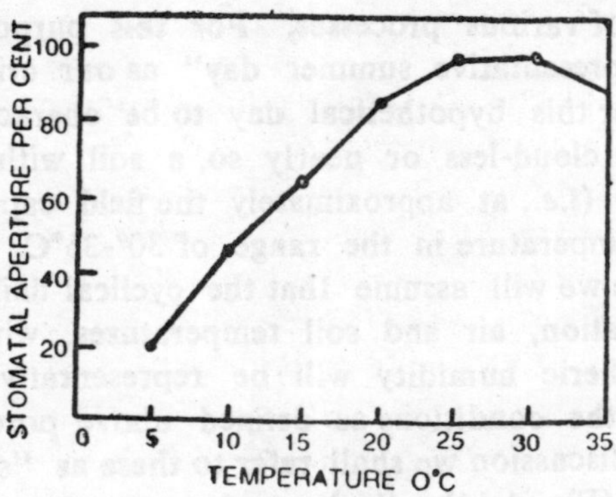


**Fig. 1.10. Relation between temperature and stomatal aperture in cotton. Data of Wilson (1948).**

environment than a rooted plant. A number of the factors in the environment of plants exhibit more or less regular daily periodicities. The most conspicuous and physiologically important factors of which this is true are solar radiation and temperature, both soil and air. Other usually less significant factors in which daily cyclical variations may occur, at least in some habitats, include wind velocity, atmospheric vapor pressure, and carbon dioxide content of the atmosphere. Since the rates of plant processes are conditioned in part by the environmental factors to which the plant is exposed, they also exhibit more or less regular daily periodicities.

Daily cycles of environmental conditions vary greatly from one plant habitat to another, depending upon latitude, altitude, directional exposure, and local climatic conditions. In a given habitat such cycles vary from season to season and during any one season with day-to-day variations in meteorological conditions. Hence the daily periodicity of any plant process may vary considerably in pattern from one day to another depending upon the prevailing cycle of environmental conditions. In the course of this book we shall have occasion to analyze the hour-to-hour variations of a number of physiological processes occurring in plants. It is desirable, therefore, to choose a definite type of diurnal cycle of environmental factors as a reference standard in terms of which to discuss daily variations in the rates of various processes. For this purpose we shall select a "representative summer day" as our criterion. We shall consider this hypothetical day to be characterized by a sky which is cloud-less or nearly so, a soil with a favorable water supply (*i.e.*, at approximately the field capacity, and a maximum temperature in the range of 30°-35°C. (86°-95°F.). Furthermore we will assume that the cyclical daily variations in solar radiation, air and soil temperatures, wind velocity, and atmospheric humidity will be representative of a day upon which the conditions as defined above prevail. In our subsequent discussion we shall refer to these as "standard day conditions" (Fig. 1.11). Such environmental conditions will actually be approximated on many summer days in moist temperate zone regions.

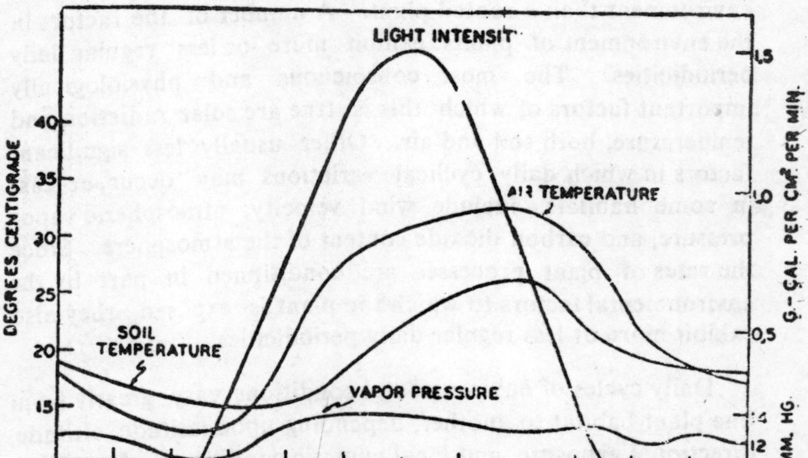


Fig. 1.11. Daily variations in certain environmental factors on a "standard day." (Light intensity measured at horizontal incidence.)

# 2

# Stomatal Control

**The striking demonstration of the importance of leaf stomata in transpiration and photosynthesis, and the ability to exercise relatively specific biochemical control of stomatal apertures, has stimulated considerable experimental work and discussion about stomatal physiology. In 1965, in *Biological Reviews*, this work was summarized with chemical inhibitors and reviewed the evidence for the role of glycolic acid metabolism in stomatal opening in the light. In the same issue of that journal, Meidner and Mansfield concluded that the $CO_2$ concentration in the intercellular spaces of leaves near the guard cells largely controls stomatal activity. Heath, Mansfield and Meidner have stressed the "complexity" of stomatal movement and disputed the suggested participation of glycolic acid in stomatal opening. Pallas also doubted that a universal mechanism could account for the observations at hand. Others have independently called attention to the error in the socalled perimeter law, which has been learned for years by students of botany, and demonstrated that, except in still air, stomatal diffusive resistance controls transpiration at all stomatal apertures. Such control may be used to alter the hydrologic cycle in nature. Most recently, and further confirming that the control of stomatal opening and closing is an exciting and lively subject, Levitt has rejected all current interpretations and has loyally returned to the ideas of a former teacher in his search for the truth.**

It is curious that the mechanism of stomatal movement should stimulate so much controversy, and one must admit at the outset that the mechanism cannot presently be given with precision. This is likely not because the process is unusually complex. It is probably simpler than many other physiological processes that are already better understood. Part of the difficulty in the interpretation of experiments with guard cells comes from the fact that guard cell function is dependent to some degree on other cells, and some of the difficulty also arises because many research workers have failed to use adequate controls and thus have observed indirect effects on stomatal movement.

Guard cells occupy a small portion of the epidermis, and either intact leaf tissue or, at best, epidermal strips must be used as experimental material. Stomatal opening depends on an increase in turgor in the guard cells relative to the adjacent epidermal cells. Thus the difference between the osmotic value in the two neighboring cells determines the degree of stomatal opening, as has been emphasized recently by different scientists. In most species, stomatal opening and increased osmotic value in the guard cells occurs in the light. This may result from photosynthetic production of organic solutes by the chloroplasts of the guard cells or from the "pumping" of ions or other solutes into the guard cells from adjacent cells. The turgor of the guard cells at any time is also dependent on the cell membranes through which water must pass. Under some conditions, stomatal opening may result from decreased turgor in adjacent epidermal cells.

Most of the factors that affect photosynthesis and the water relations in leaves also influence stomatal movement. One must take care to insure that stomatal changes observed were not caused, for example, by water deficits rather than by the factor under investigation. Here in this chapter we will examine some of the apparently conflicting results on stomatal control and attempt to sort out some of the misunderstandings in the literature. Finally, the evidence in favor of a potassium ion "pump" to account for stomatal opening in the light will be summarized.

## MEASUREMENT OF STOMATAL APERTURES

Several rapid and reliable methods for the assay of stomatal width have become available in recent years, but no single technique is best for all plant species. The casting of leaf impressions in silicone rubber, followed by the preparation of replicas of the cast with a cellulose acetate film, was invented by Sampson and has been used extensively in this laboratory for the measurement of stomatal dimensions. By this method, numerous diverse leaf samples, including leaf disks, may be observed at the same time without affecting the leaf environment. The impressions provide a permanent record, and the stomatal widths can be measured accurately at one's convenience. With care, such casts can be made repeatedly on the same area of a leaf without visible damage to the tissues. Its simplicity has no doubtled to its widespread use as a method of observing and recording stomatal movement by university undergraduates and research workers and even by school children. The major limitation of the silicone rubber technique is that it cannot be used with accuracy when the apertures are much narrower than 1 $\mu$, or with species in which the stomata are recessed in pits rather than being located on the plane of the leaf surface.

Other methods of preparing leaf replicase have been used by several groups recently. Stoddard characterized the epidermis of various leaves directly from cellulose acetate films. Others have made excellent leaf impressions with plastic films. Such materials require consider able time to set before the films can be stripped from the leaf surface, during which stomatal widths may change. They also usually cause damage to the tissue being examined.

In addition to these indirect but visual methods, some new types of porometers, in particular a convenient diffusion porometer, have been described that have advantages for certain purposes. The rapid estimation of stomatal aperture is now possible with a device that senses the humidity gradient next to a leaf, as described by Wallihan. This technique was modified by Turner and Waggoner to observe stomatal changes in needles of red pine (*Pinus resinosa*). It was not possible to do this easily by any other method in pine needles except by the relatively crude procedure of observing the degree of penetration by organic solvents. However, even the solvent infiltration method may be useful, as was shown by Moreshet and Stanhill who found a linear relation between infiltration index and porometer measurements in cotton leaves in the greenhouse and outdoors.

A viscous-flow porometer that permit measurement of the rate of air pressure drop through a leaf in a convenient manner was invented by Alvim. Waggoner showed with barley leaves that such an instrument acts according to predicted physical diffusion theory, and that the results with the porometer are well correlated with stomatal width in the range from 0.8 $\mu$ to 4.2 $\mu$, as determined from silicone rubber impressions.

Scientists considered porometers, which do not reveal stomatal dimensions directly but measure diffusive resistance, superior and have written that "detailed investigations into the mechanism of stomatal movements cannot be carried out with silicone-rubber impressions, however useful a tool for approximate studies they may be." They also incorrectly attribute to us the opinion that in the impression method "anything up to 2 $\mu$ is practically indistinguishable from a closed pore." Besides the contrary evidence with barley leaves already mentioned Shimshi had earlier found an excellent correlation ($r$=0.938) in maize leaves between stomatal apertures in the range between 0.8 $\mu$ and 6 $\mu$ obtained from silicone impressions and permeability as determined with an Alvim porometer. Another confirmation of such a relationship in the range between 1.2 $\mu$ and 3.0 $\mu$ is shown for tobacco leaf tissue in Fig. 2.1. Thus a lack of good methods for evaluating stomatal movement does not now seem to be limiting progress in understanding how guard cells function or in finding better methods of controlling their movement.

## EFFECT OF WATER RELATIONS

Since the opening of stomata depends primarily on an uptake of water by the guard cells, it is understandable that

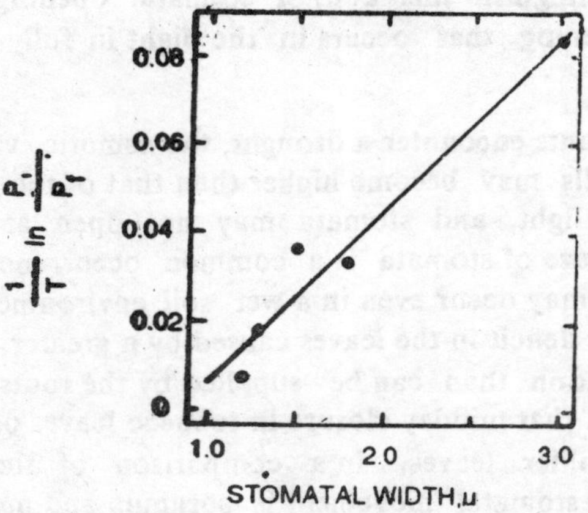


Fig. 2.1. **Comparison of leaf permeability (Alvim porometer) and mean stomatal width (silicone rubber impressions) in tobacco leaf disks. Pairs of disks from a leaf kept in darkness were floated on water at 30° and 2000 ft-c and duplicate samples were taken at various times for the two kinds of determinations. In the porometer measurements the initial pressure ($P_i$) was 200 mm Hg, and the final pressure ($P_f$) was determined after 10 sec.**

**severe water stress will inhibit stomatal opening. To avoid this pitfall, many investigators have used the technique of floating leaf disks on water, a circumstance under which water deficits cannot exist. Under such conditions stomata in tobacco leaf disks in the light in air open at about the same rate and to the same extent as is usually observed in the intact leaf. When disks on water with open stomata are placed in the dark, the stomata close more rapidly than they open, and when returned to the light a second time, the stomata open again at their initial rate. Thus the opening in the light and closing in the dark is an easily reversible process, in contrast to other types of stomatal movement that can be induced and which will be discussed later. In addition, Stalfelt has shown in leaf disks of *Vicia faba* and *Rununculus repens* that stomatal opening can occur readily in the dark in a $CO_2$-free atmosphere only if the leaf tissue has a slight water deficit. This also-**

serves to distinguish this kind of stomatal opening from the normal opening that occurs in the light in fully turgid leaf tissue.

When plants encounter a drought, the osmotic value of the epiderm cells may become higher than that of the guard cells even in the light, and stomata may not open at all. The midday closure of stomata is a common occurrence in many species. It may occur even in a wet soil environment because of the water deficit in the leaves caused by a greater water loss by transpiration than can be supplied by the roots. Macdowall showed that midday closure in tobacco leaves occurs more readily in older leaves. In a comparison of the effect of drought on stomatal movement is sorghum and maize leaves, Glover showed that after a period of one week only sorghum stomata opened. After two weeks of drought, the stomata in maize were permanently damaged, although there was still only a slight effect in sorghum stomata as evidenced by an almost complete recovery 5 days after the plants were watered. Thus the ability of stomata to recover from a severe water deficit may be an important factor in apparent drought resistance. Bannister found that the relative leaf turgidity in a given species at which stomata will close may vary with the habitat in which the plant grows. All of these studies serve to emphasize that when the stomatal mechanism is under investigation, the leaf tissue should either by turgid or at least the water status of the leaf must be kept constant while other variables are studied.

## LIGHT REQUIREMENT FOR STOMATAL OPENING

*Need for photosynthesis.* It has never been shown that guard cells devoid of chloroplasts are capable of opening in the light. Virgin observed that the development of the stomatal light response in etiolated wheat leaves, after a short light exposure needed to initiate chloroplast formation, closely followed the increase in chlorophyll-*a* content of the tissue. This suggested that the photosynthetic process must be concerned with the increase in turgor of the guard cells. Recently, Poskuta and Tomczyk carried out similar experiments with etiolated onion leaves, in which no starch is synthesized in the guard cells. Stomata did not open well until 3 or 4 days after the period of illumination that initiated chloroplast development, when the chlorophyll concentration was about 75 per cent of that of green control leaves.

Several investigations have also been carried out on the effect of wavelength of light on the opening of stomata. The most significant study of this kind was described by Kuiper who did his experiments on the maintenance of already open stomata in light with epidermal strips of *Senecio odoris* that were essentially devoid of mesophyll cells. The light was supplied by a high intensity spectrograph. Maximal opening occurred at 432 m$\mu$ and at 675 m$\mu$, the same maxima as were found in the absorption spectrum of spinach chloroplasts. This work provides strong evidence that photosynthesis in the guard cells themselves is responsible for the maintenance of stomatal opening in the light. With these epidermal strips, 2(3,4-dichlorophenyl)-1,1-dimethyl urea (DCMU) at $10^{-5}$M, a well known inhibitor of Photosystem II and noncyclic photophosphorylation closed the stomata completely in the light. This observation raises the question of whether photosynthesis is needed to produce carbon compounds in the guard cells or is primarily required for the production of ATP in the light.

*Stomata up or down.* Stalfelt described experiments from which he concluded that net $CO_2$ uptake (the "production effect") played an important part in light opening. He measured stomatal opening in leaf sections from three species which contain stomata only on their lower surface, either when the stomata were floated down against the water or up facing the atmosphere. When stomata were down there could be little net photosynthesis, and indeed he found that stomata opened two to three times as wide in from 3 to 6 hr when the stomata were up.

Table 2.1, Experiment 1, shows a similar result with leaf disks from still another species, the rose. However, when the stomata were up (Experiments 2 and 3), they opened nearly as well in flowing $CO_2$-free air as in normal air, showing that

Table 2.1. Effect of changes in the Gas Phase on stomatal opening in rose leaf disks in light.

| *Expt. No.* | *Time in Light, min* | *Gas Phase* | *Increase in Stomatal Width, μ* | |
|---|---|---|---|---|
| | | | *Stomata Up* | *Stomata Down* |
| 1 | 30 | Still air | 0.4 | 0.2 |
| | 90 | Still air | 1.1 | 0.3 |
| | 240 | Still air | 4.5 | 2.5 |
| 2 | 90 | Flowing air | 2 5 | 1.5 |
| | 90 | Flowing $CO_2$-free air | 2.2 | 1.8 |
| 3 | 120 | Flowing air | 2.8 | |
| | 120 | Flowing $CO_2$-free air | 2.5 | |

Pentafoliate rose leaves (stomata are present only on the lower surface in this species) were excised and kept with their petioles in water in the dark for 1 hr. Leaf disks (1.6 cm diameter) were cut from the three terminal leaflets and floated in pairs in 2 ml of water with the stomata either facing the gas phase (stomata up) or against the water (stomata down). At the end of the times shown at 30° and 1000 ft-c (a mirror reflected approximately the same light to the lower surface), impressions of the lower epidermis were cast in silicone rubber to determine stomatal widths.

In Experiment 1, the disks were floated in partly-covered beakers. In Experiments 2 and 3, moistened air or $CO_2$-free air was passed through the atmosphere over the disks in 25 ml Erlenmeyer flasks at a rate of several flask volumes per minute starting 5 min before the light was turned on.

lack of net photosynthesis did not greatly inhibit stomatal opening in this tissue. This suggests that sufficient photosynthesis can occur in guard cells from internally derived $CO_2$ to permit good rates of stomatal opening in the light, and the poorer opening when the stomata were down must have been for a different reason, such as the poorer availability of oxygen rather than from a lack of net $CO_2$ fixation.

*Opening of stomata in the dark.* Wide opening of stomatal pores in the dark is characteristic of crassulacean plants, whose metabolism is geared to net $CO_2$ fixation in the dark, as was shown by Nishida. Even in such species, dark opening occurred only after a light period, and the magnitude of the opening was related to the duration of illumination in the preceding day. A slight "endogenous" diurnal opening of stomata takes place in the dark in normal air in some species, as

Stalfelt found in intact leaves of *Vicia faba*. When the leaves were excised, however, opening in the dark occurred only in an atmosphere of $CO_2$-free air, suggesting that this kind of opening differs from the usual opening in the light that takes place readily in excised leaves. Another difference was that, the dark opening was also very much smaller after the first 24-hr period. The "endogenous" dark opening was less than 20 per cent of that in the light, and in the absence of a water deficit was only 2 per cent as great. Dark opening occurs best in the presence of a water deficit providing another distinguishing characteristic from the light opening which takes place so readily in fully turgid tissue.

Mansfield has described dark stomatal opening in $CO_2$-free air in *Xanthium* leaves and showed that the pores were always smaller than in normal air in the light. He also indicated that in continuous darkness, stomata closed partially several hours after reaching a maximal aperture. In contrast, stomata that open in the light remain fully open under continuous illumination for at least 10 hr. Thus it would appear that in all cases where opening in the dark has been observed, its characteristics are sufficiently different from the normal light opening to suggest that a different mechanism operates in the dark, and it may bear little relation to the light-requiring process that is common to all but plants with crassulacean type metabolism.

*Effects of blue light*. Mansfield and Meidner examined stomatal opening in *Xanthium* at low light intensities in a closed system at the $CO_2$, compensation point. There was little difference between red and blue light on the compensation point (110 to 115 ppm $CO_2$ at 21°), but stomatal opening was greater in blue light. Blue light was also more effective in *Allium cepa*. Raschke investigated opening in *Vicia faba* in $CO_2$-free air in blue light. When the pores were open maximally, a change to an atmosphere of nitrogen produced a much increased stomatal opening. When the gas phase was changed back to $CO_2$-free air, they returned to their former opening, but then a second exposure to nitrogen produced very little increase in width. Since opening in the light is inhibited in an atmosphere of nitrogen and opening in light and closing in darkness is freely reversible the blue light effect may also be related to the dark opening processes, but does not appear to be important in the normal opening mechanism in the light.

## $CO_2$ CONCENTRATION IN THE INTERCELLULAR SPACES

*Stomatal opening in $CO_2$-free air in darkness.* During the last 20 years, considerable attention has been paid to the hypothesis that opening in the light results from the lowering of the $CO_2$ concentration in the substomatal spaces, and closing in the dark is caused by the increase in $CO_2$ from dark respiration. The greatest support for this view comes from the observations that, in some species, stomatal opening occurs in the dark and is sometimes accelerated in a $CO_2$-free atmosphere. As indicated earlier, such dark opening appears to differ in a number of respects from the opening normally observed in the light, and in any case no reasonable biochemical explanation for such a $CO_2$ hypothesis has yet been suggested.

A slow (7-hr) stomatal opening by leaf disks of *Vicia faba* in the dark in $CO_2$-free air was described by Stalfelt who also found that the increase in the surplus in the osmotic value of the guard cells over that in the epidermal cells was well correlated with the increase in pore width. Longuet showed how critical the $CO_2$ concentration is in order to obtain dark opening of *Pelargonium zonale* stomata. At 0.03 per cent $CO_2$ in air the stomata were closed, but when lower than 0.02 per cent or greater than 2 per cent to 4 per cent $CO_2$, the stomata opened. These experiments further emphasize difficulties in accepting a hypothesis of stomatal movement based largely on the $CO_2$ concentration.

*Stomatal closure by high $CO_2$ concentration in light.* In spite of the uncertainties about the role of internal $CO_2$ concentration in the normal range on opening in the dark (or light), the effect of high concentrations of $CO_2$ on stomatal closure in the light is an easily observable phenomenon. Pallas studied the effect of increasing $CO_2$ levels on transpiration of several species, and found, based on the concentration at which the transpiration rate remained constant with further increases of $CO_2$, that stomata were closed at the following concentrations : tomato, 0.1 per cent ; cotton, 0.1 per cent ; maize, 0.2 per cent ; and sorghum, 0.3 per cent. In tobacco leaf disks floated on water in the light, $CO_2$ levels above 0.2 per cent in air also inhibit stomatal opening. High concentrations of $CO_2$ such as these are known to greatly alter the nature of the products formed in photosynthesis particularly the synthesis of glycolic acid and hence its metabolic products, and this could be responsible for stomatal closure.

Raschke used leaf sections of *Zea mays* to show that the linear light intensity requirement for stomatal opening is the same in atmospheres of air, $CO_2$-free air, and nitrogen. He therefore, concluded that opening is not dependent on $CO_2$ assimilation. Such an interpretation is uncertain because there is a constant supply of $CO_2$ present from respiration (and including photorespiration in most species) inside leaves. Moreover, the $CO_2$ concentration inside a leaf would be considerable at the beginning of a period in the light no matter what the atmosphere might be outside. It was also observed with maize and tobacco leaf disks floating on water that the rate and extent of stomatal opening in the light is the same in air as in $CO_2$-free air. In intact maize leaves, however, Moss found that stomatal opening is inhibited as the $CO_2$ concentration is raised from zero to 400 ppm, and the inhibition is greatest at higher leaf temperatures and disappeared at 14°. Thus the closing at warmer temperatures in intact leaves may also have been related to leaf water deficits.

*Changing $CO_2$ concentration inside leaves in light.* Stomatal opening can be studied in leaves of different species that vary greatly in their $CO_2$ compensation point, the steady state concentration of $CO_2$ in a closed, illuminated system. In leaves or leaf disks of maize, the $CO_2$ compensation point is less than 10 ppm, while in most species such as tobacco leaf tissue this figure is at least 50 ppm under usual conditions. The $CO_2$

compensation point characteristically increases greatly with rise in leaf temperature in species like tobacco. Since the concentration of $CO_2$ is the same inside and outside the leaf at the $CO_2$ compensation point, and it is easily measured by comparing stomatal opening in an open system (in flowing air) with that in a closed system (at the $CO_2$ compensation point), one can observe if there is any effect at widely differing internal $CO_2$ concentrations.

Estimates of the $CO_2$ levels inside maize and tobacco leaves in an open and closed system have been presented and a more detailed analysis is available in Table 2.2. These results were obtained from a rational model of a leaf, similar to tobacco,

**Table 2.2.** $CO_2$ Concentration at mid-stoma in a Simulated Tobacco leaf at 30° (from Waggoner, 51).

| *$CO_2$ Conc. in Atmosphere, ppm* | *$CO_2$ Concentration at Mid-Stoma, ppm* | | |
|---|---|---|---|
| | *Light Intensity, ly $min^{-1}$* | | |
| | *0.1* | *0.4* | *1.2* |
| 100 | 95 | 93 | 95 |
| 300 | 260 | 252 | 253 |
| 600 | 516 | 491 | 488 |
| | $CO_2$ Compensation Point, *ppm* | | |
| | 71 | 61 | 76 |
| | Net Photosynthesis at 300 ppm $CO_2$ *mg $dm^{-2}$ $hr^{-1}$* | | |
| | 7.9 | 14.6 | 15.9 |

which was constructed by Waggoner to account for net photosynthesis by incorporating the effects of such factors as light, temperature, $CO_2$ concentration, respiration, etc. This model simulates the activity of real leaves in a number of ways. It also shows that the $CO_2$ concentration at mid-stoma is always considerably greater in normal air than at the $CO_2$ compensation point at light intensities from one-tenth to that of full sunlight. In maize, these differences would be still greater. Nevertheless it was found that stomata in leaf disks open at the same rate and to the same extent in moving air as at the

$CO_2$ compensation point in both species. Stomata also open wider at warmer temperatures in a closed system in spite of the higher internal concentration of $CO_2$. Hence, the $CO_2$ concentration inside leaves in the normal range need not have a pronounced effect on stomatal opening in the light.

If stomatal closing in the dark resulted from an increased $CO_2$ concentration in the substomatal spaces, closing in the dark should be inhibited by $CO_2$-free air. However, it was found that the initial rate of closing in *Xanthium* in the same in both atmospheres, although the final extent of closing is less in the absence of $CO_2$. These experiments also indicate that $CO_2$ concentration does not have an important role in the normal opening and closing of stomata.

## INHIBITORS OF STOMATAL MOVEMENT

Experiments with biochemical inhibitors of stomatal movement have provided crucial evidence of the relation between stomatal aperture and rates of net photosynthesis and transpiration, and they have provided some insight into the biochemical mechanism responsible for this movement. It was realized in the beginning that since guard cells are located on the outer surfaces of leaves, the greatest chance of achieving specificity in controlling stomata with inhibitors would likely be by supplying such compounds as a spray on leaves or by floating leaf tissue in solutions. Allowing substances to reach guard cells through roots, stems, or petioles, and hence making it more likely that inhibitors will affect mesophyll cells as well as guard cells, introduces unnecessary complications in studies on the mechanism of stomatal action. It also seems obvious that, in biochemical studies with inhibitors, one should use a compound at the lowest possible effective concentration, for the shortest time necessary. One should then confirm by using independent chemical measurements that the biochemical changes expected are indeed being carried out. It is unfortunate that in a number of recently published papers few of these precautions were considered, and hence there is serious doubt about the conclusions derived from such work.

If a substance acts specifically on the guard cells of a leaf to close stomata, it can be shown from diffusion theory that the transpiration rate should be inhibited to a greater extent than net photosynthesis. On the other hand, if photosynthesis in the mesophyll cells is also seriously interfered with by an inhibitor that also closes stomata, the decrease in photosynthesis would be greater than that of transpiration. Thus a comparison of these two processes provides a simple test of the specificity of a substance on guard cell metabolism.

One group of workers has been dismayed by are array of substances that interfere with stomatal movement and considers this broad range a part of the evidence that there is no primary mechanism of stomatal opening. However, oxygen uptake in dark respiration is carried out by the cytochrome, oxidase reaction, but respiration may be interfered with by many different kinds of inhibitors that may affect glycolysis, the citric acid cycle, or the cytochrome oxidase reaction itself. Inhibitors, when used judiciously, should help to unravel the stomatal mechanism just as they have the respiratory process.

Inhibitors are known that interfere with stomatal opening, prevent closing of already open stomata in the dark, and substances are known that even stimulate opening. The action of many of these compounds has already been summarized. Only those inhibitors about whose action a great deal is known, or whose unusual properties have only recently been described, will be discussed here.

*Inhibitors of photosynthesis and phosphorylation.* The substituted ureas DCMU and CMU, which are herbicides and inhibit the Hill reaction at low concentrations, close stomata in the light in epidermal strips and in leaf disks at concentrations less than $5 \times 10^{-5}$ M. Allaway and Mansfield supplied CMU at $1 \times 10^{-4}$ M through the petiole to excised leaves of *Rumex* and *Vicia faba* after the stomata were opened in light. The leaves continuously absorbed the CMU for at least several hours, and the stomata closed to various extents in normal air. When $CO_2$-free air was then passed under pressure through the leaves treated with CMU, the stomata opened somewhat in comparison to those of leaves provided with normal air. However, no leaf was treated with $CO_2$-free air in the light in the absence of CMU for comparison. If opening was in fact greatly stimulated by $CO_2$-free air under conditions that appear to have been so adverse for photosynthesis and photophosphorylation by all leaf cells, it might indicate that the enhancement of opening by $CO_2$-free air bears little connection with the normal process of stomatal opening in the light and might be caused by indirect effects on cells other than guard cells.

Stomatal closure in the light is also caused by atrazine (2-chloro-4-ethylamine-6-isopropylamino-s-triazine), another herbicide that inhibits the oxygen-evolving reactions of photosynthesis. Smith and Buchholts showed that this substance inhibits transpiration and closes stomata when sprayed on bean leaves. This inhibitor diminished transpiration in cotton, maize, and soybean plants when supplied in the nutrient solution, and practically stopped transpiration within several hours when sprayed on leaves of *Vicia faba*. When applied to the soil at herbicidal concentration immediately after planting, stomata of maize were unaffected by atrazine. No publications indicate whether transpiration was inhibited more than photosynthesis when stomata were closed by treatment of leaves with atrazine.

*Growth regulators.* Another growth regulator, 2, 4, 5-trichlorophenoxyacetic acid, may be added to the list of those already known to inhibit stomatal opening. When sprayed on apricot leaves at low concentrations, stomata were closed, and the effect persisted for about two weeks. Kozinka showed that stomata were closed and transpiration was decreased in barley seedlings sprayed with the potassium salt of 2-methyl-4-chlorophenoxyacetic acid at concentrations from $10^{-3}$ M to $10^{-6}$ M.

Livne and Vaadia made the interesting observation that when excised barley seedlings were placed in the light in solution of $10^{-5}$ M kinetin, stomatal opening was greatly enhanced within 2 hr. These results were confirmed by Meidner and by Luke and Freeman. The latter group indicated that wheat and oat leaves were more sensitive than barley. The wider stomatal opening induced by kinetin resulted in an increase in net photosynthesis of 12 per cent, as might be expected from the decreased stomatal diffusive resistance.

*Inhibitors acting directly on membranes.* Water must pass through cell membranes in order to make the guard cells turgid, and cell organelles also play an essential part in the opening of stomata. Alkenylsuccinic acids and their derivatives, $CH_3—(CH_2)_n—CH=CH—CH_2—CH(COOH)—CH_2COOH$, are surface-active agents that effectively close stomata in leaf disks. Kuiper has shown that these compounds at low concentrations enhance the permeability of bean roots to water, and the same chemical structures required to demonstrate this activity are needed to inhibit stomatal opening. These inhibitors probably act on membranes through the activity of their hydrocarbon chain of appropriate length on the lipid portion of membranes. Siegenthaler and Packer oberved that the monomethyl ester of decenylsuccinic acid, between $10^{-3}$ M and $10^{-4}$ M inhibits electron transport and photophosphorylation as well as changes in the volume of chloroplast membranes in isolated spinach chloroplasts in the light. Baddeley and Hanson observed that the same compound at $10^{-4}$ M and $10^{-5}$ M acts as an uncoupler of oxidative phosphorylation in isolated mitochondria from maize shoots and is similar in its mode of action to linoleic acid. Thus these inhibitors function on cellular organelles, as would be expected if they interferred with the behaviour of the lipids of membranes.

The inhibitor most used in studies on decreasing transpiration through stomatal control is phenylmercuric acetate. It probably forms mercaptides with the sulfhydryl groups of the protein in membranes, and this may account for the longer lasting inhibition than is usually observed with alkenylsuccinic acids. In isolated spinach chloroplast preparations, phenylmercuric acetate inhibits NADP reduction and accompanying noncyclic photophosphorylation at concentrations between $10^{-4}$ M and $10^{-6}$ M. Rao, Fallin and Fang observed with $Hg^{203}$-labeled phenylmercuric acetate supplied to pea roots that this substance is concentrated in the mitochondrial fraction in comparison with similarly labeled mercuric acetate. Mercuric acetate, in comparison with phenylmercuric acetate, has little effect on stomata.

Mansfield apparently still retains the view that even substances like phenylmercuric acetate act on guard cells by inhibiting photosynthetic $CO_2$ uptake and thus increasing the $CO_2$ concentration inside the leaf. To test this, he carried out some experiments in which inhibitors were supplied through the cut petiole, a hazardous technique with reactive inhibitors for reasons already discussed. Other experiments were done with epidermal strips of *Vicia faba*. In an atmosphere of $CO_2$-free air, the stomata closed more slowly in $10^{-4}$ M phenylmercuric acetate than in normal air. Evidence, to be discussed later under antitranspirants, makes it highly unlikely that when phenylmercuric acetate is sprayed on leaves, the mechanism by which it closes stomata can be related to that suggested by Mansfield. Instead, its action must be primarily on the guard cells themselves, and these relatively few cells could hardly provide any large contribution to the total release or uptake of $CO_2$ inside a leaf.

Graniti discovered that the fungus *Fusicoccum amygdali*, a pathogen causing almond and peach canker, when grown in submerged culture produces a substance that alters the water relations of plants This compound, Fusicoccin A, has been characterized and its structure is reminiscent of a substituted sterol glycoside (MW 680). When supplied to cut tomato leaves through the stem at $10^{-5}$ M, this natural product doubled the transpiration rate and increased stomatal opening much as has been mentioned for kinetin. It also slowed stomatal closure in the dark (A. Graniti and N. Turner, personal communication), in a manner similar to higher concentrations of azide.

Mansfield and Heath indicated that natural "smog" had a small inhibitory effect on stomatal opening *Xanthium* leaves which was not permanent. Todd and Propst found that ozone in the air reduced transpiration in bean, tomato, and *Coleus* leaves. It is possible that ozone will react with unsaturated groups in the lipid portion of membranes to induce a temporary stomatal closure.

## GLYCOLIC ACID METABOLISM IN STOMATAL OPENING

Since α-hydroxysulfonates, which are effective competitive inhibitors of the glycolate oxidase reaction, were found to inhibit stomatal opening in the light or to close already open stomata, it seemed important to investigate the possible role of glycolate metabolism in stomatal movement. Experiments with these inhibitors, as well as a comparison of several environmental factors that affect the synthesis of glycolate and its oxidation, were first carried out by Zelitch and Walker and have been extended more recently in order to examine the relation of these factors to stomatal opening in the light. The results are consistent in every instance with the postulated importance of glycolate metabolism and may be summarized as follows :

(*a*) The concentration of α-hydroxy-2 pyridinemethanesulfonic acid required to inhibit by 50 per cent the stomatal opening in tobacco leaf disks was about the same as that needed to inhibit glycolate synthesis 50 per cent under the same conditions.

(*b*) After stomata were opened in the light in leaf disks, $10^{-3}$ M sodium 1-hydroxy-1-decanesulfonate effectively closed stomata (in sodium tartrate buffer-at pH 4.6), and this closure was reversed in a competitive manner by sodium glycolate (also buffered at pH 4.6). On the other hand, solutes like glucose, malate, or potassium chloride did not reverse the effect of the inhibitor under the same conditions. Glycolate also did not reverse stomatal closure induced by another inhibitor, phenylmercuric acetate, again emphasizing the specific nature of the action of this α-hydroxysulfonate and its connection with glycolate metabolism.

(*c*) The rate of stomatal opening by leaf disks in light was greater in air than in an atmosphere of nitrogen, and the synthesis of glycolate by the same tissue was likewise greatly stimulated in air in comparison with nitrogen.

(*d*) The rate and extent of stomatal opening in the light in leaf disks floating on water was the same in $CO_2$-free air as in normal air, while the rate of glycolate synthesis was only slightly less (27 per cent) in $CO_2$-free air compared with normal air. Stomatal opening was much less at $CO_2$ concentrations in the gas phase greater than about 0.2 per cent and marked inhibition of glycolate biosynthesis also occurred at concentrations of $CO_2$ above 0.2 per cent.

(*e*) The closure of stomata in the light induced by 1.8 per cent $CO_2$ was reversed only by sodium glycolate in a comparison with a large number of metabolites and salts.

A number of lines of evidence thus indicate that glycolate metabolism plays an important part in the opening of stomata. This may result because glycolate metabolism serves to reoxidize NADPH, especially at low $CO_2$ concentrations, and hence to simulate noncyclic phosphorylation needed for a "pump" where glycolate may function in catalytic concentrations, or perhaps because glycolate acts as an intermediate in the synthesis of carbohydrate. These interpretations of the role of glycolate have been criticized, however, and in order to clear up some misunderstandings in the literature, we will discuss some of the objections that have been raised.

Meidner and Mansfield stated that there was no evidence that glycolate synthesis is better at low (zero) $CO_2$ concentrations and, to bolster their argument, cited work on the excretion rather than the synthesis of glycolate in *Chlorella* cells. These latter workers did not examine the excretion of glycolate at "zero" $CO_2$ but did in fact observe a severe inhibition of 0.25 per cent $CO_2$. In tobacco leaf tissue, the rate of synthesis of glycolate in light is not much less at "zero" $CO_2$ than in normal air. They also cited an apparent paradox since glycolic acid ($10^{-2}$ M) inhibited stomatal opening slightly, whereas it was used by us in reversal experiments to bring about opening. Glycolic acid is a fairly strong acid, $p$K=3.83, hence there is a decided difference between the use of the free acid at a low pH and the conditions used in the reversal experiments (pH 4.6).

These workers repeated the objections already mentioned, and further stated that "at zero carbon dioxide there can be little if any photosynthetic formation of glycolate, and on Zelitch's hypothesis illuminated stomata should therefore show closure in carbon dioxide-free air." This is in error, since glycolate synthesis in tobacco leaf tissue is little affected by the absence of $CO_2$ in the atmosphere, because under such environmental conditions there is a considerable turnover of $CO_2$ inside the leaf arising from dark respiration and photorespiration. In fact, this turnover of carbon may be so great in "zero" $CO_2$ that only one-third of the $CO_2$ generated by photorespiration may be released into a rapid stream of $CO_2$-free air outside the leaf even when the stomata are open.

Meidner and Mansfield continuously supplied excised tobacco and *Xanthium* leaves through the petioles with $10^{-2}$ M $\alpha$-hydroxy-2-pyridine methanesulfonic acid in the light for 90 min. With stomata closed, the $CO_2$ compensation point in *Xanthlum* increased from 36 ppm to 121 ppm because of inhibition of photosynthesis. In $CO_2$-free air, stomata closed by the inhibitor opened, and in 0.2 per cent $CO_2$ they closed. The investigators concluded that stomated closing is not related to inhibition of glycolate oxidation, but results from the increase in $CO_2$ concentration in the intercellular spaces. However, Scientists have found that when leaf disks were floated on the same inhibitor solution for periods up to 20 min or longer, depending on the temperature, net photosynthesis is not affected and is even enhanced under certain conditions in tobacco, although glycolate oxidation was clearly interfered with since glycolate accumulated. Scientists have also recently found that when an $\alpha$-hydroxysulfonate is supplied to tobacco leaves in a manner more likely specifically to affect guard cells, and not through the petiole, stomata closed with no change in the $CO_2$ compensation point, indicating that the $CO_2$ concentration inside the leaf is not involved. Thus they conclude that the results described by Meidner and Mansfield were not necessarily caused by an increase in the $CO_2$ concentration inside the leaf, but that coincidentally all chloroplasts, including those in the guard cells, were adversely affected by the manner in which the inhibitor was supplied. Thus effects on stomatal widths induced by lowering the normal $CO_2$ concentration still appear to be unrelated to the primary mechanism of stomatal opening in the light.

Mouravieff found that when leaf tissue of *Leucanthemum* was floated on a solution of α-hydroxy-2-pyridinemethanesulfonic acid, the stomata closed, but the inhibitor failed to close stomata in *Veronica* tissue. An attempt to reverse the closing with glycolic acid also failed. However, in the original reversal experiment, a different inhibitor was used and the pH was controlled.

Levitt repeated some of the misunderstandings about glycolate metabolism already discussed. He also expressed skepticism that glycolate could even penetrate cells in commenting on the reversal experiments, although there are many experiments in the literature showing how readily sodium or potassium glycolate is metabolized when supplied to leaf tissues.

## USE OF INHIBITORS OF STOMATAL OPENING AS ANTITRANSPIRANTS

A leaf disk assay has been described that facilitates the search for compounds that close stomata, and methods of selecting inhibitors that are highly specific on guard cells have also been discussed by Waggoner and Zelitch. The most reliable inhibitor used thus far is phenylmercuric acetate. Zelitch and Waggoner first used this substance on excised tobacco leaves in experiments designed to test whether stomatal diffusive resistance was a greater part of the total diffusive pathway for transpiration than for net photosynthesis, as was predicted by Fick's law of diffusion. Several properties make phenylmercuric acetate useful in model studies. It closes stomata when sprayed on leaves at low concentrations ($10^{-4}$M); it affects only the stomata in the area where it is sprayed, hence it is not likely to be translocated and to affect cells other than guard cells ; and one treatment inhibits stomatal opening for at least 14 days without apparent toxicity in many species. The poor translocation of this inhibitor was confirmed in studies on transpiration in *Vicia faba* leaves.

Scientists found that closure of stomata in excised leaves resulted in a greater inhibition of transpiration than of photosynthesis, indicating that the inhibitor affected primarily stomatal diffusive resistance. Also when sprayed on intact tobacco plants in the greenhouse, phenylmercuric acetate diminished transpiration from 16 per cent to 43 per cent for 16 days, depending on the concentration of the inhibitor used, and the effect on growth was small. These results are consistent with the considerable evidence that, except in still air, diffusion of water vapor and $CO_2$ through stomatal affects transpiration and photosynthesis at all stomats widths in leaves as well as in artificial pores whose dimensions simulate those of natural stomatal cavities. Hence experiments with phenylmercuric acetate indicate that transpiration and $CO_2$ uptake must be negligible through the cuticle in comparison with stomatal diffusion.

Botalish also found that, when sprayed on maize leaves, phenylmercuric acetate closed stomata and more adversely affected transpiration than photosynthesis in a chamber. A single treatment on sunflower plants outdoors closed stomata for 37 days, transpiration was decreased by 14 per cent, and growth was essentially unaffected. Slatyer and Bierhuizen showed that dipping the foliage of cotton plants in a greenhouse into $10^{-4}$ to $10^{-5}$ M phenylmercuric acetate closed stomata and inhibited transpiration proportionately more than photosynthesis.

Waggoner, Monteith and Szeicz first demonstrated the role of stomatal diffusive resistance under field conditions with barley plants that were sprayed with the monomethyl ester of nonenylsuccinic acid ($10^{-3}$ M). A reduction in transpiration of 20 per cent to 30 per cent was observed that persisted for several days, only as long as the stomata were closed by the inhibitor. Fulton was able to close stomata in potato leaves growing outdoors by spraying them with $10^{-3}$ M monomethyl ester of decenylsuccinic acid. He claimed no effect on transpiration. However, a comparison of the stomatal widths in treated and control leaves with their respective rates of evapotranspiration shows that water loss was inhibited when the stomatal differences were greatest.

Granger and Edgerton observed that spraying phenylmercuric acetate on apple leaves closed stomata, but there was variable damage to the leaves. Toxicity was also found by Keller, who sprayed this inhibitor an spruce foliage outdoors at $10^{-4}$ M. Transpiration of 30 per cent less for 30 days, but there was a greater inhibition of photosynthesis, suggesting that mesophyll cells were also adversely affected in this tissue in addition to the guard cells.

Davenport sprayed $10^{-3}$ M decenylsuccinic acid or phenylmercuric acetate on creeping red fescue (*Festuca rubra*) growing in a lysimeter in the field, and found transpiration was inhibited 5 per cent and 10 per cent respectively without measurable affect on growth of the grass. Finally, the potential importance of controlling stomata on increasing stream flow in watersheds was clearly shown by Waggoner and Bravdo in one summer season, and this was confirmed the following season by Turner and Waggoner. They sprayed the foliage of 16 m tall red pine trees in a forest with $10^{-3}$ M phenylmercuric acetate at the beginning of the summer, observed increased stomatal diffusive resistance in the treated trees during daylight hours, and found a significantly higher soil moisture content around the treated trees at the end of the summer.

## GENETIC CONTROL

An example of a drought-resistant grass has been cited in which the stomata failed to open even in a wet environment. Brown and Pratt observed inactive stomata in many of the 69 species of grasses examined, although it is uncertain whether this was a genetic or an environmental effect. Waggoner and Simmonds studied a potato mutant ("Droopy") in which the stomata remained open even when the leaves wilted. When sprayed with $10^{-3}$ M phenylmercuric acetate, however, the stomata closed, and transpiration was inhibited to the same level as in normal leaves treated in a similar manner. Tal likewise described three mutants of tomato in which excessive wilting occurred because transpiration was greater and stomata remained open. With these plants, $10^{-4}$ M phenylmercuric acetate closed stomata very poorly in comparison with normal tomato leaf disks.

Zelitch and Day noted that stomata of a yellow mutant of tobacco, first described by Burk and Menser, open in the light at about twice the rate of wild type tobacco. Thus plants are known in which both the rate and extent of stomatal opening are genetically controlled, although no biochemical explanations for these interesting aberrations are yet known.

## IONS AND A POTASSIUM ION "PUMP" IN LIGHT OPENING

Several mineral deficiencies have recently been shown to retard stomatal opening, presumably because of adverse effects on photosynthesis and phosphorylation. Shimshi showed that withholding nitrogen and iron caused reduced stomatal apertures in several species, and these inhibitions were well correlated with a decrease in chlorophyll concentration. Wallace and Frolich found poor stomatal opening in phosphorous-deficient tobacco leaves, and Buczek observed decreased transpiration in nitrogen-deficient tomato plants. Peaslee and Moss showed a more specific effect of potassium deficiency on stomata in maize leaves ; normal leaves opened about 6.5 $\mu$, while the widths in deficient stomata were less than 1 $\mu$. Moreover, the resulting lower net photosynthesis was similar to that found when normal stomata were closed with a spray of phenylmercuric acetate. Hence it appeared that guard cells could not compete very well with mesophyll cells in obtaining potassium ions.

Fujino, with use of cytochemical methods, made the important observation that guard cells of several species contain large concentrations of potassium when they are open in the light and relatively small quantities when they are closed in the dark. In epidermal strips of *Commelina communis*, which were floated on phosphate buffer at pH 5.6 containing $5 \times 10^{-3}$ M potassium chloride, stomata opened 7 μ in 120 min in light, and the potassium content of the guard cells increased. When he added $10^{-2}$ M ATP to the medium, the stomata opened 14 μ in the same time, and a still greater quantity of potassium accumulated in the guard cells. The possibility that the increase in potassium ions in the guard cells is related to the functioning of an ATP-operated pump that increased the osmotic value was suggested by these experiments.

Fischer independently described experiments on the uptake of potassium ions (or $^{86}Rb^+$ as a substitute) in epidermal strips of *Vicia faba* floated on a medium containing buffer and $10^{-2}$ M potassium chloride. Although this epidermal tissue was a good choice since the stomatal opening was stimulated in light, it was also an unfortunate system because stomata in this tissue atypically open to some extent in the dark, and this caused high blank values for potassium uptake. Opening in the light in $CO_2$-free air in this system was accompanied by a decrease in solute potential (increase in solute concentration), and maximal stomatal opening in the dark required a ten-fold higher potassium concentration in the medium. From measurement of the $^{86}Rb^+$ uptake in the light, on the assumption that it took place in the guard cells, there would have been an increase in potassium concentration of 0.3 M, enough to account for the opening observed.

In these interesting experiments with *Vicia faba* epidermal strips, an atmosphere of $CO_2$-free air in the light was compared with one of normal air in the dark, although it appeared that $CO_2$-free air offered very little advantage. More recently, these experiments have been extended by Humble and Hsiao, who found that rubidium and potassium ions are much more effective in light opening than lithium, sodium or calcium ions, and preliminary evidence was obtained that potassium ions move out of the guard cells when the stomata close.

Pallaghy measured the potential between the guard cells and the external medium in epidermal strips of tobacco in the light in open stomata. When potassium chloride was in the medium, the greatest change in the electrochemical potential was observed indicating that potassium ions were accumulating. However, the stomata in this system did not close well in the dark or in the presence of high concentrations of $CO_2$, hence this is a limitation in work with this tissue.

Thus, in spite of the absence of as precise quantitative data as one would wish for, there is good evidence that stomatal opening in the light is mediated by an inward potassium ion pump that concentrates this ion in guard cells. If this hypothesis holds true, then stomatal opening in the light would require only small amounts of photosynthetic $CO_2$ fixation in the guard cells—just enough perhaps to maintain a catalytic concentration of glycolate so that its metabolism could stimulate noncyclic photophosphorylation and thus provide the energy of ATP for the pump. The potassium ions might come from the adjoining epidermal cells, or perheps from the chloroplasts in the mesophyll, or even to some extent from the chloroplasts in the guard cells themselves. It is also likely that carbon compounds produced in photosynthesis would contribute to the increase in osmotic value in the guard cells, especially at the later stages of stomatal opening, since there are clear differences with respect to the requirement for oxygen in the atmosphere between the initiation and the maintenance of stomatal opening in the light.

# 3

# Biological Nitrogen Fixation

Only bacteria and blue-green algae have evolved the capacity to fix nitrogen, a process governed by a set of genes called the nitrogen fixation (Nif) genes. The Nif genes have been defined in detail in the free-living bacterium *Klebsiella pneumoniae*.

Free-living bacteria such as *Klebsiella* couple the $NH_4^+$ produced by nitrogenase to cellular biosynthesis and do not export any detectable $NH_4^+$ into the growth medium ; nitrogenase is not produced by the organism if combined nitrogen (*e.g.* $NH_4^+$ or $NO_3^-$) is present in the environment. In contrast, the nitrogen fixation (Nif) genes of symbiotic bacteria such as *Rhizobium* spp. forming root nodules on leguminous plants are normally found to be "derepressed" (producing nitrogenase activity even in the presence of $NH_4^+$) leading to the synthesis and export of fixed nitrogen into the outside environment (cytoplasm of the host plant cell, in the case of *Rhizobium* spp.). Thus the derepression of the Nif genes resulting in the production of enormous quantities of fixed nitrogen from atmospheric nitrogen plays a crucial role in man's food chain on a global scale.

There is now considerable evidence that genetic as well as biochemical mechanisms may function to integrate production of fixed nitrogen with utilization for celluler needs. The purpose of this review is to discuss current concepts of the regulation of nitrogen fixation in both free-living and symbiotic bacteria.

## DEREPRESSION OF THE NIF GENES

Recent advances in the control and expression of Nif genes in *K. pneumoniae* will be discussed in this section with emphasis on nitrogenase-derepressed mutant strains which export large quantities of fixed nitrogen.

Ammonia produced by nitrogenase is assimilated by the bacterium to the level of glutamate by the enzymes glutamine synthetase and glutamate synthase. Biosynthesis of nitrogenase was found to be repressed by $NH_4^+$ or amino acids. It was difficult to distinguish the repressive effect of amino acids from that of $NH_4^+$ produced by the cell (by internal metabolism). This problem was eliminated by the use of mutant strains which produced nitrogenase even in the presence of $NH_4^+$ (nitrogenase-derepressed mutants). Studies with these strains showed that the repressive nature of $NH_4^+$ was connected to the cellular ability to convert the $NH_4^+$ to the level of glutamine and glutamate. In essence, these experiments show that nitrogenase was produced under conditions in which $NH_4^+$ was not assimilated into amino acids, and nitrogenase production was repressed whenever $NH_4^+$ was assimilated to the level of amino acids.

Three classes of nitrogenase-derepressed mutant strains were isolated and described in *K. pneumoniae*, starting with a mutant strain which lacked glutamate synthase activity ($Asm^-$) as the parent (Table 3.1). Biochemical analysis of these strains showed that one class of mutants produced glutamine synthetase activity but not glutamate dehydrogenase activity (*e.g.* strains SK-24, SK-28, and SK-29). The other class of mutants produced no detectable glutamine synthetase activity (*gln* $A^-$). The second class was further divided into two groups : 1. strains that produce catalytically inactive glutamine synthetase protein (*e.g.* strains SK-27, SK-55, and SK-56) ; 2. strains with no detectable glutamine synthetase protein

**Table 3.1. Properties of nitrogen fixation (Nif) derepressed mutants of *K. pneumoniae**.**

| *Strain* | *Nitrogenase (μmoles/hr per mg protein)* $-NH_4^+$ | $+NH_4^+$ | *Glutamine synthetase* (*nanomoles/min per mg protein*) $-NH_4^+$ | $+NH_4^+$ | *Glutamate dehydrogenase* (*nanomoles/min per mg protein*) $-NH_4^+$ | $+NH_4^+$ | $NH_4^+$ *excreted (μmoles/mg protein)* |
|---|---|---|---|---|---|---|---|
| M5A1 | 4.04 | 0.00 | 878 | 229 | <5 | 126 | — |
| Asm-1 | 3.14 | 0.00 | 653 | 147 | 18 | 116 | — |
| SK-24 | 2.80 | 2.42 | 684 | 866 | <5 | <5 | 92 |
| SK-28 | 3.27 | 3 05 | 1006 | 1055 | <5 | <5 | 280 |
| SK-29 | 2.82 | 2.01 | 739 | 627 | <5 | <5 | 220 |
| SK-25 | 4.47 | 3.70 | 0 | 0 | 12 | 16 | 450 |
| SK-26 | 3.82 | 2.91 | 0 | 0 | 17 | 18 | — |
| SK-27 | 2.88 | 1.81 | 0 | 0 | 17 | 15 | 270 |
| SK-55 | 2.53 | 1.92 | 6 | 0 | 20 | 20 | 230 |
| SK-56 | 3.19 | 2.29 | 14 | 21 | 22 | 24 | 190 |

* rom ( ) and (50).

determined immunologically (*e.g.* strains SK-25 and SK-26). The mutant strains of all three classes excreted large quantities of $NH_4^+$ from fixed $N_2$. Some of the strains excreted $NH_4^+$ for more than one week while being in a nondividing state.

A class of glutamine requiring mutants (isolated from an Asm + strain of *K. pneumoniae*) was described by Streicher *et al.* These mutants, in contrast to the other *gln*⁻ strains described before (for example, strain SK-27) failed to produce any detectable nitrogenase activity under conditions in which the parental strains as well as strain SK-27 produce nitrogenase activity. On further analysis these strains were found to produce nitrogenase activity under conditions in which the glutamine concentration in the medium (a necessary growth supplement) was depleted to very low levels.

It has been found recently that certain combinations of amino acids, such as glutamine and aspartate, completely repressed nitrogenase biosynthesis in strains derepressed for nitrogenase biosynthesis in the presence of $NH_4^-$. Also $NH_4^+$ in combination with high concentrations of glutamine completely repressed the biosynthesis of nitrogenase in the glutamine requiring strains. These strains assimilated $NH_4^+$ in the presence of high concentrations of glutamine (1 mg/ml). A strong correlation has been established between the repression of nitrogenase biosynthesis by $NH_4^+$ and subsequent assimilation of $NH_4^+$, presumably to the level of amino acids. The simplest interpretation of these results is that the overall mechanism of regulation of nitrogenase biosynthesis involves the conversion of $NH_4^+$ to the level of amino acids.

The type of control system described above for nitrogenase biosynthesis in *Klebsiella* appears to operate in other free-living $N_2$-fixing organisms as well. This is borne out by the fact that nitrogenase biosynthesis is derepressed in the presence of $NH_4^+$ in a culture medium which contains L-methionine-D, L-sulfoximine (MSX), an inhibitor of glutamine synthetase (Table 3.2).

MSX, a structural analog for glutamate (Fig. 3.1), was first used by Gordon and Brill to study the regulation of nitrogenase production in both *K. pneumoniae* and *Azotobacter vinelandii*. These workers observed that the nitrogenase was derepressed in these organisms in the presence of MSX, and the $NH_4^+$ produced from $N_2$ by *A. vinelandii* under these conditions was excreted into the medium. Similar experiments were carried out by various investigators using a variety of organisms (*Rhodospirillum rubrum*, *Anabaena cylindrica*, *Spirillum lipoferum*, Table 3.2). In the presence of MSX, both *R. rubrum* and *A. cylindrica* failed to assimilate $NH_4^+$ produced by nitrogenase and instead excreted it into the medium. In both these organisms, MSX has been shown to inhibit glutamine synthetase activity. Since glutamine synthetase is the first enzyme in the conversion of $NH_4^+$ to the level of glutamate and further into cellular material, it is reasonable to assume that the depression of nitrogenase production is related to the inhibition of $NH_4^+$ assimilation by MSX.

Based on preliminary experiments, it has been suggested that glutamine synthetase acts as a controlling element in the transcription of the Nif genes. However, in view of the results discussed above, it seems important to determine the nature of other pleiotrophic effects of these regulatory mutations, such as the transport of various metabolites into the cell, the rate of assimilation of $NH_4^+$, etc. Additional information is therefore required before the component(s) of the cell which controls the production of nitrogenase can be identified.

## ENERGY COST OF $N_2$ FIXATION

Experiments carried out in a number of laboratories have established that biological nitrogen fixation requires a large input of metabolic energy. Evidence has been presented that the supply of energy may often be a rate-limiting step in symbiotic $N_2$ fixation. Thus, indirectly, energy supply may play a crucial role in controlling $N_2$ fixation. Therefore, a complete picture of nitrogenase regulation will have to include the control of supplies of ATP and reductant, essential substrates for the nitrogenase reaction.

### Nitrogenase Catalyzed $H_2$ Evolution

All nitrogenase preparations isolated so far have been found to catalyze an ATP dependent $H_2$ evolution, even in the presence of $N_2$. Evidence has been presented indicating that this reaction also occurs in vivo in various nitrogen fixing organisms. Schubert and Evans reported that loss of energy as ATP and reducing power, in soybean and other legume root nodules, attributed to nitrogenase catalyzed $H_2$ evolution may be as high as 40-60% of the total energy flow through nitrogenase. Quantitative in vivo measurements are, however, complicated by the presence, in most nitrogen fixing organisms, of other hydrogenase systems which mediate evolution or uptake of $H_2$.

Mutant strains of *K. pneumoniae* which were blocked in the conventional hydrogen producing system have been isolated as chlorate-resistant derivatives of the nitrogenase derepressed mutant strains SK-24 and SK-25. It was determined that the only source of $H_2$ evolution in these chlorate-resistant mutant strains was via nitrogenase. Therefore, these strains provided a convenient tool for determination of nitrogenase-catalyzed $H_2$ evolution *in vivo*. The results indicated that nitrogenase

**Table 3.2** Relationship between levels of $NH_4^+$ assimilation enzymes and nitrogenase repression or derepression (by $NH_4^+$) in various nitrogen-fixing organisms.

| Organism | Repression by $NH_4^+$ | Glutamine synthetase | Glutamate synthase NADH | Glutamate synthase NADPH | Glutamate dehydrogenase NADH | Glutamate dehydrogenase NADPH | Derepression of $N_2$-ase | Ref. |
|---|---|---|---|---|---|---|---|---|
| *Klebsiella pneumoniae* | 100% | 26% | UD | 58 | UD | 126 | − | 50 |
| *Klebsiella* mutants | 0 | 100% | UD | <5 | UD | <5 | + | 50 |
| *Klebsiella pneumoniae* in presence of MSX | 0 | ND | ND | ND | ND | ND | + | 20 |
| *Azotobacter vinelandii* in presence of MSX | 0 | ND | ND | ND | ND | ND | + | 20 |
| *Spirillum lipoferum* | 100% | 13% | <5 | 52 | 27 | <5 | − | 38 |
| *Spirillum lipoferum* in presence of MSX | 0 | ND | ND | ND | ND | ND | + | 38 |
| Blue-green algae in presence of MSX | 0 | 9% | ND | ND | ND | ND | + | 55 |
| Symbiotic blue-green algae | 50% | 6% | ND | ND | ND | <5 | + | 56 |
| *Rhodospirillum rubrum* in presence of MSX | 0 | 5% | ND | ND | ND | ND | + | 65 |

| | | | | | | | | |
|---|---|---|---|---|---|---|---|---|
| ***Rhizobium trifolii*** | | | | | | | | |
| (free-living. aerobic) | | | | | | | | |
| *Glutamate | | 100% | 18 | 8 | 18 | <5 | | 35,36 |
| *Leucine | | 64% | 6 | 20 | 6 | ND | | 51 |
| ***Rhizobium trifolii*** | | | | | | | | |
| (free-living $N_2$-ase | | | | | | | | |
| inducing) | 50-70% | 13% | <5 | <5 | <5 | ND | + | 37 |
| Bacteroids-*R. japonicum* | ND | 14% | 55 | <5 | 19 | <5 | + | 13 |
| *R. leguminosarum-* | | | | | | | | |
| *Vicia-faba* | ND | 9% | <5 | 12 | 15 | <5 | + | 13 |
| *R. leguminosarum-* | | | | | | | | |
| *Pisum sativum* | ND | 1% | <5 | <5 | 33 | <5 | + | 13 |

*in vivo*, even under optimum conditions (25°C, pH 7.3), catalyzed the production of about 1.3 moles $H_2$ per mole $N_2$ reduced. It seems probable that nitrogenase-catalyzed $H_2$ evolution is an important factor contributing to the high energy requirement for $N_2$ fixation. The data ($H_2/NH_4^+$ molar

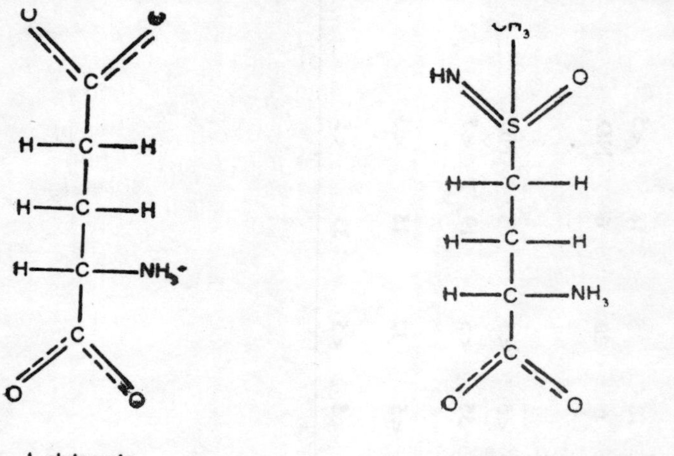


Fig. 3.1. Structure of L-glutamate and its structural analog L-methionine-D, L-sulfoximine (MSX).

ratio of 0.65) suggests that almost one-third of the energy flow through nitrogenase is lost as $H_2$ evolved. Some of the energy lost through nitrogenase-catalyzed $H_2$ evolution may be recovered by recycling $H_2$ through uptake hydrogenase. It has been estimated that 2-3 ATR may be obtained per $H_2$ oxidized in hydrogen-bacteria.

## Apparent ATP Requirement

Reduction of $N_2$ or other substrates by nitrogenase requires ATP in addition to a suitable reductant. Most recent in vitro studies with nitrogenase from various organisms have reported a minimal requirement of 12-15 moles ATP per mole $N_2$ reduced, but the values vary considerably with the experimental conditions. Estimates of the in vivo energy requirement for various organisms vary from 4-5 to 29 ATP/$N_2$ (Table 3.3).

**Table 3.3. In vivo ATP requirement for $N_2$ fixation reported for various organisms.**

| *Organism* | *Method of estimation* | *ATP/$N_2$* | *Reference* |
|---|---|---|---|
| *Azotobacter chroococcum* | By comparing growth yields on $N_2$ and $NH_4^+$ | 4-5 | 17 |
| *Clostridium pasteurianum* | By comparing growth yields on $N_2$ and $NH_4^+$ | 20 | 16 |
| *Klebsiella pneumoniae* | By comparing growth yields on $N_2$ and $NH_4^+$ | 29 | 24 |
| *Klebsiella pneumoniae* | From $NH_4^+$ excretion by Nif derepressed mutant strains | 14-16 | 1 |

These estimates are based on comparisons of growth yields with $N_2$ or $NH_4^+$ as N source.

The low value reported for *Azotobacter chroococcum* may in part be caused by recycing of the $H_2$ evolved through nitrogenase sincc $H_2$ can support oxidative phosphorylation in *A. chroococcum*. The estimated value of 4-5 ATP/$N_2$ also involved a rather large extrapolation to eliminate respiration loss attributed to the respiratory protection" of nitrogenase. The ure of nitrogenase-derepressed mutant strains of *K. pneumoniae* allowed the products $NH_4^+$ and $H_2$ to be determined directly.

Nongrowing cells of the derepressed mutant strains consumed 7.5-8.5 glucose per $N_2$ fixed under optimum conditions. The ATP requirement for $N_2$ fixation may be estimated when the fermentation pathways are known. *Klebsiella pneumoniae* ferments glucose through the Embden-Meyerhof-Parnas pathway to pyruvate which yields 2 ATP per glucose. Additional ATP can be obtained through the clastic cleavage of pyruvate. Acetate production may be used as a measure of this additional ATP. An apparent ATP requirement of 21-25 ATP/$N_2$ was obtained for five different *K. pneumoniae* mutant strains under optimum conditions (25°C, pH 7.3). These values were calculated from the total glucose consumption rates with no correction for basal levels of glucose catabolism. These calculations, therefore, also include the cells' "maintenance" energy requirement. The apparent total ATP requirement increased when the pH was lowered below 6.0 or when the temperature was increased above 25°C, and also after long incubation periods.

An estimate of the "maintenance" energy requirement may be obtained from the rate of glucose consumption by Nif$^-$ minus mutant strains. This correction gave an apparent ATP requirement for the nitrogenase reaction of 14-16 ATP/$N_2$. This is considerably lower than the value of 29 ATP/$N_2$ estimated for growing cells of *K. pneumoniae* by comparing growth yields with $N_2$ of $NH_4^+$ as N source (Table 3.3).

If the observed nitrogenase-catalyzed $H_2$ evolution was taken into consideration (a $H_2/NH_4^+$ molar ratio of 0.65), then an apparent ATP/2e value of about 4 was calculated. This value is in good agreement with most recent values reported for purified nitrogenase preparations.

The reduction of $N_2$ to $NH_4^+$ also requires energy in the form of low potential electrons. The observed minimal $H_2/NH_4^+$ ratio of 0.65 implies that at least 4.3 electron pairs were consumed per $N_2$ reduced. Low potential electrons and ATP are partially interconvertible energy forms in respiring organisms, and 4.3 low potential electron pairs could have given perhaps 13 ATP if used in oxidative phosphorylation (assuming an ATP/2e value of 3). This implies that the total in vivo energy cost for $N_2$ fixation in terms of ATP equivalents may be close to 30 ATP/$N_2$ for the nitrogenase reaction.

When considering the energy cost of $N_2$ fixation by any biological system, it is essential to take into account the cells' basal metabolic rate ("maintenance" energy requirement). In order to maintain a high level of nitrogenase activity over a longer period of time, a certain amount of protein turnover in the cell appears to be essential. Inhibition of cellular protein synthesis by chloramphenicol leads to a decay of nitrogenase activity. Taking this into consideration, the minimal energy requirement for $N_2$ fixation may be as high as 35—40 ATP equivalents per $N_2$ reduced.

The contribution of the "maintenance" energy requirement in the overall energy cost in probably increasing with decreasing rates of $N_2$ fixation. It may also vary from organism to organism. Substantial amounts of energy may be lost in aerobic $N_2$-fixing bacteria like *Azotobacter* in "respiratory protection" of nitrogenase.

In conclusion *in vivo* experiments indicate that $N_2$ fixation requires a very large input of metabolic energy. The evolution of very sophisticated genetic control systems for nitrogen fixation may be understood in this context. Because of the high energy requirement, the organisms fixing $N_2$ when other N sources are available for growth would be expected to be at a disadvantage.

## SYMBIOTIC NITROGEN FIXATION

Based on the discovery that certain species of *Rhizobium* are capable of producing nitrogenase in a free-living state, it is clear that the root nodule bacteria (*Rhizobium* spp.) harbor the *nif* genes. The levels of nitrogenase activity produced by free-living *Rhizobium* sp. have been reported to approximate that of isolated bacteroids. However, it is important to note that none of these organisms are capable of growing in the free-living state at the expense of fixed $N_2$ as sole N source. Using $^{15}N_2$, more than 94% of fixed $N_2$ was found to be exported by free-living *R. japonicum*. Similar results were reported for isolated bacteroids. The $N_2$ fixation system of free-living *Rhizobium* spp. presumably consists of nitrogenase (both the Fe-Mo protein and Fe protein subunits) as well as a variety of enzymes for supplying both ATP and reductant (including reduced pyridine nucleotide-ferredoxin reductase, ferredoxins, flavodoxin, and perhaps other electron coupling factors.) A major shift in the pattern of synthesis of cytochromes and other components of the electron transport chain may also occur, leading to the formation of a new respiratory chain that is functional under low $O_2$ concentrations. Differential gene expression may account for these marked shifts in metabolism. In other words, the commitment to $N_2$ fixation by the free-living *Rhizobium* sp. not only involves the expression of the nitrogenase structural genes, but also a whole series of other supporting pathways and enzymes. It has been observed that the hydrogenase system which takes up hydrogen from the environment, as well as dissimilatory nitrate reductase, are simultaneously induced in $N_2$-fixing cultures of free-living *Rhizobium* sp. These enzyme systems are capable of contributing energy to $N_2$ fixation. Some of these data suggest that $N_2$-fixing cultures of free-living rhizobia may resemble the bacteroids in the root nodule. Thus it is of considerable interest to investigate the gene products which are produced along (or only) with nitrogenase (as well as their control) in free-living *Rhizobium* spp.

In this section, a key set of enzymes which appear to be switched off during $N_2$ fixation in *Rhizobium* spp. are described, the $NH_4^+$ assimilation enzymes (glutamine synthetase and glutamate synthase), the enzymes responsible for converting the $NH_4^+$ derived from $N_2$ to the level of amino acids.

Using $^{15}N$ analysis, it has been found that free-living rhizobia excrete a majority of their fixed nitrogen, presumably as $NH_4^+$. For example, as much as 94% of the $^{15}N_2$ fixed by free-living *R. japonicum* was recovered as $NH_4^+$ from the cell supernatant (Table 3.4). As summarized in Table 3.4 the $^{15}N$-enrichment in the $NH_4^+$ fraction recovered from the cell supernatant was 65% to 80% higher for cowpea strain 32H1 and 94% higher for *R. japonicum* strain CB1809 than the enrichment detected in the $NH_4^+$ fraction obtained from the total cellular nitrogen. The demonstration of the relatively higher $^{15}N$ enrichment in the cell supernatants indicates that

**Table 3.4.** Export of fixed nitrogen by free-living *Rhizobium* spp.

| *Rhizobium species* | *Experiment* | *Atoms % $^{15}N$ excess** *Cell nitrogen* | *Cell supernatant fraction* | *nmoles N fixed/mg protein per hr* | *%$^{15}N$ in supernatant fraction* |
|---|---|---|---|---|---|
| "cowpea" | I | 0.150 | 0.423 | 0.17 | 65 |
| strain 32H1 | II | 0.040 | 0.202 | 0.09 | 80 |
| *R. japonicum* | | 0.020 | 0.341 | 0.09 | 94 |

*Referred to control cell and supernatant fractions grown in the presence of unlabeled $N_2$.

a major portion of the fixed nitrogen was exported by these cultures.

This corroborates well with a number of studies carried out with isolated bacteroids in which the activities of $NH_4^+$-assimilatory enzymes leading to biosynthesis of the glutamate-glutamine family of amino acids were found to be extremely low (13, 31, 43). Bergersen and Turner reported earlier that $^{15}N_2$ was fixed and released as $^{15}NH_4^+$ using isolated bacteroids.

The mechanism of modulation of the $NH_4^+$ assimilation enzymes of *Rhizobium* is of considerable interest because the levels of these enzymes may in large part determine the fate of fixed $N_2$, permitting the bulk to flow out of the bacteroids to the plant. In other words, the partitioning of fixed $N_2$ between the symbiont and its host may be dictated by the activities of these enzymes.

There is indirect evidence that glutamate may play a role in the regulation of the $NH_4^+$-assimilatory enzymes leading to export of fixed nitrogen as $NH_4^+$. For example, when *R. trifolii* (free-living) was presented simultaneously with both glutamate and $NH_4^+$, it showed a marked preference for glutamate as nitrogen source. The culture failed to utilize any detectable $NH_4^+$ as long as glutamate was present in the medium. In contrast, when glutamate in the medium was replaced by L-asparate or L-leucine, *R. trifolii* utilized $NH_4^+$ for cellular growth. Glutamate synthase activity in the cells was also found to be low when the culture medium contained L-glutamate. The presence of either asparate or leucine in the medium (replacing glutamate) allowed the organism to induce higher levels of glutamate synthase activity. The presence or absence of glutamate synthase corresponded well with the $NH_4^+$ utilization pattern of the cell. $NH_4^+$ also repressed the levels of glutamine synthetase activity. It is possible that the block in the conversion of $NH_4^+$ to glutamate (due to low levels of glutamate synthase activity) is responsible for the derepression of nitrogenase, even in the presence of $NH_4^+$.

## A UNIFIED CONCEPT OF SYMBIOTIC NITROGEN FIXATION

Only three types of $N_2$-fixing bacteria have evolved beneficial symbiotic associations with higher plants—*Rhizobium* spp. (root nodule bacteria) of leguminous plants, *Frankia* (an unidentified, actinomycete-like soil organism) forming root nodules on *Alnus*, *Purshia*, etc., and heterocystous blue-green algae, the most promiscuous of the symbionts, with a variety of suitable hosts. Since the rhizobial symbiosis has been studied in greater detail, it will be the center of our attention in this section. The focus will be on the requirements for symbiotic $N_2$ fixation which may be shared by all forms of symbiotic $N_2$-fixing microorganisms (Figure 3.2).

The sole purpose of the symbiont in the symbiotic $N_2$ fixation is the production of $NH_4^+$ for the host. It is of interest to note in this connection that the bacteroids in the nodule may resemble the energy-producing organelles of the plant cell, the mitochondria. The bacteria in the nodule (bacteroids) seem to lack a rigid cell wall and are osmotically labile. Bacteroids are known to metabolize various carbon compounds supplied by the plant cell. The literature is scanty regarding the major carbon compound(s) supplied by the plant cell. It is possible that more than one carbon compound might be made available by the plant. The presence or absence of specific permeases in the bacteroid membrane may control the transport of organic compounds into the bacteroid cytoplasm. Once inside the bacteroid, the carbon compounds are transformed and metabolized via an active TCA-cycle coupled to an energy-yielding respiratory chain capable of functioning under low $O_2$ concentrations. Higher $O_2$ concentrations would inactivate the $O_2$-sensitive nitrogenase activity. Leghemoglobin is believed to function as an oxygen carrier maintaining a sufficient and constant $O_2$ concentration near the bacteroid membrane.

Two additional systems, probably membrane-bound, may play an important role in energy metabolism—the $H_2$ uptake system and energy-linked nitrate reductase. Since efficiency of energy metabolism is increasingly recognized as a crucial aspect of symbiotic $N_2$ fixation, these two systems are worth describing in some detail. Most $H_2$-utilizing bacteria were found to induce a membrane-bound $H_2$ uptake system, presumably coupled to the major respiratory chain. Dixon observed that hydrogenase activity of rhizobial bacteroids, broken by sonic disruption, was readily centrifuged down and was not detectable in the supernatant. Schubert and Evans have provided evidence that the $H_2$ uptake system of nodules is a bacterial function since net $H_2$ production varied when different bacterial strains were used to inoculate the same variety of host plant.

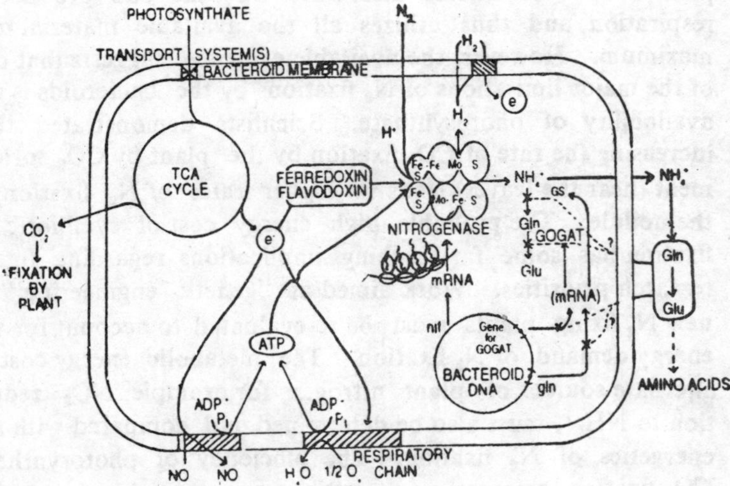


**Fig. 3.2.** A model for symbiotic $N_2$-fixation by bacteria. See text for details. GS-glutamine synthetase; GOGAT-glutamate synthase; *nif* and *gln* are the genes for nitrogenase and glutamine synthetase respectively.

Since nitrogenase produces certain amounts of $H_2$ during the reduction of $N_2$, the strains which are capable of utilizing $H_2$ as an energy source would evolve very little free $H_2$. On the other hand, plants inoculated with bacterial strains incapable of utilizing $H_2$ would evolve $H_2$. Based on the measurements of $H_2$ and $C_2H_4$ produced by nodules derived from a given strain of *Rhizobium* in association with the plant, Schubert and Evans developed an index called "relative energy efficiency" to measure the efficiency of energy used in the nodule. The question of whether the $H_2$ uptake systems of the bacteroids contribute significantly to the "relative energy efficiency" can only be answered by the use of isogenic strains of *Rhizobium* sp. which lack the $H_2$ utilization reactions. Bacteroids seem to have evolved additional energy-yielding reactions. For example, it has been reported that the energy production coupled to anaerobic $NO_3^-$ reduction can support $N_2$ fixation in isolated bacteroids. The contribution of this pathway to the whole plant remains to be determined. The plant cell in the nodule also fixes $CO_2$, the end product of respiration, and thus utilizes all the available material to a maximum. However, the available evidence suggests that one of the major limitations of $N_2$ fixation by the bacteroids is the availability of photosynthate. Scienlists demonstrated that increasing the rate of $CO_2$ fixation by the plant by $CO_2$ enrichment (near the leaves) leads to higher rates of $N_2$ fixation by the nodule. The probable high energy cost of symbiotic $N_2$ fixation has some far-reaching implications regarding future research priorities. Work aimed at "genetic engineering" of new $N_2$ fixing plants must be re-evaluated to account for the energy demand of $N_2$ fixation. The metabolic energy cost of alternate sources of plant nitrogen, for example $NO_3^-$ reduction to $NH_4^+$, must also be determined and compared with the energetics of $N_2$ fixation. The efficiency of photosynthetic $CO_2$ fixation emerges as a crucial step requiring increasing attention.

During symbiotic nitrogen fixation, the genes and operons necessary for nitrogen fixation, energy producing systems, etc. are expressed. However, some bacteroid genes seem to be repressed. Nothing is known at this stage about the extent as well as identity of the various genes which are repressed under symbiotic conditions. One of the important series of reactions, the $NH_4^+$ assimilation pathway, seems to be either severely or completely repressed (inhibited ?). In isolated bacteroids, the levels of the enzymes glutamine synthetase and glutamate synthase were found to be low (Table 3.2). This finding provides a biochemical explanation for the export of fixed $N_2$ as $NH_4^+$ by the bacteroids. Research is now centered on the mechanism of regulation of these pathways. Plant enzymes rather than the bacteroid enzymes have been proposed to be physiologically important in the assimilation of the $NH_4^+$ produced by nitrogenase. During the development of the nodule, glutamine synthetase and glutamate synthase activities increased considerably in the plant fraction but not in the bacteroid fraction of the nodule. The first step in the assimilation of $NH_4^+$ (Fig. 3.2) is proposed to be the synthesis of glutamine by the plant glutamine synthetase. The amide nitrogen of glutamine is transferred to $\alpha$-ketoglutarate to produce glutamate by glutamate synthase. The build-up of glutamate (and glutamine), known to inhibit the assimilation of $NH_4^+$ in free-living *Rhizobium* sp., may function to repress the $NH_4^+$ assimilation pathways of the bacteroid.

A summary of experiments on the mechanism of nitrogenase regulation in free-living and symbiotic organisms is presented in Table 3.2. Note that in all symbiotic systems, including *Rhizobium* spp., as well as blue-green algae (lichen system), both glutamine synthetase and glutamate synthase activities are found to be low. Since *Rhizobium* spp. produce higher levels of both glutamine synthetase and glutamate synthase activities under free-living conditions, the question of how the host regulates the functioning of these genes becomes an important one. Undoubtedly the regulation of the activity and biosynthesis of both the bacteroid and the plant enzymes play a crucial role in symbiotic $N_2$ fixation. These are studies for the future.

There is rapid progress in the genetics of *Rhizobium* with systems for gene transfer and recombination recently reported for two organisms. This provides a fresh approach to studies of the molecular biology of $N_2$ fixation by *Rhizobium*, including the key symbiotic genes and their regulation and function.

# 4

# Nitrate Reduction in Higher Plants

Nitrogen constitutes approximately 2 per cent and carbon approximately 40 per cent of the dry weight of plants. On a world-wide scale it is estimated that 200 billion tons of carbon are fixed annually by photosynthetic processes which on a basis of approximate analysis would require the incorporation of 10 billion tons of nitrogen. Apart from those species that have a symbiotic association with nitrogen fixing bacteria, the bulk of the plant nitrogen arises from the reduction of nitrate taken up from the soil. It is of interest to note that 6 million tons of nitrogen, in various fertilizer forms, are applied annually in the United States alone. Since the reduction of nitrate to the usable (ammoniacal) form requires the donation of 8 electrons, while carbon requires but 4 electrons, the reduction of nitrate to ammonia utilizes a massive amount of solar energy.

The sequence of reviews in the *Annual Review* series chronicles the development of knowledge of the general aspects of nitrogen metabolism in plants. In addition, a large number of reviews are more directly concerned with the metabolism of inorganic nitrogen.

The current review is directed primarily on the developments since 1962, and confined principally to the literature relating to higher plants.

## ENZYMATIC CONVERSION OF NITRATE TO NITRITE

*General characteristics.* Although the enzymatic reduction of nitrate to nitrite has been inferred for over half a century and the existence of nitrate reductase conclusively established for 15 years the enzyme has not, as yet, been fully purified or characterized from higher plants. From studies with partially purified preparations, the following characteristics have been reported : (*a*) a molecular weight of 500,000 to 600,000 ; (*b*) the enzyme contains molybdenum $Fe^{++}$ and FAD ; (*c*) Michaelis constants ($K_m$) for $NO_3^-$, $FMNH_2$, and NADH are in the order of $2\times10^{-4}$ M 3 to $6\times10^{-4}$ M and $2.5\times10^{-4}$ M ; (*d*) a pH optimum of 7.5 when NADH is the electron donor and 6.0 to 6.25 when NADPH is the electron donor ; (*e*) maximum activities of 700 and 600 m$\mu$ moles $NO_2^-$ produced $min^{-1}$, mg $protein^{-1}$ for the enzyme from corn and spinach leaves respectively ; (*f*) the enzyme is adaptive in that the activity is inducible by nitrate and by molybdenum ; and (*g*) the enzyme is unstable both *in vitro* and *in vivo* [the enzyme exhibits a half-life of 4 hr in detached corn leaves at 35° and a diurnal variation].

*Electron donors.* There is a preponderance of evidence that nitrate reductase from leaves of most plant species has a specific requirement for NADH as electron donor. The earlier and commonly held belief that NADPH was the preferred cofactor had its origin in the initial work on nitrate reductase obtained from the leaves from young soybean seedlings. It is ironic that the enzyme from this tissue, when extracted in the presence of cysteine, is twice as active with NADH as with NADPH and that the leaves from mature plants lose the capability of utilizing NADPH. Recently, Sims *et al.* reported that *Lemna minor* plants exhibited a 20-fold increase in NADPH-specific nitrate reductase when sucrose was added to the culture medium. They indicate that these plants contain two enzymes, each with specific cofactor requirement, and that the primary energy source determines the cofactor specificity of the nitrate reductases.

It would appear that the enzyme from young soybeans is exceptional in that it is capable of directly utilizing either cofactor. The enzyme from young leaves of several other plant species show only traces of activity when NADPH is the added electron donor. It is not known whether the traces of activity obtained with NADPH in these instances represent direct utilization or transformation of NADPH to NADH by phosphatase or transhydrogenase.

In some instances, the addition of FMN or FAD to the nitrate reductase assay mixture stimulated activity when pyridine nucleotides were the electron donor. Furthermore, Stoy demonstrated that photoreduced riboflavin under anaerobic conditions was more effective than NADH as an electron donor for nitrate reductase from wheat. This suggests that the pyridine nucleotide may not be the primary electron donor to the enzyme.

Hageman *et al.* found that partially reduced benzyl viologen could also serve as an electron donor for nitrate reductase extracted from may row leaves. Subsequently, Del Campo *et al.* reported that illuminate broken chloroplasts were able to reduce nitrate to nitrite in the presence of benzyl viologen. Although ferredoxin or ferredoxin plus NADP would not substitute for the dye, the physiological cofactors FMN or FAD could. Since pyridine nucleotides were not required for nitrate reduction it was proposed that the well-known pyridine nucleotide dependent nitrate reductase was, in fact, two separate enzymes, a NADP reductase and nitrate reductase. An extensively purified preparation of nitrate reductase from spinach required only the addition of nitrate and reduced FMN or FAD to effect reduction. The FAD could be reduced in the light by grana or in the dark by NADPH and a NADP : reductase. Based on this evidence, it was proposed that the enzyme be reclassified as $FMNH_2$ ($FADH_2$) : nitrate oxidoreductase. However, apart from the nature of the electron donor, the properties of the enzyme described by Paneque *et al.* are similar ($K_m$'s for $FMNH_2$—$1.5 \times 10^{-5}$ M, $NO_3^-$—$1.8 \times 10^{-4}$ M, and pH optimum between 7.3 and 7.8) to the pyridine nucleotide dependent enzyme.

Several following lines of evidence suggest that there is no substantial basis for renaming the enzyme. In fact, these same investigators later found that their highly (800-to 1000-fold) purified enzyme could utilize NADH as the electron donor for nitrate reduction. No additional enzyme or cofactor was required. In this respect the spinach enzyme is comparable to that present in most other plant species. However, it was postulated that the reduction of nitrate by NADH may require a separate NAD reductase and an unknown cofactor to transfer electrons to the proposed $FMNH_2$ $(FADH_2)$ : nitrate oxidoreductase. More recently, these workers have shown the occurrence in chloroplasts of a NADH-specific diaphorase capable of supplying electrons to FMN for nitrate reduction.

Schrader *et al.* investigated the possibility that nitrate reduction in plant leaves is achieved by : (*a*) two separate enzymes each with different cofactor requirements ; (*b*) a single enzyme, possibly composed of subunits that can utilize both reduced NAD or flavins as electron donors ; or (*c*) $FMNH_2$ $(FADH_2)$ : nitrate oxidoreductase that is capable of utilizing reduced pyridine nucleotides indirectly *via* separate pyridine nucleotide reductases and flavin. Based on the observations that the $FMNH_2$ requirement for optimal activity was 40- to 100-fold higher than for NADH and that all attempts to separate the NADH and $FMNH_2$ dependent activities by purification, differential induction, and differential decay failed, it was concluded that alternative (*b*) was most likely. This opinion is also supported by the recent crystallization of a pyridine nucleotide dependent nitrate reductase from *Neurospora*.

Thus it appears that nitrate reductase is versatile in its use of electron donors. However, all these observations can be accommodated in an electron transport scheme (Fig. 4.1). A similar scheme described by Hewitt and Nicholas is based on direct evidence obtained with microbial systems ; however, it is equally applicable to higher plants.

Nitrate reductase most commonly found in plant leaves is depicted as a molybdoflavoprotein that is capable of accepting electrons directly from NADH, or $FMNH_2$ $(FADH_2)$ but not from NADPH. An alteration in binding site for the pyridine nucleotide would explain the NADPH specific enzyme reported by Sims *et al.* or for the enzyme from young soybean leaves. As stated in 1956 by Stoy the *in vivo* role of reduced flavins as electron donors is unknown.

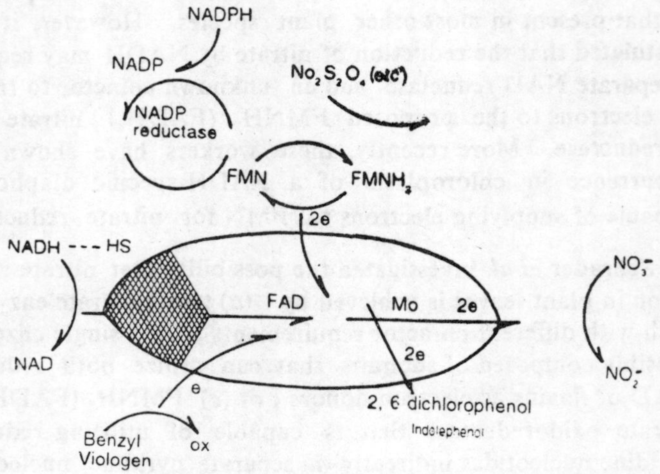


Fig. 4.1. Electron Transport Scheme.

For the NAD-specific nitrate reductase from plants the electrons appartatly can be accepted by the enzyme directly from NADH, free $FMNH_2$ ($FADH_2$), or reduced dyes such as benzyl viologen. NADPH can serve as an electron donor only indirectly *via* NADP ; reductase and FMN. It appears that the component (subunit) binding the NADH is more labile during purification and storage than that part of the enzyme binding FMNH. A sulphydryl group is implicated in the binding of NADH in that the addition of sulphydryl reagents restores activity lost during purification and alleviates the inhibitory effects of *p*-chloromercuribenzoate. For the plant enzyme no clear delineation can be given as to electron entry sites. It has been shown that when oxidized 2,6-dichlorophenolindolephenol (DCPIP) was added to the plant enzyme (standard NADH nitrate reductase assay), the dye accepted the electrons and excluded the reduction of nitrate. It does not appear that DCPIP accepts electrons from the molybdenum as DCPIP can still be reduced when nitrate reduction is blocked by cyanide. The identification of molybdenum as the metal prosthetic group in nitrate reductase has been demonstrated only in the case of soy-beans and wheat. However, physiological observations indicate that molybdenum is involved in nitrate metabolism in higher plants. Attempts to use reduced molybdenum as an electron donor for nitrate reductase from plants were unsuccessful in contrast to its successful use with the enzyme from *Neurospora*. Garrett and Nason have recently shown that the enzyme from *Neurospora* contains cytochrome *b*-55. They have tentatively positioned the cytochrome between FAD and molybdenum in the electron flow pattern, depicted in the preceding diagram.

## ENZYMATIC REDUCTION OF NITRITE

*Characteristics of nitrite reductase*. The enzyme has not been purified or extensively studied ; however, evidence is available to support the following statements : (*a*) the enzyme is probably not a flavoprotein, but receives electrons from ferredoxin when the latter has been reduced by illuminated chloroplasts or by NADPH and a diaphorase enzyme (*b*) the Michaelis constant ($K_m$) for $NO_2^-$ is between $10^{-4}$ M and $10^{-6}$ M but the cause of this wide spread of values is not known ; (*c*) the $K_m$ value for ferredoxin is 0.01 mM ; (*d*) the enzyme is located in the chloroplasts ; (*e*) it is induced by nitrite and appears to be more stable *in vivo* than nitrate reductase.

*Electron donors*. Early reports of the reduction of hydroxylamine and nitrite by extracts from plants supplied with exogenous reduced pyridine nucleotides were not conclusive and were not verified by subsequent work. Cresswell and Hewitt found that the failure to detect ammonia in assay systems that showed loss of hydroxylamine and oxidation of reduced pyridine nucleotides reported by Roussos and Nason could be explained by oxidation of the hydroxylamine. Both Cresswell and Sanderson and Cocking concluded that reduced pyridine nucleotides did not function as direct electron donors for nitrite reductase.

A new insight into nitrite metabolism was provided by the demonstration of a nitrite reductase in cell-free extracts of higher plants capable of reducing nitrite and hydroxylamine to ammonia. In the initial work, benzyl viologen reduced chemically by hydrogen or enzymatically by NADPH was used as the electron donor for the enzyme. Concurrently, Huzisige and Satoh demonstrated that nitrite was reduced by illuminated spinach chloroplasts fortified with an additional soluble protein fraction. Sanderson and Cocking identified a similar enzyme-grana system from tomatoes and suggested that the cofactor involved in nitrite reduction is present in the grana. Nason suggested that the electron carrier in the photochemical system might be the photosynthetic pyridine nucleotide reductase described by San Pietro and Lang. Following the demonstration that ferredoxin was the electron transporting component of photosynthetic pyridine nucleotide reductase it was independently demonstrated by a number of groups that ferredoxin could replace benzyl viologen as the electron carrier in nitrite metabolism and was apparently the physiological carrier for the reduction of nitrite to ammonia. Reduced ferredoxin for the enzymatic reduction of nitrite can be supplied by light and a chloroplast system or by NADPH *via* ferredoxin reductase as shown in (Fig. 4.2).

Subsequent studies demonstrated that nitrite reductase partially purified from corn, spinach, and marrow leaves cannot accept electrons directly from NADPH in the dark, but requires ferredoxin or nonphysiological dyes as electron carriers for reduction of nitrite to ammonia. In the light, ferredoxin is apparently the physiological cofactor and nitrite can be a recipient of electrons from the noncyclic photophosphorylation process. For each mole of nitrite reduced, 1 mole of ammonia, 1.5 moles of oxygen, and 3 moles of ATP are formed. Electrons, which would normally be transferred from ferredoxin to NADP *via* ferredoxin-NADP reductase, become rerouted in some circumstances to nitrite *via* the nitrite reductase enzyme. Addition of NADP to a chloroplast system which is reducing nitrite suppresses nitrite reduction in favor of the reduction of NADP. The mechanism of this adjustment is not yet clear. The initial electron donor, water, can be replaced by ascorbate-dichlorophenolindolephenol couple after inhibition of photosystem II by treatment with heat or *p*-chlorophenyldimethylurea.

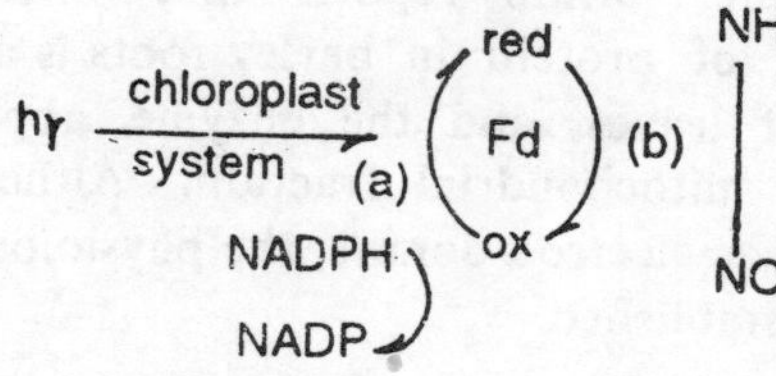


(a) Ferredoxin - NADP : reductase
(b) Nitrite reductase

Fig. 4.2. Reduction of ferredoxin.

The findings that ferredoxin, which could be reduced photochemically functions as the electron donor in nitrite reduction agrees with the physiological observation of accelerated nitrite utilization in the light. However, it has been reported that wavelengths of light other than those which promote photosynthesis function in stimulating nitrite metabolism. Further physiological evidence of the function of ferredoxin in nitrite assimilation is suggested by the experiments of Kessler and Czygan and Cresswell *et al.* that iron-deficient plants exhibit a decreased capacity to reduce nitrite. Presumably the iron deficiency would cause a reduction in the level of ferredoxin. The fact that nongreen tissues can metabolize nitrite suggests that electron donors other than ferredoxin might function in these systems, since to date the presence of ferredoxin in plants has been reported only in photosynthetic tissues. However, scientist report that a factor required for nitrite reduction in extracts from tomatoes is present in cellular particles from organs other than leaves. If ferredoxin is present in nongreen tissues, then nitrite reduction could proceed via NADPH and an appropriate diaphorase or enzyme system. Such a system is consistent with the observation of Butt and Beevers of an increased respiratory metabolism of carbohydrate via the pentose phosphate pathway in maize roots in the presence of nitrite. It was proposed that nitrite reduction utilized the

NADPH generated by the oxidation of glucose-6-phosphate, thus constituting a coupled enzyme system. However, it was not demonstrated that nitrite had been reduced. It is of interest that the addition of nitrate did not stimulate the oxidation of glucose-6-phosphate. Miflin reports that nitrite reductase activity per unit of protein in barley roots is approximately equal to that of leaves, and the enzyme appeared to be associated with a mitochondrial fraction. Although viologen dyes could serve as electron donors, the physiological reducing system was not established.

In view of experimental data showing that nitrite reduction is inhibited by DNP, it has been proposed that this step of nitrite metabolism required the utilization of high energy phosphate. However, the demonstration by Wessels that DNP could serve as a Hill oxidant, and the finding of Del Campo *et al.* that ferredoxin was able to catalyze the reduction of DNP, indicate that such arguments may not be valid. It is possible that the decreased nitrite reduction in the presence of DNP is the result of a diversion of electrons for DNP reduction rather than associated with the uncoupling action of DNP.

Ahmed and Morris have subsequently indicated that their earlier suggestion that reduction of DNP might account for its inhibitory effect on nitrite metabolism may not be valid. Furthermore, such an explanation for the mode of action of DNP in inhibiting nitrite metabolism is difficult to reconcile with the observation that arsenate, another uncoupler, also inhibits nitrite metabolism.

While the conversion of nitrite to ammonium by nitrite reductase in the presence of ferredoxin has been established, the intermediate reduction stages have not been determined. Ferredoxin is a single electron donor and thus reduction might proceed by a series of one-electron stages ; such a scheme has been proposed by Fewson and Nicholas. If such intermediates do occur, it appears unlikely that they are released in the free form. Hageman *et al.* originally pointed out that the rate of conversion of nitrite to ammonia was more rapid than the reduction of hydroxylamine to ammonia, indicating that hydroxylamine was not a free intermediate. In a detailed survey on the comparative rates of nitrite and hydroxylamine reduction, Hewitt *et al.* indicated that the nitrite reductase from marrow and spinach also reduced hydroxylamine, but the $K_m$ values for hydroxylamine exceeded those for nitrite. Nitrite severely inhibited hydroxylamine reduction, whereas hydroxylamine had a negligible effect on nitrite reduction. It was concluded that free hydroxylamine is not an intermediate in nitrite reduction in higher plants. Thus the conversion of nitrite to ammonia requires a series of single electron steps with all the reactions occurring on a single enzyme. These findings are in agreement with the investigations of Lazzarini and Atkinson and Kemp *et al.* who earlier had concluded that in *E. coli* the reduction of nitrite to ammonia was carried only by a single enzyme without the production of free intermediates.

Hewitt *et al.* showed two major hydroxylamine reductase activities in leaves of marrow (*Cucurbita pepo*) and spinach. One, which was associated with nitrite reductase activity, could accept electrons from ferredoxin. The other, which did not reduce nitrite, accepted electrons from reduced benzyl viologen but not from physiological donors. Neither of these hydroxylamine reductases seems to be obligate for the reduction of nitrite to ammonia, and their physiological role is obscure. Hydroxylamine produced endogenously or supplied in the culture solutions should be reduced and utilized by the plant. Bundel *et al.* have demonstrated that various plants have the capacity to incorporate $N^{15}$ from $N^{15}$-labeled hydroxylamine, and this process is stimulated by light. This hydroxylamine could be reduced to ammonia by one of the reductases described by Hewin *et al.* Bundel has reported that a hydroxylamine reductase can be obtained from a variety of plant tissues that utilizes NADH as the electron donor. It is difficult to reconcile this work with the other investigations reported.

## GENERATION OF ELECTRON DONORS FOR NITRATE ASSIMILATION

The current biochemical evidence shows that the enzyme involved in the reduction of nitrate requires reduced pyridine nucleotides but that the reduction of nitrate in leaves in the light involves ferredoxin. In the dark, it would appear that reduced ferredoxin can also be generated by NADHP it leaf tissue. The nature of the cofactor for nitrite reduction in nonphotosynthetic tissue has not been established.

*Nitrate reductase.* The question of the source of reduced pyridine necleotides for nitrate reduction has been an enigma. In nonphotosynthesis tissue, the observations that nitrate influences oxygen consumption and the existence of a NADH-dependent nitrate reductase suggest that respiratory metabolism supplies the electron donors for nitrate reduction. Although the commonly held view has been that light is rather closely linked to nitrate reduction in photosynthetic tissue the same respiratory metabolism could be providing the reductant in both light and dark. Moreover, the fact that leaves will reduce nitrate in the dark indicates the involvement of respiration. Currently, it is not clear whether the influence of light on nitrate metabolism resides in its effect on nitrate reductase level per se, availability of nitrate, the generation of reductant or a combination of all.

If the influence of light is mediated through the generation of reductant, the reductant could be produced directly by the photosynthetic system or derived indirectly from the carbon products. Since the chloroplast is the prime source of reduced pyridine nucleotide (primarily NADHP), and since nitrate reduction can be coupled to light via NADP and a chloroplast preparation it has been accepted that photosynthetically generated NADPH was the electron donor for nitrate reduction in leaves.

With time this pathway did not appear to satisfy all experimental observations, as nitrate reductase from green leaves was NADH- rather than NADPH-specific, and the enzyme appears to be localized in the cytoplasm rather than in the chloroplasts.

There is evidence that chloroplasts can photoreduce NAD as well as NADP *in vitro* however, the $K_m$ is much larger

($3.8 \times 10^{-3}$ M versus $9.8 \times 10^{-6}$ M) and the rate of reduction slower (25 to 60 per cent). It is obvious that if NADH is generated in the chloroplasts and supplied to the enzyme in adequate amounts, nitrate could be reduced. However, Ogren and Krogmann have shown that illumination of leaves converts much of the NAD present in the chloroplasts to NADP and markedly increases the amount of NADPH. No increase in NADH was observed upon illumination.

In proposing the direct photosynthetic reduction of nitrate, Ramirez *et al*- have shown that nitrate can serve as the terminal electron acceptor for a noncyclic photophosphorylation. If this is the case, it would be expected that nitrate would be reduced at an accelerated rate in the absence of $CO_2$. However, it has been shown that plants exposed to light but deprived to $CO_2$ reduce nitrate at a very show rate. The requirement of $CO_2$ for nitrate reduction may be explained by the observation of Walker *et al.* These workers found that oxygen production by isolated intact chloroplasts is dependent upon an adequate supply of $CO_2$. Apparently, *in vivo* $CO_2$ is required for the continued flow of electrons in noncyclic photophosphorylation. Alternatively, the requirement of $CO_2$ for rapid nitrate reduction may indicate that the reductant utilized for nitrate reduction is generated in conjunction with or from the carbon products formed during $CO_2$ fixation.

In this context, Klepper vacuum infiltrated leaf discs from morning-glory and corn with water, pyruvate, phosphyglycerate (PGA), 3-phosphoglyceraldehyde (3 PGAld), glucose-6-phosphate (G-6-P) and fructose-1-6-diphosphate (FDP), in the presence of nitrate. The tissue was then incubated in the dark in the infiltration medium and rate of nitrite production determined. The least and equal amounts of nitrite were found in the control PGA and pyruvate treatments. The tissue infiltrated with the three sugar phosphates produced more (up to two fold) nitrite than the control. The amount of nitrite formed was related to the concentration of the sugar phosphates added to the medium.

When nitrite was substituted for nitrate in a comparable set of experiments, a slow but steady decrease in nitrite content was observed. There was no effect of the various substrates on nitrite utilization. However, if these experiments were conducted in the light (2,500 ft-c), nitrite was rapidly lost. This pronounced effect of light on nitrite utilization is consistent with previous observations.

In confirming experiments, Klepper found that clarified homogenates (30,000×*g* for 15 min) when supplied with (PGA), ATP, NADPH, and NAD could reduce appreciable amounts of nitrate to nitrite. No nitrite was produced if any of the four metabolites were omitted. The addition of phosphoglyeraldehyde (3-PGAld) and NAD gave maximum reduction of nitrite and was as effective as NADH alone. The addition of NADPH alone was ineffective.

These processes per se constitute no more than the coupling of several enzymes that are present in the homogenates. However, they do permit the development of a basic concept of the influence of light on nitrate reduction.

It can be assumed that the chloroplast, when illuminated *in vivo* and supplied with $CO_2$ would be capable of producing the gamut of phosphorylated trioses, hexoses, sugars, and starches. As the phosphorylated intermediates or carbohydrates move into the cytoplasm, they are metabolized by the glycolytic enzymes and thus provide the energy source for nitrate reduction via the NAD-dependent phosphoglyceralde dehydrogenase (PGD). Heber *et al.* have estimated that approximately 70 per cent of the total NAD-dependent PGD is localized in the cytoplasm. Furthur, the NADH-dependent nitrate reductase has been found to be exo-chloroplastically located.

This hypothesis is attractive in that it : (*a*) explains the marked stimulation of light on nitrate reduction dependent upon $CO_2$ by increasing the supply of 3-PGAld which in turn generates NADH ; (*b*) concurs with Kessler's view on the involvement of ATP, NAD, and carbon compounds ; and (*c*) provides a means of recycling the NADH, generated by the appreciable amounts of NAD-dependent PGD localized in the cytoplast of leaf tissues. This data permits the conclusion that 3-PGAld and the cytoplasmically located NAD-dependent PGD is the electron generating system for nitrate reductase in both light and dark.

In contrast, the work of Mendel and Visser based only on the incorporation of $^{15}NO_3^-$ into reduced nitrogen, suggests a mitochondrial source of electrons in the dark and a photosynthetically produced reductant in the light. Their conclusions are based on enhanced nitrate assimilation obtained with substrates for mitochondrial respiration or with illumination and the inhibition of nitrate assimilation in the dark by addition of iodoacetate ($10^{-4}$ M). In this vein, Klepper found that $10^{-4}$ M iodoacetate did not suppress reduction of nitrate to nitrite, while $10^{-3}$ M gave a 70 per cent inhibition. Since Mendel and Visser did not measure nitrite accumulated in the media or tissue, it can be argued that the enhanced incorporation which they observed in the light is more pertinent to nitrite assimlation than to the actual reduction of nitrate to nitrite.

*Nitrite reductase.* It is significant that in studies of nitrate metabolism of leaf tissues in the dark that respiration is able to support the reduction of nitrate to nitrite, but further reduction of nitrite appears to be closely coupled to light in leaf tissues. Such observations are consistent with a requirement for reduced ferredoxin for nitrit assimilation. In nonphotosynthetic tissues it has been demonstrated that the conversion of nitrate to nitrite, apparently supported by respiration, proceeds rapidly with secretion of nitrite into the surrounding media. However, the subsequent utilization of nitrite is dependent upon adequate aeration. Thus nitrate can be converted to nitrite under partially anaerobic conditions, but the subsequent reduction of nitrite appears to depend on oxidative reactions. If nitrite functions as a terminal electron acceptor, as respiratory studies indicate it is difficult to envisage why nitrite utilization should be stimulated by oxygen.

Currently, the electron donor for nitrite reduction in non-photosynthetic tissue is not known.

## LOCATION

The majority of recent studies on nitrate metabolism in higher plants have been made with leaf tissues, because of the abundance of enzymes and availability of material. There is adequate evidence, however, to show that the capacity for nitrate metabolism is not solely associated with leaves. The most convincing evidence that all living cells have the capability for assimilating nitrate are those studies in which tissues are grown in a medium in which nitrate serves as the sole nitrogen source. The initial work indicated that while nitrate supported growth of excised tomato roots ammonia failed to do so. This was a surprising observation in view of the predicted conversion of nitrate to ammonia prier to the incorporation of ammonia into an organic form. However, it was indicated by Sheat *et al.* that ammonia was an effective nitrogen source for excised roots if suitable precautions were taken to maintain pH. Growth on nitrate was optimal at acidic conditions, while neutral to slightly alkaline conditions were optimal for ammonium nitrogen. Roots could also be cultured effectively with nitrite serving as the sole nitrogen source ; however, the concentration of nitrite had to be maintained at a constant low level, or else toxicity symptoms developed. It was further demonstrated that growth of roots in nitrate was dependent upon the presence of an adequate level of molybdenum. In contrast, roots grown on nitrite are capable of growth in the absence of molybdenum, although it stimulated growth.

In the absence of photosynthetically derived reductant to be utilized for nitrate reduction in roots or nongreen tissues, it appears that the reducing potential is furnished by the respiration of carbohydrates. Extensive surveys in this area have been performed by Yemm and co-workers, who demonstrated that incubation of excised barley roots with nitrate, nitrite, or ammonia resulted in a changed respiratory activity as indicated by changes in $CO_2$ output or $O_2$ consumption. Characteristically, it was found that incubation of the roots in nitrate or nitrite resulted in a stimulated carbon dioxide production and a less marked increase in oxygen consumption. Thus the net effect of these treatments was to increase the respiratory quotient (RQ) from 0.9 to 1.15 and 1.25 for nitrite and nitrate, respectively, clearly indicating that the respiratory mechanism had been modified. It was suggested that nitrate and nitrite might be replacing oxygen as the hydrogen acceptor. It is significant that the different forms of nitrogen supplied to the roots had distinctive effects on respiration. Thus the change in RQ produced by nitrate was slower than that caused by nitrite. Furthermore, although nitrate accumulated in roots incubated in nitrate, the same was not true with nitrite. It appeared that nitrite was taken up and metabolized immediately. The data could be taken to indicate an adaptive development of nitrate reductase, whereas nitrite reductase (or at least the capacity to metabolize nitrite) is constitutive. In this respect it is interesting that Czygan reports a similar situation in the alga *Aukistrodesmus*, nitrate reductase being a soluble adaptive enzyme and nitrite reductase constitutive and particle bound. Detailed analyses by Yemm and Willis indicated that in addition to altering the respiratory activity, the nitrogenous compounds were assimilated by the roots in a manner consistent with a conversion to ammonia prior to their incorporation into glutamine.

The mechanism by which nitrate is reduced to ammonia in nongreen tissues is not well established. Eckerson described a scheme in which aqueous extracts of tissues of fresh plant organs were employed to measure the nitrate reducing ability of the structure involved. In this scheme it was demonstrated that the capacity of the extracts to convert nitrate to nitrite was dependent upon the maintenance of pH of 7.2 and an adequate supply of glucose and oxygen. Using these methods, it was consistently demonstrated that root extracts contained nitrate reducase. Eckerson demonstrated that reducase activity (*i.e.*, the capacity to convert nitrate to nitrite under defined conditions) varied with plant part, species, and age. Although investigations were thorough and detailed, it appears that the conversions being measured were nonenzymatic because it was shown that boiled plant extracts were effective in converting nitrate to nitrite.

More convincing evidence for the enzymatic reduction of nitrate in root extracts is provided in the pioneer paper of Evans and Nason however, nitrate reductase activity of the extracts was very low. Candela *et al.* Scientists reported only low levels of nitrate reductase in extracts of roots from cauliflower, corn, and sugar cane, respectively.

Sanderson and Cocking demonstrated appreciable nitrate reductase in extracts from roots of several plant species; however, the activity of the root extracts was consistently lower than that in leaf extracts. The properties of nitrate reductase from the root extracts were very similar to those of the leaf enzyme, and it was concluded that they were identical. It was also demonstrated that the level of nitrate reductase activity in excised to mato roots grown in sterile nutrient culture was comparable to that found in intact roots, thus precluding the possibility that microbial enzymes were being measured. It was interesting that Sanderson and Cocking were unable to detect nitrate reductase activity in leaves or roots of apples; it had been postulated that since nitrate level was low in xylem sap, then the reduction of nitrate and amino acid assimilation occurred predominantly in roots. Grasmanis and Nicholas reported that reduced benzyl viologen functioned as the electron donor for an apparently atypical nitrate reductase in apple roots. More recently, however, Klepper and Hageman have demonstrated that by using adequate levels of polyvinylpyrrolidone in the extraction medium it is possible to extract a typical NADH-dependent nitrate reductase from apple leaves, stems, petioles, and roots. Thus it appears that, given adequate nitrate levels, all parts of the apple plant have the capacity to utilize nitrate in a seemingly conventional manner (NADH as electron donor). Using similar criteria to those of Bollard Wallace and Pate indicate that plants may be categorized as those which reduce nitrate in their roots (field pea) and those which do not (*Xanthium*). In support of this view, they report that all of the nitrate is metabolized in the roots of peas and only reduced nitrogen is found in the exuded sap, but only if the plants are cultured on low levels of nitrate. However, if the plants are supplied with nitrate levels higher than 10 ppm, nitrate is transported to the shoot and nitrate reductase induced. In contrast, *Xanthium* transports 95 per cent of the supplied nitrate to the shoots and negligible amounts are metabolized in the root.

Boutard has reported that the distribution of nitrate reductase activity in etiolated barley seedlings cultured in the absence of nitrate is similar to that in field peas on minimal nitrate. When nitrate was supplied to illuminated barley, nitrate reductase activity was increased in the leaves and decreased in the roots. The nitrate reductase detected in pea roots was dependent on NADH and thus appeared characteristic of that found in most plants. The capacity for nitrate reduction in pea roots appears to be confined to the elongating zone, *i e.*, the region of most active respiration. In contrast, no nitrate reductase was detected in the roots of *Xanthium.*

Boutard and Coupe *et al.* extracted nitrate reductase from barley roots and concluded that roots are more efficient than leaves in nitrate assimilation. Miflin has reported on the occurrence of nitrate and nitrite reductase in extracts of roots of young barley seedlings, and the level of nitrate reductase activity on a per unit protein basis was comparable to that from leaf tissue. There appear to be two systems capable of metabolizing nitrate in barley root extracts, a soluble component utilizing NADH as electron donor and a particulate fraction which can utilize succinate more efficiently than NADH. A similar succinate stimulated nitrate reduction by particulate fractions from different organs of tomatoes is reported by Sanderson and Cocking. Particulate fractions of barley roots were able to reduce nitrite in the presence of reduced benzyl viologen, but the physiological cofactor was not demonstrated.

Since the concept that roots play an important if not a major role in reducing nitrate is widely accepted, it is worthwhile to consider the relative abundance of nitrate reductase in roots and shoots (leaves). Miflin's work shows that the specific activity of nitrate reductase is slightly higher in roots than in leaves. However, if activity is computed on the basis of fresh weight or plant part, the shoots have a much greater amount of the enzyme. This is illustrated by work with field peas showing that shoots contain about the same amount of enzyme as the roots only when nitrate levels of the nutrient medium were maintained at 10 ppm. Even when the plants were grown on low (500 ppm) levels of nitrate, the shoots had from 3 to 12 times more enzyme than the root.

Roots from young corn seedlings contained but 20 per cent of the activity found in the shoots. The capacity of corn leaves to reduce nitrate (estimated by nitrate reductase levels found in composite samples of all leaves, sampled twice weekly over the growing season) exceeded by 3- to 4-fold the total reduced nitrogen found in the total shoot and ear at the end of the growing season. Leaves of corn and wheat maintain high 10 to 15 $\mu$mole $NO_2^- \times hr^{-1} \times$ (grams fresh weight) average levels of nitrate reductase activity throughout the growing season.

From an agronomic standpoint, and especially in view of current fertilizer practices, this work as well as that of Wallace and Pate would suggest that except for the early seedling stages, the bulk of the nitrate is reduced in the leaves of crop plants, excluding legumes and paddy rice.

The only report found on the occurrence of nitrite reductase in roots was that of Miflin. He showed that a particulate fraction from barley roots was able to reduce nitrite using reduced benzyl viologen as the electron donor. The presence of a physiological cofactor was not demonstrated. Klepper and Hageman using the assay system of Joy and Hageman were able to detect nitrite reductase activity in extracts from leaves but failed to detect nitrite reductase activity in extracts from roots stem, or petiole tissue.

Other reports have been made of nitrate reductase activity in extracts from nongreen tissue. Spencer was able to isolate and purify nitrate reductase from the embryos of germinating wheat seed ; characteristically, the reduction of nitrate was coupled specifically to reduced NAD. Varner has recently reported that nitrate reductase activity can be detected in extracts of barley aleurone cells which have been incubated with nitrate. Elsner and Hageman have also demonstrated the engyme in extracts of corn scutella preincubated with nitrate. Filner's studies with tobacco pith cultures clearly demonstrate the capacity of this tissue to utilize nitrate as principal nitrogen source.

Thus it appears that most plant cells and tissues have the capacity to utilize nitrate, and given an adequate supply of nitrate, the presence of nitrate reductase specific for reduced pyridine nucleotide can be demonstrated in these tissues.

Although it is obvious that nonphotosynthetic tissue can reduce and assimilate nitrate, more work is needed to establish the mechanism for reduction of nitrite to the ammonium ion.

*Subcellular location.* The observations that nitrate metabolism in green tissues is intimately associated with photosynthesis, and in nongreen tissue appears to be related to respiratory metabolism, suggest that the enzymes required for nitrate metabolism might be directly associated with organelles responsible for photosynthesis and respiration. There is strong evidence for the association of at least part of the machinery for nitrate utilization being associated with the chloroplasts ; however, it is debatable whether nitrate reductase is located within this organelle. Evans and Nason originally demonstrated that illuminated chloroplast grana could provide the electrons for nitrate reduction, but the chloroplast grana themselves were unable to reduce nitrate unless fortified with a purified nitrate reductase. Del Campo *et al.* reported that illuminated isolated chloroplasts had the capacity to reduce nitrate. However, the system had to be supplied with benzyl viologen as electron carrier, and it appeared that the reductase was present in a ferredoxin free-chloroplast extract. In another report purporting to show nitrate reduction in chloroplasts, it was demonstrated that nitrite was not formed by illuminated washed broken chloroplasts unless nitrate reductase (prepared from a $27,000 \times g$ supernatant of a spinach leaf homogenate) was added. Thus, although it is clearly demonstrated that chloroplasts have the capacity to generate a reductant for nitrate reduction, it is not conclusively demonstrated that nitrate reductase is located in the chloroplast. On the contrary, the extraction techniques used for the preparation of nitrate reductase seem to indicate a cytoplasmic location. It is significant that the same workers were to able demonstrate that nitrite reduction could proceed at an appreciable rate in a system consisting only of washed chloroplasts and ferredoxin. While the addition of a ferredoxin-free chloroplast extract increased nitrite reduction, it was not an absolute requirement, as in the case for nitrate reduction by the chloroplast system.

In a critical study aimed at establishing the intracellular location of enzymes of nitrate metabolism in leaf tissue, Ritenour *et al.* compared nonaqueous and aqueous techniques and concluded that nitrite reductase is located within the chloroplast, but that nitrate reductase and glutamic acid dehydrogenase are not. Glutamic acid dehydrogenase was shown to be localized within the mitochondria. Nitrate reductase was stated to be an exo-chloroplastic enzyme, but the possibility was not eliminated that the enzyme might be associated with the external chloroplast membrane which is removed during aqueous and nonaqueous isolations.

Coupe *et al*, also using nonaqueous techniques, but using density gradient centrifugation separation rather than stepwise fractionation, concluded that nitrate reductase of barley was a chloroplastic enzyme. The apparently conflicting results may be related to the method of interpretation of data. Ritenour *et al.* based their conclusions on the distribution of marker enzymes, whereas Coupe *et al.* determined activity in relation to distribution of chlorophyll and protein. It view of the possibility that nitrate reductase could be associated with the chloroplast membrane, it should be worthwhile to use the more functional chloroplasts isolated by the methods of Jensen and Bassham and Walker to re-evaluate the current views of localization of nitrate and nitrite reductase. In work of this type, recognition should be given to Stocking's warning that enzymes can be adsorbed by, as well as lost from, the chloroplasts.

Further indirect evidence for the cytoplasmic as opposed to the chloroplastic location of nitrate reductase is provided by the observation of Schrader *et al.* which demonstrated that chloramphenicol inhibited the synthesis of nitrite reductase but not of nitrate reductase. Anderson and Smillie had previously shown that at appropriate levels chloramphenicol binds with chloroplast ribosomes and thus inhibits protein synthesis at that site, but is without effect on cytoplasmic ribosomes.

The distribution of enzymes involved in nitrate metabolism in roots and other nongreen tissues—*e.g.*, barley aleurone layers, corn scutellum, wheat embryos—is similarly open to speculation. The observed influence of nitrate in root respiration might indicate an association of nitrate metabolism with the mitochondria. However, Spencer stated that nitrate reductase from etiolated wheat embryos was soluble, and the extraction procedures utilized by other investigators suggest a cytoplasmic location. In contrast, Sanderson and Cocking and Miflin reported that particulates from a range of nongreen tissue were able to produce nitrite from nitrate in the presence of succinate. Miflin's work also indicates that nitrite reduction with reduced dyes as the electron donor occurred with particulate preparations. It is interesting that the crude homogenates of barley roots used NADH more effectively than succinate and that NAD activity was retained predominantly in the supernatant fraction, while the succinate mediated reduction was confined to the particulate fraction. It has been possible to demonstrate the alteration of respiration of mitochondrial preparations from corn roots similar to those reported by Willis and Yemm for intact barley roots. The utilization of nitrate and nitrite could not be estimated due to the fact that nitrite disappeared nonenzymatically when reaction mixes were acidified with trichloroacetic acid as the termination of the reaction. Caution is also needed in interpreting experiments supposedly demonstrating metabolism of nitrate by particulates because of the interference by microorganisms. In this respect, Dubinins reports that nitrate and nitrite respiration of roots occurs readily under nonsterile conditions, but the induction of nitrate respiration in roots themselves after the beginning of anaerobiosis is slow.

## REGULATION

A characteristic feature of the nitrate metabolism in higher plants is its susceptibility to a range of environmental conditions. It has been established that light, drought, mineral nutrition, hormonal treatment, plant age, and genetic composition all influence the capacity of nitrate reduction. In most of these cases it appears that the control of nitrate reduction is mediated through a regulation of the enzyme nitrate reductase.

Nitrate reductase is the logical point to effect regulation of the input of reduced nitrogen for the plant because it is (*a*) the first enzyme in the pathway ; (*b*) the rate limiting step ; (*c*) substrate inducible ; and (*d*) relatively unstable (high turnover rate). Further, the toxic effects of excess levels of nitrite and ammonium ions also indicate the desirability of regulating their production.

*Substrate induction, Nitrate.* Evidence that nitrate reductase is inducible by nitrate has been provided by several workers. A feature of these observations is that different levels of nitrate are required for optimum induction in different species and in some cases high concentrations of nitrate are needed. These differences probably indicate differences in rate of uptake between species because with a given species the level of enzyme is dependent, within limits, upon the concentration and rate of supply of nitrate to the tissue.

The induction of nitrate reductase in various tissue in response to inducers is related to maturation and age. In general, the ability to induce nitrate reductase is related to the capacity for protein synthesis, and increasing maturity is associated with a decreased nitrate reductase inducibility. It is not clear whether the change in inducibility with age is directly associated with a decreased capacity for protein synthesis or due to an accumulation of soluble nitrogenous products which specifically repress the synthesis of nitrate reductase. The reduced capacity for induction which occurs with increasing age is not an irreversible process. Wallace and Carr and Pate have shown that there is a resurgence of nitrate reductase activity in old leaves remaining on plants after partial defoliation or detopping. Such leaves frequently contain high nitrate before defoliation and thus the increased induction cannot be interpreted as being associated with an increased delivery of inducer following defoliation or detopping. It may be that the acceleration of synthesis of nitrate reductase is associated with an increased availability of soluble precursors which facilitates the synthesis of proteins. There is certainly a requirement for an adequate level of precursors for protein and nucleic acid synthesis before induction of nitrate reductase can occur. Schrader has shown with corn seedlings that appreciable induction of nitrate reductase by nitrate, in short-term experiments, can only be shown if the seedlings have been grown in ammonia containing medium. The ammonia grown seedlings, although lacking nitrate reductase, had a higher nucleic acid, protein, soluble nucleotide, and amino acid content than the water controls and induced readily on transfer to nitrate. Boutard using nitrogen deficient etiolated barley seedlings, indicated that nitrate reductase activity was present in seedlings cultured in distilled water and the activity was predominantly associated with roots. Addition of nitrate caused an increase in nitrate reductase activity in the leaves and a decline in the roots. The increase in nitrate reductase, which was much smaller than that observed in corn, was greater in illuminated barley seedlings than in those maintained in darkness. The data of Boutard is analogous to that reported by Wallace and Pate for pea seedlings receiving minimal nitrate.

Ammonium or nitrite ions did not induce nitrate reductase in cauliflower or corn seedlings; however, these ions were reported to induce (directly or indirectly) the enzyme in radish cotyledons. Paradoxically, negligible levels of nitrate reductase were found in cotyledons from radish seedlings cultured on nitrite for 7 days.

Induction of nitrate reductase by nitrate is followed (3 hrs) by the appearance of nitrite reductase, whereas no lag period is noted if nitrite is used as the inducer. This work, in conjunction with the independent induction and regulation of nitrate and nitrite reductases in *Lemna*, suggests that synthesis of the two enzymes is not coordinate.

*Molybdenum.* It has been demonstrated that in molybdenum-deficient plants grown in the presence of nitrate, molybdenum can serve as an inducer for nitrate reductase. Since no increase in enzyme activity could be induced by adding molybdenum to cell-free extracts of deficient plants, it appears that the metal is involved in the induction process and is not merely activating pre-existing protein. Furthermore, the induction of nitrate reductase by molybdenum shows the same sensitivity to inhibitors of protein synthesis as does induction by nitrate.

*Light.* The level of nitrate reductase in leaf tissue varies diurnally and is influenced by intensity of illumination. The mode of operation of light in controlling nitrate reductase is complex. However the data of Candela *et al.* and Hageman and Flesher clearly indicate that light influences the level of the enzyme per se. Thus, extracts from plants exposed to increasing periods of darkness show a progressive decrease in nitrate reducing ability (even though adequate electron donors present in the assay medium) which is restored on subsequent illumination. It is difficult to delineate the action of light in controlling the level of nitrate reductase. Beevers *et al.* demonstrated that light was not an absolute requirement for induction of nitrate reductase in green leaf tissue as long as sufficient nitrate was present in the induction medium. Furthermore, the induction of nitrate reductase in nongreen tissue indicates that light is not a prerequisite for induction. Beevers *et al.* indicated that illumination increased the uptake of nitrate into corsseedlings and suggested that this stimulated uptake could account for the effect of light on induction. A similar conclusion for the effect of light or nitrate utilization had been made by Warburg and Negelein. This conclusion is supported by the observations of Hope and Walker, MacRobbie and Dainty, and

Nagai and Tazawa of changed permeability of cell membranes following illumination. However, it is recognized that some of the stimulating effect of light an induction of nitrate reductase may be associated with the increased protein synthesis occurring after illumination. In contrast to the observations with corn and radish, Kannangara and Woolhouse indicate that there is an absolute and direct requirement for photosynthetic $CO_2$ fixation for the induction of nitrate reductase in leaves of *Perilla*. It will be of interest to see it *Perilla* tissue can induce nitrate reductase in the dark in the presence of sufficient nitrate. Another possible function of light in induction may lie in its influence on redox potential in leaf tissue. Showe and DeMoss have demonstrated that induction of nitrate reductase in *Escherichia coli* is dependent upon a reductive environment in the cell. A generally more reduced state of leaf cells following illumination may account for the observed greater induction following illumination.

*Regulation of nitrate reductase by endogenous metabolites.* The failure to demonstrate the occurrence of appreciable nitrate reductase in extracts from tissues containing nitrate e g., plants grown under shaded conditions or mature leaves can be attributed to a variety of causes. It is possible that the nitrate which is accumulated is maintained in the vacuoles and in this location cannot function as an inducer. A lowered capacity for protein synthesis in shaded or more mature tissue could also contribute to a failure of induction. Alternatively, under shaded conditions or in older tissue nitrate metabolism may be impaired due to a depletion of photo-synthetically produced electron donors and ATP. Thus intermediate products of nitrate metabolism might accumulate and repress induction of nitrate reductase. End product repression of nitrate reductase has been demonstrated with *Neurospora* and *Chlorella*. However, the primary metabolites of nitrate reduction, nitrite or ammonia, do not inhibit or suppress the induction of nitrate reductase in higher plants.

Filner demonstrated that casein hydrolysates and 11 amino acids, added individually, would repress the synthesis of nitrate reductase of cultured tobacco pith cells. Arginine or lysine would counteract the repressive action of any of the 11 amino acids, when tested in matched pairs. In like manner, cysteine and isoleucine were effective "derepressors" except when added with methionine or alanine. He also reported that since the casein hydrolysate repressed induction in proportion to its ability to meet the nitrogen needs of the cells, the repression by the individual amino acids does not appear to be consequence of toxic growth effects. Smith and Thompson recently reported that when 20 protein amino acids, orinithine, citrulline, urea, and ammonium ion were tested as regulators of nitrate reductase in *Chlorella*, only arginine, lysine, leucine, glutamate and the compounds named above were effective in suppressing induction. While it is possible that the first five compounds were providing ammonium ions, which would effect repression, leucine and glutamate would appear to be unique. None of these compounds inhibited nitrate reductase activity.

In contrast, most of the compounds used by Filner and Smith and Thompson, as well as other compounds, were not effective in inhibiting nitrate reductase induction in excised corn seedlings or radish cotyledons. Of several secondary (phenylpropanes and derivatives) metabolities of nitrogen metabolism tested, only coumarin, *trans*-cinnamate, and *trans*-o-hydroxycinnamate inhibited induction of nitrate reductase. Since these compounds inhibited protein synthesis in general, they cannot be considered, as specific co-repressors of nitrate reductase activity. Carbamyl phosphate was shown to inhibit nitrate reductase in *Chlorella*. Other works demonstrated that carbamyl phosphate and its chemical derivative cyanate were shown to be competitive inhibitors of nitrate reductase and not co-represssors.

*Induction, de novo synthesis, and decay of nitrate reductase.* The rapid fluctuations of nitrate reductase activity that occur daily and seasonally, when plants are placed in the dark, or exposed to heat or drought, indicate an effective *in vivo* regulatory mechanism. Obviously, fluctuations in activity could be brought about by changes in relative rates of synthesis or activation and breakdown or deactivation. Thus an increased activity could be due to a decreased breakdown or to an increased activation or synthesis; conversely, a decreased activity could be associated with an increased breakdown or deactivation or with decreased synthesis. The inhibition of induction of nitrate reductase by actinomycin D, cycloheximide, L-azetidine-2-carboxylate, and a range of substituted amino acids and bases constitutes the best evidence that increase in activity during induction is due to *de novo* synthesis of the enzyme rather than activation. These results should be interpreted with some caution. While these studies show that a prerequisite for induction of nitrate reductase is a normal RNA and protein synthesis, they do not demonstrate conclusively *de novo* synthesis of the enzyme. If fact, thus far all attempts to demonstrate *de novo* synthesis have been inconclusive. These failures are attributable to (*a*) the inability to isolate the enzyme in pure form which precludes finger printing and demonstration of uniform incorporation of label; (*b*) the possibility that nitrate is not a specific inducer for nitrate reductase but rather stimulates a gamut of activities; and (*c*) the fact that during the normal course (3 to 5 hr) of induction, especially with chlorophyllous tissue, the newly synthesized amino acids dilute and mask incorporation of exogenously labeled amino acids.

It is not known whether the enzyme is only inactivated, completely destroyed, or both, with changes in environment that lead to a loss of nitrate reductase activity. For example, when plants or leaves containing nitrate are transferred to the dark, there is a rapid decline in nitrate reductase activity. In contrast, nitrite reductase activity shows a much less rapid decline. A similar decline in nitrate reductase activity occurs if nitrate grown seedlings are transferred to a nitrate free medium. Schrader *et al.* demonstrated that in the presence of nitrate and cycloheximide (to inhibit protein synthesis), nitrate reductase activity declined at a rate which permitted an estimated enzyme half-life of about 4 hr. A similar rate of enzyme depletion was observed when seedlings preinduced with nitrate were transferred to a nitrate free medium, but without cycloheximide. Kannangara and Woolhouse found that the rate of decline in nitrate reductase activity in the light in the absence of $CO_2$ (a condition which they state prevents nitrate reductase induction) was similar to that observed when plants were transferred to darkness. Thus it appears that the enzyme is continuously declining in activity, but in the presence of light or sufficient nitrate or both, this decline is more than counter-balanced by the apparent synthesis of new enzyme. It is not known whether the decreased enzyme activity of extracts from plants placed in the dark or depleted or nitrate is due to enzyme degradation, or partial inactivation due to change in enzyme configuration in the absence of substrate, or loss of essential prosthetic groups.

*Genetic regulation*. It has been shown that corn inbreds and hybrids differed widely (up to five fold) in their seasonal level of nitrate reductase and that the level of activity was under genetic control. Commercial corn inbreds were used and all genotypes used had the capacity to produce nitrate reductase. Thus it should be emphasized that in corn the genetic control studied is over the level of activity and not concerned with the number of structural genes involved in the synthesis of the enzyme as reported by others for microorganisms. A subsequent study demonstrated that by proper combination of inbred lines, followed by selection based on enzyme assay, a hybrid could be developed with a high, medium, or low level of nitrate reductase activity. One other aspect of the corn studies was the finding that certain $F_1$ hybrids obtained by crossing inbreds with low (relative) levels possessed a heterotic level of nitrate reductase activity. This superiority in ability to reduce nitrate existed from seedling stage to maturity. Further investigations of this phenomenon with two corn inbreds and their progeny ($F_1$, $F_2$, $F_3$, $F_4$ generations) indicate that the level of activity of this material was under control of a two-gene system. Whether this control of level of activity is due to different rates of synthesis or decay cannot be stated at this time. In these experiments, inducer ($NO_3^-$) levels in the tissue were not considered to be the causal factor in the regulation of the enzyme level. Inducer level could be a regulating factor since as Griffith has found that different grass strains possess different nitrate levels ; unfortunately, nitrate reductase levels were not measured in these grasses.

Marked variations in levels of nitrate reductase have also been observed among several varieties of cauliflower and with various wheat genotypes. The work with wheat also indicates a relatively simple genetic control system.

That this genetic control is real and of practical significance was established with wheat by showing that the seasonal input of reduced nitrogen to the plant, computed on an acreage basis from the enzyme assay, was related in a linear manner to the protein accumulated in the vegetation and grain or grain alone. This work with wheat and corn provides an initial step in the establishment of guidelines for breeding plants with genetically controlled potential for metabolic activities and their products.

*Hormonal effects*. As early as 1950, studies revealed that several broad-leaf weed species and sugar beets accumulated nitrate when treated with herbicidal levels of 2, 4-dichlorophenoxyacetate (2,4-D). In contrast, no record of nitrate accumulation has been reported for plants of the Gramineae family, similarly treated. An explanation for this divergence in response was reported by Beevers *et al.* They found that nitrate reductase activity was decreased in cucumber plants sprayed with 2,4-D solutions, while the opposite effect was noted with corn seedlings. The changes in enzyme activity were positively correlated with changes in protein content for both species. Nitrate content varied inversely with enzyme activity. The results obtained with growth chamber plants were reproducible, while much greater variability was encountered when the plants were grown under field conditions.

Taylor was among the first to find that sub-herbicidal levels of 2,4-D would increase dry weight accumulation of plants. Subsequently, Wort applied mixtures of 2,4-D and micronutrients as foliar dusts and observed increases in growth rates and yields for several crop plants. These treatments have not been widely adopted commercially because of the variability of results obtained from season to season and with location.

Huffaker *et al.* found that the isopropyl ester of 2,4-D, supplemented with iron salts or chelates, would increase grain yields of barley and wheat when applied at appropriate stages of plant development. With these treatments, the percentage of protein in the grain was increased in barley but not in wheat. However, if the isooctyl ester was used in conjunction with Fe-diethylenetriaminepentaacetate as the spray treatment for wheat, grain yields were depressed slightly, but protein content was significantly increased. Treatment of field beans with the isopropyl ester and micronutrients increased bean yields significantly. The major problem encountered in this work, especially with beans, was predicting the proper concentration, time of application, and the interaction with environment that would insure beneficial results. Similar problems have been encountered in effecting efficient weed control with herbicides.

Recently, Tweedy *et al.* observed that simazin (2-chloro-4,6-bisethylamino-*s*-triazine) increased growth rates and nitrogen content of tolerant plants. They found the effects were more pronounced if the plants were cultured on suboptimal levels of nitrate or grown at low temperatures. These changes were associated with increased nitrate reductase activity, and plants cultured on ammonium salts did not respond in a similar manner with treatment. Further studies with peas and rye grown in growth chambers confirmed and extended these observations. Field trials where six different plant species were treated with subherbicidal levels of simazin were reported to show that appropriate treatments increased total crude protein production per acre from 10 to 52 per cent. It was stated that the simazin treatment enhanced uptake of nitrate from the media. The data indicated that in some instances the nitrate content of treated tissue was higher. Since protein content was measured by a Kjeldahl procedure which could reduce a portion of the nitrate present in the tissue, some of the protein values, especially for the forages, are questionable. Again, as with 2,4-D treatments, the effects on growth, yield, and protein content were variable for a given crop and simazin level applied, depending on environment.

It is of interest that in induction studies with seedlings or mature corn leaves and radish cotyledons no enhancement has been observed by additions of indoleacetic acid, various cytokinins and gibberellins, or 2,4-D to the induction medium. Varner reported that gibberellic acid did not stimulate the induction of nitrate reductase by nitrate in barley aleurone layers.

The major conclusions for this section are that (*a*) the enhancement of yields and protein content achieved with sublethal levels of herbicides emphasizes the role of nitrate reductase in maximizing grain yields with high protein content ; (*b*) nitrate reductase activity is interwoven with metabolism as a whole ; (*c*) hormonal and herbicidal effects on nitrate reductase are probably indirect through changes in overall metabolism ; (*d*) the application of herbicides and related compounds to achieve enhanced yields and protein production offer promise but are developing slowly because of the complexity of the system. Since the appearance of Kessler's review in 1964, additional information on the enzymes and mechanisms of nitrate assimilation has been published. Most interesting in this respect has been the demonstration of the existence of nitrite reductase and its physiological cofactor, ferredoxin, in leaf tissue. Thus, there is now evidence for the complete sequence of enzymes capable of converting nitrate to ammonia. The nature of the precise mechanism of reduction of nitrite remains to be elucidated. Current data suggest that nitrite reductase affects the transfer of six electrons at one enzyme surface in converting nitrite to ammonia, a disturbing phenomenon from the standpoint of thermodynamics and kinetics. In contrast, knowledge concerning nitrite metabolism in nonphotosynthetic tissues is meager, and future investigations into this area are warranted.

# 5

# The Role of Chloride in Oxygen Evolution

A brief review on the role of chloride in oxygen evolution is presented in this chapter. This chapter deals with the discovery of the phenomenon, the site of action, specificity and effectiveness, possible binding sites and a possible mechanism of action. The phenomenon was discovered by Otto Warburg and has been studied extensively since the pioneering work of Sei Izawa and coworkers. The site of action is on the electron donor side of the Photosystem II (PSII) between the so-called water oxidation complex (WOC) and the electron donation sites of external donors such as $NH_2OH$, diphenylcarbazide and catechol. The effectiveness of the anions in supporting water oxidation, protecting $O_2$ evolution against mild heating and in preventing the dissociation of the extrinsic polypeptides of WOC, follows the series : $Cl^- > Br^- > NO_3^- > I^- > ClO_4^-$. Bicarbonate ($HCO_3^-$), that appears to be required on the electron acceptor side of PSII, is ineffective in replacing $Cl^-$. Sulphate ($SO_4^{2-}$) and $PO_4^{3-}$ are totally ineffective, and $F^-$ and $OH^-$ might even be inhibitory. The size, ionic field and hydration energies of the anions seem to play significant roles in their effectiveness. It appears that several (4 to 40) $Cl^-$ ions may be weakly and reversibly bound per WOC. In all likelihood, there are two binding domains : intrinsic : on the PSII reaction centre proteins $D_1$ and $D_2$ (perhaps, on the lumenal histidines): and extrinsic : on the extrinsic 33 kDa polypeptide (perhaps, on lysines and/or arginines). Removal of the extrinsic polypeptides seems to increase the [Cl$^-$] requirement for water oxidation activity. It is suggested that the intrinsic sites are the catalytic sites and the extrinsic sites are the regulatory sites. In the absence of Cl$^-$, the WOC is stuck in an abnormal state, and there is no $O_2$ evolution activity. Our hypothesis for Cl$^-$ action is : it facilitates the abstraction of H$^+$s from $H_2O$ during its oxidation to $O_2$ ($2H_2O=4H^++e^-+O_2$). Scientists have suggested that Cl$^-$ binds to a positively charged amino acid (N$^+$) which is close to a negatively charged amino acid (B$^-$) in the neighbourhood of the substrate $H_2O$ molecules. Binding to N$^+$ changes the pKa of B$^-$ such that its affinity to $H_2O$ protons increases. This facilitates H$^+$ abstraction from $H_2O$ during its oxidation to $O_2$. In this hypothesis, release of H$^+$s is accompanied by the release of Cl$^-$ ions.

**Discovery**

Warburg and Luttgens discovered that various anions (Cl$^-$, Br$^-$, I$^-$ and $NO_3^-$) stimulate oxygen evolution during Hill reaction in water-washed broken chloroplasts. Chloride was the most effective anion, but rhodanide, $SO_4^{2-}$ and $PO_4^{3-}$ were ineffective in stimulating the Hill reaction. Gorham and Clendening provided the first clear demonstration that Cl$^-$ is specifically involved in $O_2$ evolution during the Hill reaction, and that the Cl$^-$ effect is not just the reversal of injurious effects of exposure to light. Gorham and Clendenning further noted that (*a*) Cl$^-$ shifts the *p*H optimum of the Hill reaction from 6.5 to 7.5 ; and (*b*) this Cl$^-$ effect has similarity to that in dialyzed α-amylase. It is now generally accepted that Cl$^-$ is a necessary cofactor for $O_2$ evolution, based mainly on the pioneering work of S. Izawa and coworkers, Hind *et al.* ; Izawa *et al.* ; Kelley and Izawa ; Muallem *et al.* and Izawa *et al.* After 44 years of research, that occurred mostly in spurts, the molecular mechanism by which Cl$^-$ activates $O_2$ evolution remains unknown. Readers are encouraged to consult earlier reviews by Homann *et al.* Govindjee *et al.* Critchley. Homann, Coleman and Govindjee and Coleman.

**Site of Action**

Figure 5.1 shows a current Z-scheme for electron flow from $H_2O$ to NADP$^+$. This figure is a modification of that pre-

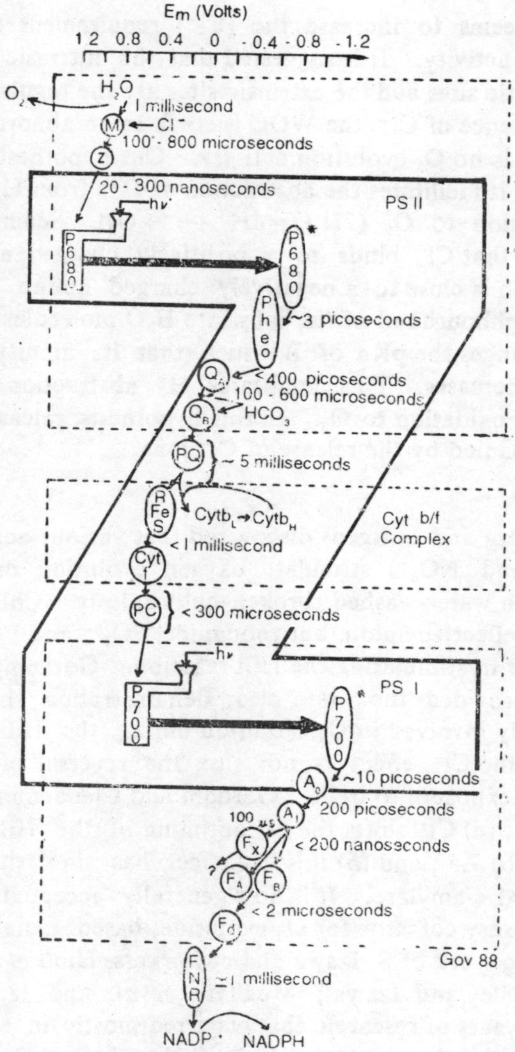


Fig 5.1. **The 'Z' scheme of electron transport in photosynthesis.**

sented earlier by Govindjee and Eaton Rye ; all the symbols are defined in the legend of the figure. The site of $Cl^-$ action is shown here to be on the electron donor side of photosystem II (PSII) between water oxidation and the electron donation by Z of PSII (see left top of Figure 5.1 ; this site is suggested to involve the charge accumulator "M", of the water oxidation complex (WOC).

Figure 5.2 shows a current model of the thylakoid membrane with four major protein complexes : photosystem II, plastoquinol-plastocyanin oxidoreductase or cytochrome $b_6/f$ complex, photosystem I, and the coupling factor ($CF_1$—$CF_0$) or ATPase ; this figure is also a modification of that presented earlier. The site of $Cl^-$ action is on the lumenal portion of photosystem II. The diagram shows $Cl^-$ ions bound to the extrinsic 33 kilodalton polypeptides : $Cl^-$ ions bound to the intrinsic polypeptides are not shown due to lack of space.

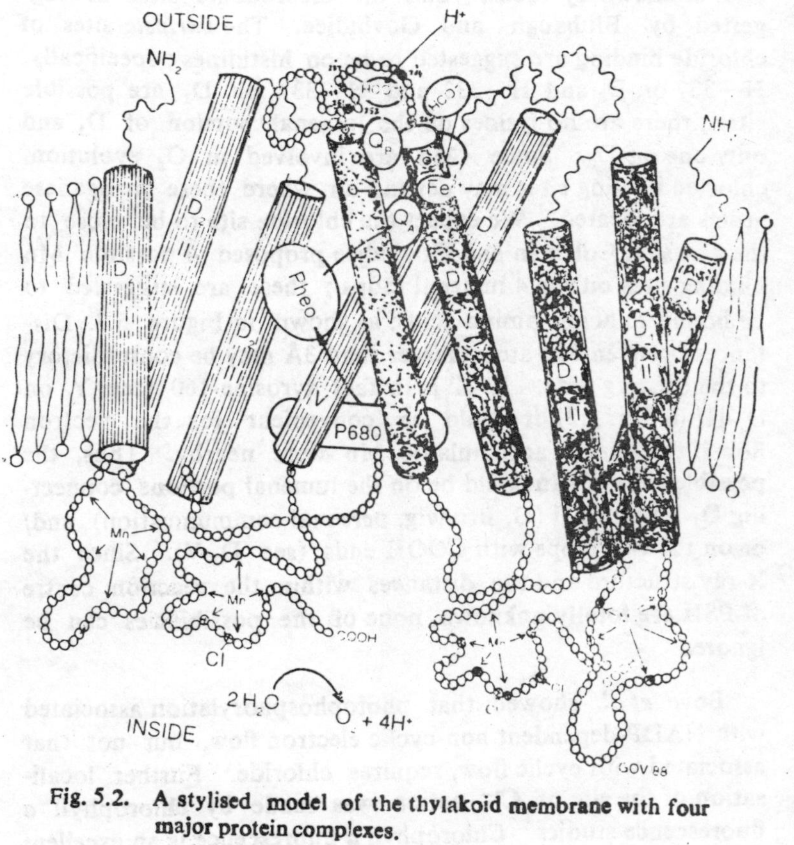


Fig. 5.2. A stylised model of the thylakoid membrane with four major protein complexes.

Figure 5.3 shows a scheme of the reaction centre of PSII, except that cytochrome *b*559 is not shown because its function is unknown. The reaction center of PSII was first isolated by Nanba and Satoh and shown to contain two polypeptides of 32-34 kDa, lebelled as $D_1$ and $D_2$, and the two subunits of cytochrome *b*559 (4 and 9 kDa). This preparation contains 4 chlorophyll and 2 pheophytin molecules, but lacks plastoquinones $Q_A$ and $Q_B$. The diagram shows the membrane portion of $D_1$ and $D_2$ polypeptides arranged in the same fashion as the L and M polypeptides of bacterial reaction centres ; the lumenal portions of $D_1$ and $D_2$ as suggested by Coleman and Govindjee, the herbicide niche (see amino acids, as solid circles) as shown by Trebst ; and the bicarbonate sites as suggested by Blubaugh and Govindjee. The *intrinsic* sites of chloride binding are suggested to be on histidines ; specifically, H—337 on $D_2$ and H—332 and H—337 on $D_1$ are possible sites ; there are no lysines on the lumenal portion of $D_1$ and only one on $D_2$. Since 4Mn are involved in $O_2$ evolution, chloride binding sites may depend on where these manganese atoms are located. We expect the chloride site to be close to the Mn site, Coleman and Govindjee proposed 4 possible Mn binding sites on the 4 lumenal loops ; these are suggested to be bound to acidic amino acids, as shown in Figure 5.3. Distances between Mn atoms larger than 3Å may be contradictory to the existing data. If 'Z' is in fact tyrosine-160 (see Y on $D_1$-III in Fig. 5.2), it would be 'convenient' for the electron flow if the charge accumulator Mn were nearby. Thus, the possible sites of Mn could be on the lumenal portions connecting $D_1$-I and $D_1$-II (G. Brudvig, personal communication) and/or on the two loops with COOH ends (see $D_1$-V). Since the X-ray structure and the distances within the reaction centre of PSII are totally unknown, none of the possibilities can be ignored.

Bove *et al.* showed that photophosphorylation associated with NADP-dependent non-cyclic electron flow, but not that associated with cyclic flow, requires chloride. Further localisation of the site of $Cl^-$ action was made by chlorophyll *a* fluorescence studies. Chlorophyll *a* fluorescence is an excellent

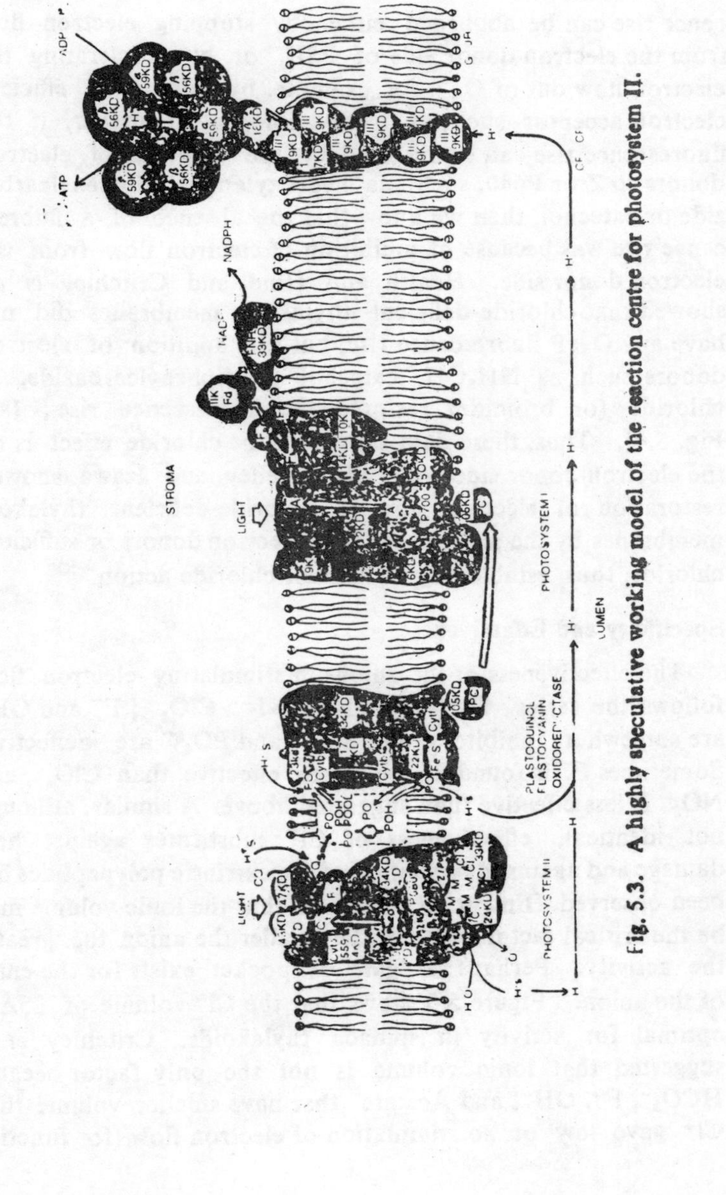


**Fig. 5.3. A highly speculative working model of the reaction centre for photosystem II.**

monitor of PSII reactions. Upon illumination of a darkadapted sample, fluorescence rises instantly to O(Fo) level, and then further increases to a 'P' (F maximum or Fp) level. This O→P rise is a monitor of the reduction of $Q_A$ to $Q_A^-$. This fluorescence rise can be abolished either by stopping electron flow from the electron donor side of PSII, or by accelerating the electron flow out of $Q_A^-$, for example, by adding an efficient electron acceptor such as methylviologen. However, if the fluorescence rise can be restored by the addition of electron donors to Z or P680, such as hydroxylamine, diphenylcarbazide or catechol, then we know that the absence of a fluorescence rise was because of inhibition of electron flow from the electron donor side. Health and Hind and Critchley *et al.* showed that chloride-deficient thylakoid membranes did not have any O→P fluorescence rise, but the addition of electron donors such as $NH_2OH$, catechol or diphenylcarbazide, or chloride (or bromide) restored the fluorescence rise; (see Fig. 5.4). Thus, these data show that the chloride effect is on the electron donor side of PSII. Kelley and Izawa showed restoration of electron flow in chloride-deficient thylakoid membranes by the addition of such electron donors or sufficient chloride, thus, establishing the site of chloride action.

**Specificity and Effectiveness**

The effectiveness of the anions in stimulating electron flow follows the series : $Cl^- > Br^- > NO_3^- > I^- > ClO_4^-$ ; $F^-$ and $OH^-$ are somewhat inhibitory, and $SO_4^{2-}$ and $PO_4^{3-}$ are ineffective. Sometimes $F^-$ is found to be more effective than $ClO_4^-$, and $NO_3^-$ is less effective than suggested above. A similar, although not identical, effectiveness of $Cl^-$ substitutes against heat damage and against dissociation of the extrinsic polypeptides has been observed. Hind *et. al.* suggested that the ionic volume may be the critical factor ; that is, the smaller the anion the greater the activity. Perhaps a 'gate' or 'pocket' exists for the entry of the anion. Figure 5.5 shows that the $Cl^-$ volume of 25Å is optimal for activity in spinach thylakoids. Critchley *et al.* suggested that ionic volume is not the only factor because $HCO_3^-$, $F^-$, $OH^-$, and Acetate$^-$ that have smaller volume than $Cl^+$ gave low or no stimulation of electron flow (for function

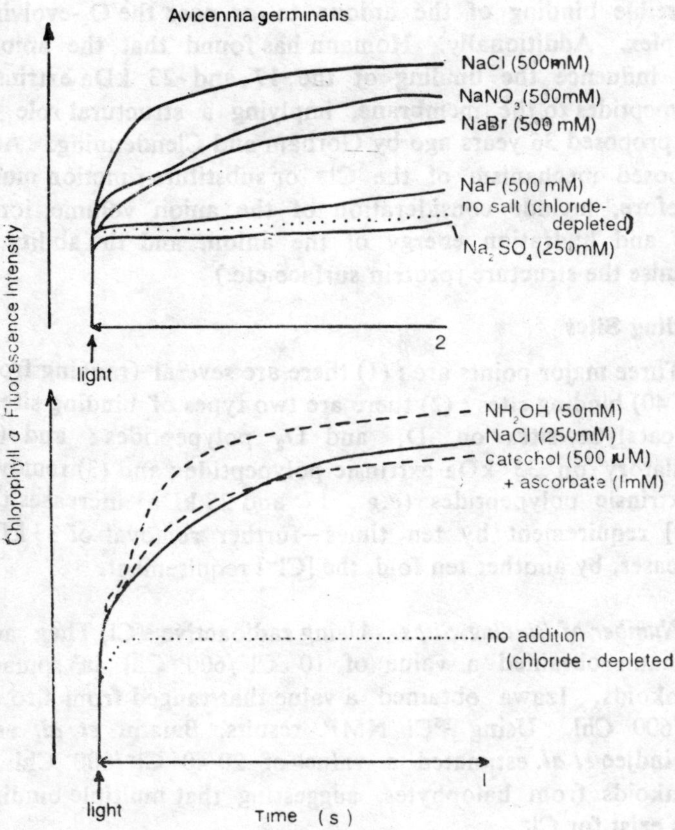


**Fig. 5.4. Chlorophyll a fluorescence transients in halophyte thylakoids.**

**of $HCO_3^-$ on electron acceptor side. A plot of the effectiveness of the anion as a function of ionic field (= $Q_A/e_s r_A^2$, where $Q_A$ is the anion charge, $r_A$ is the Pauling radius and $e_s$ is the differential dielectric constant of water) showed that the effectiveness of anions decreases with decreasing ionic field. However, anions with large ionic field ($HCO_3^-$, $SO_4^{2-}$, $Ac^-$, $PO_4^{3-}$, $OH^-$ and $F^-$) were also relatively ineffective in stimulating electron flow.**

**Coleman *et al.* had suggested that the similarity of the specificity of the anions for stimulating electron flow and**

protection against mild heat damage is due to similar selective reversible binding of the anions to or near the $O_2$-evolving complex. Additionally. Homann has found that the anions also influence the binding of the 17 and 23 kDa extrinsic polypeptides to the membrane, implying a structural role of $Cl^-$ proposed 36 years ago by Gorham and Clendenning. Any proposed mechanism of the $Cl^-$ or substitute function must, therefore, include consideration of the anion volume, ionic field and hydration energy of the anion, and the ability to organise the structure (protein surface etc.)

**Binding Sites**

Three major points are : (1) there are several (ranging from 4 to 40) binding sites ; (2) there are two types of binding sites : (*a*) catalytic sites on $D_1$ and $D_2$ polypeptides : and (*b*) regulatory on 33 kDa extrinsic polypeptide : and (3) removal of extrinsic polypeptides (*e.g.*, 17 and 23 kDa) increases the [$Cl^-$] requirement by ten times—further removal of 33 kDa increases, by another ten fold, the [$Cl^-$] requirement.

*Number of Binding Sites.* Using radioactive $^{36}Cl$, Theg and Homann obtained a value of 10 $Cl^-$/600 Chl in spinach thylakoids. Izawa obtained a value that ranged from 4 to 40 $Cl^-$/600 Chl. Using $^{35}Cl$ NMR results, Baianu *et al.* and Govindjee *et al.* estimated a value of 20-40 $Cl^-$/600 Chl in thylakoids from halophytes, suggesting that multiple binding sites exist for $Cl^-$.

*Multiple Binding Sites.* The existence of multiple sites of $Cl^-$ was most dramatically demonstrated by $^{35}Cl$-NMR measurements on photosystem II and thylakoid membranes from spinach. Plots of $^{35}Cl$ NMR linewidth, corrected for that in chloride solution, show several maxima that occur at roughly the same [$Cl^-$] as intermediary plateaus in Hill activity (at low light intensities) versus [$Cl^-$] plot, indicating a possible correlation between the two measurements.

These maxima were obtained from data collected by an NMR instrument at the University of Illinois at Urbana (NSF-250 ; 250 MHZ homebuilt NMR spectrometer : spectra

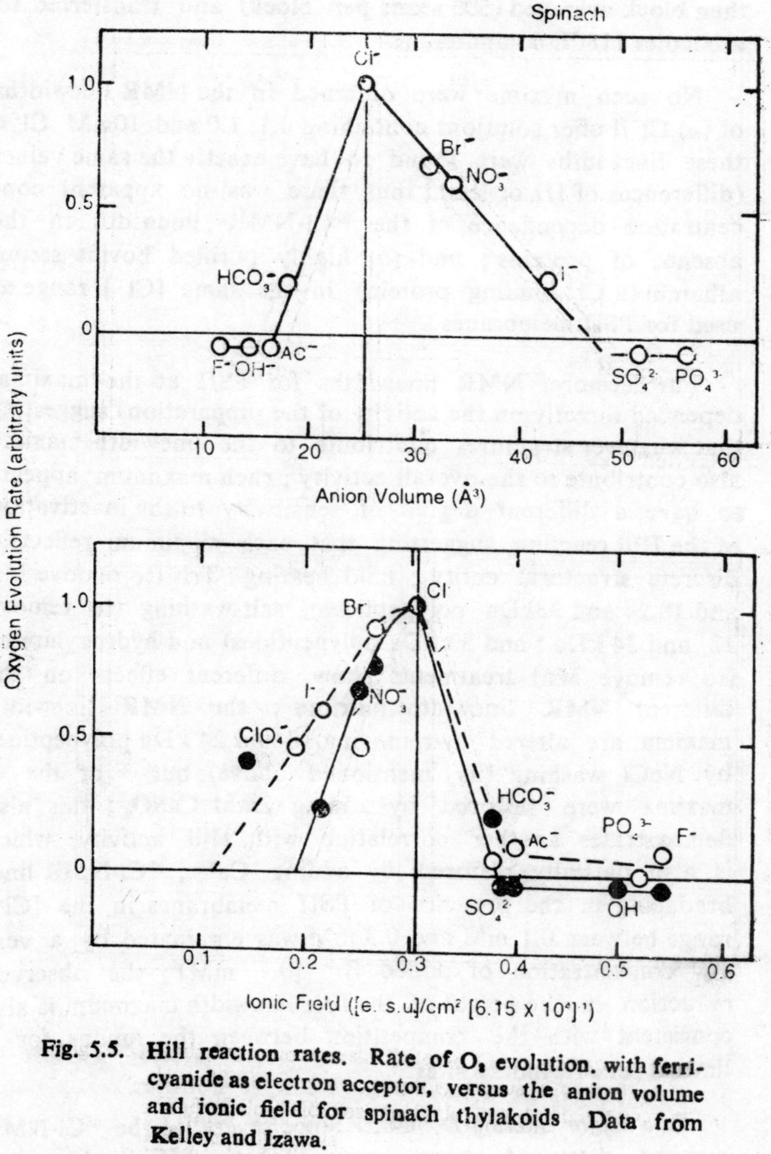


**Fig. 5.5. Hill reaction rates. Rate of $O_2$ evolution, with ferricyanide as electron acceptor, versus the anion volume and ionic field for spinach thylakoids Data from Kelley and Izawa.**

**obtained at 24,508 KHz using a 33$\mu s$ 90° pulse and 360 ms recycle time; signals detected in quadrature with 32K data points and a spectral width of ± 25,000 Hz; signals were**

then block averaged (500 scans per block) and transferred to a Nicolet 1180E computer.

No such maxima were obtained in the NMR linewidths of (*a*) $Cl^-$/buffer solutions containing 0.1, 1.0 and 10mM $Cl^-$; these linewidths were found to have exactly the same values (differences of Hz or less); thus, there was no apparent concentration dependence of the $^{35}$Cl-NMR linewidth in the absence of proteins; and (*b*) highly purified bovine serum albumin (a $Cl^-$ binding protein) in the same [$Cl^-$] range as used for PSII membranes.

Furthermore, NMR linewidths for PSII at the 'maxima' depended directly on the activity of the preparations suggesting that whatever structures contribute to the linewidth maxima also contribute to the overall activity; each maximum appears to have a different degree of sensitivity to the inactivation of the Hill reaction, suggesting that each maximum reflects a discrete structural entity; mild heating: Tris (to remove Mn and 18,24 and 33kDa polypeptides), salt-washing (to remove 18 and 24 kDa; and 33 kDa polypeptides) and hydroxylamine (to remove Mn) treatments show different effects on the different NMR linewidth maxima; the NMR linewidth maxima are altered by removing 18 and 24 kDa polypeptides by NaCl washing (as mentioned above) but 3 of the 4 maxima were restored by adding 2mM $CaSO_4$; this also demonstrates another correlation with Hill activity, which is also partially restored by adding $Ca^{2+}$; $^{35}$Cl-NMR linebradening in the presence of PSII membranes in the [$Cl^-$] range between 0.1 mM and 0.3 mM was eliminated by a very low concentration of added $Br^-$ (0.1 mM); the observed reduction in the height of the first linewidth maximum is also consistent with the competition between the anions for a limited set of binding sites.

*Two Major Binding Domains.* Since several of the $^{35}$Cl-NMR maxima, discussed above, occur at higher [$Cl^-$] than that needed for activating $O_2$ evolution, it is suggested that these may be related to the *regulatory*, rather than the catalytic, function of $Cl^-$. Mild heating (3 min. 30°C) eliminates the

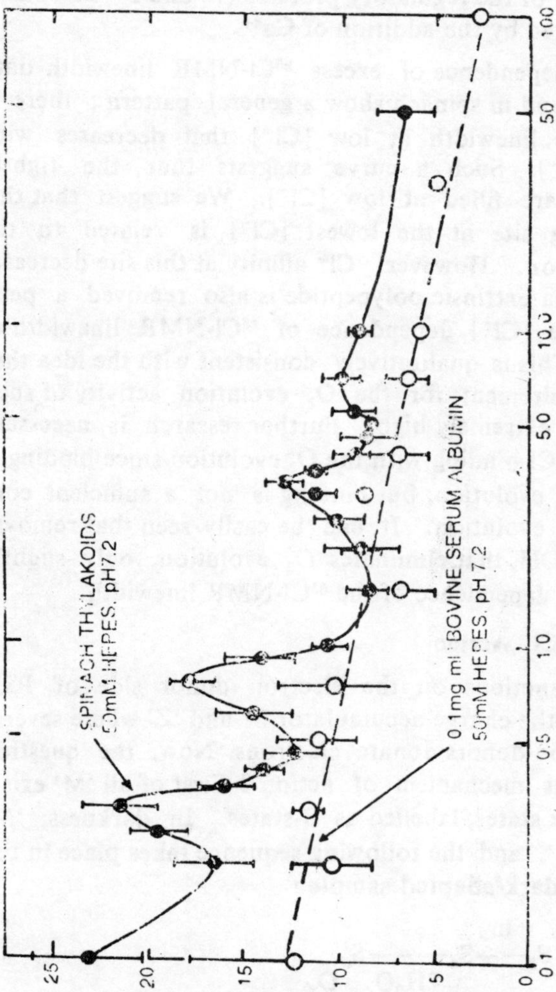


**Fig. 5.6. $^{35}$Cl-NMR binding curve for thylakoid membranes and for bovine serum albumin (BSA).**

high [$Cl^-$] 'kinks' of maxima and minima in both the [$Cl^-$] dependence of the $^{35}Cl^-$NMR linewidth and the Hill reaction. Furthermore, these are the ones that are dramatically altered by the removal of the regulatory proteins (18 and 24 kDa) and partially restored by the addition of $Ca^{2+}$.

The [$Cl^-$] dependence of excess $^{35}Cl$-NMR linewidth data in halophytes and in spinach show a general pattern : there is a large excess linewidth at low [$Cl^-$] that decreases with increasing [$Cl^-$]. Such a curve suggests that the tightly binding sites are filled at low [$Cl^-$]. We suggest that this tightly binding site at the lowest [$Cl^-$] is related to the catalytic function. However. $Cl^-$ affinity at this site decreases when the 33kDa extrinsic polypeptide is also removed a peak appears in the [$Cl^-$] dependence of $^{35}Cl$-NMR linewidth at $\sim$ 0.5 mM. This is qualitatively consistent with the idea that the [$Cl^-$] requirement for the $O_2$ evolution activity of such preparations is extremely high. Further research is necessary to correlate $^{35}Cl$ binding with the $O_2$ evolution since binding is required for $O_2$ evolution, but binding is not a sufficient condition for $O_2$ evolution. It can be easily seen that removal of Mn by $NH_2OH$, that eliminates $O_2$ evolution, only slightly alters the [$Cl^-$] dependence of the $^{35}Cl$-NMR linewidth.

**Mechanism of $Cl^-$ Action**

The $Cl^-$ functions on the electron donor side of PSII between or at the charge accumulator M and 'Z' where several artificial electron donors donate electrons. Now, the question is : What is its mechanism of action ? First of all 'M' exists in several redox states, labelled as S-states. In darkness, 'M' is in state '$S_1$', and the following sequence takes place in the first cycle in a dark-adapted sample :

$$S_1 \xrightarrow{h\nu_1} S_2 \xrightarrow{h\nu_2} S_3 \xrightarrow{h\nu_3} S_4 \xrightarrow[2H_2O \quad O_2]{} S_0$$

And, the following sequence occurs in the subsequent cycle(s) :

$$S_0 \xrightarrow{h\nu_4} S_1 \xrightarrow{h\nu_5} S_2 \xrightarrow{h\nu_6} S_3 \xrightarrow{h\nu_7} S_4 \xrightarrow[2H_2O \quad O_2]{} S_0$$

Here, $h\nu_n$ stands for a light flash, and $S_n$ for a redox state of 'M', the higher n's representing higher oxidation states. Thermoluminescence is a powerful tool to monitor the S-states because it represents the back reaction of PSII involving the S-states and the electron acceptor side. Thermoluminescence is suggested to appers as follows :

$$S_2.Z.P680.Pheo.Q_A.Q_B^- \text{ (created by light reaction)} \xrightarrow{\text{warm}}$$

$$S_1.Z^+.P680.Pheo.Q_A^-.Q_B \xrightarrow{\text{warm}} S_1Z.P680^+.Pheo^-Q_A.Q_B \xrightarrow{\text{warm}}$$

$$S_1.Z.P680^*.Pheo.Q_A.Q_B \xrightarrow{\text{spontaneous}} S_1.Z.P680.Pheo.Q_A.Q_B + \text{light}$$

This light appears as a glow peak 'B' in thermoluminescence curves and is said to originate from an $S_2Q_B$ back reaction. On the basis of their thermoluminescence results. Homann *et al.* and Vass *et al.* suggested that $Cl^-$ depletion generates an abnormal '$S_2$' state. It was, however, Muallem *et al.* who had first suggested that $Cl^-$ depletion causes the formation of defective oxidant storage. Several investigators, using a variety of experimental approaches, have confirmed this conclusion. Flash illumination of dark-adapted $Cl^-$ depleted preparations generates an abnormal $S_2$-state (labelled as '$\Sigma_2$ state') which has an abnormally long lifetime and has a lower than normal oxidation potential. This $\Sigma_2$ state, unlike the normal $S_2$ state, does not produce the normal multiline ESR signal for manganese. However, the $g=4.1$ signal of the $S_2$ state is still present in $Cl^-$ depleted samples. Addition of $Cl^-$ restores the normal $S_2$ signal. $^{35}Cl$-NMR linewidth broadening, suggestive of Cl-binding, appears to be dependent upon the S-state transitions. Preston and Pace showed its relationship to the $S_2+S_3$ states.

In order to formulate a theory for the mechanism of $Cl^-$ action, it is useful to keep in mind the following points : (1) Inactivation by $Cl^-$ depletion is reversible ; it appears that only electrostatic interactions are involved ; (2) $Cl^-$ depletion is accelerated by incubating the thylakoids at high pH ; (3) $Cl^-$ binding is pH dependent and reversible (4) activation of the Hill reaction by added $Cl^-$ shows hyperbolic kinetics, indicating saturation ; (5) activation of the Hill reaction by anions is relatively, but not exclusively, specific for $Cl^-$, the selectivity does not resemble the lyotropic Hofmeister series ; and there is a competitive interaction of anions ; (6) the pH optimum of the Hill reaction is shifted to more alkaline pH by $Cl^-$ binding : this suggests that $Cl^-$ addition is equivalent to an acidification ; $Cl^-$ may be involved in splitting of $H_2O$ and protonation steps ; and (7) an interaction of $Ca^{2+}$ with the $Cl^-$ effect exists.

To this check list, we may add the following unique properties of $Cl^-$ in the water oxidising steps : there are two types of $Cl^-$ binding/interactions ; $Cl^-$ binding is controlled by a group with rather low pKa : there is modulation of $Cl^-$ binding by the extrinsic polypeptides prevention of binding of other anions and chemicals by $Cl^-$ : effect of sulphate ions on $Cl^-$ action suggesting that its interaction involves deprotonation, and the variation of $Cl^-$ binding during S-state transitions.

There are two major effects of $Cl^-$ : a regulatory (a structural) role ; and a catalytic role. We shall discuss here only the latter. Govindjee *et al.* suggested that the function of $Cl^-$ was to act as a counterion to positive charges on the oxygen-evolving complex. Since these charges arrive after each light flash, it was suggested that $Cl^-$ binding occurred after each S-state transition. The release of $Cl^-$ was associated with the release of a positive charge, *i.e.*, a proton ($H^+$). Since there is no proton release during $S_1 \rightarrow S_2$ transition, the unbinding of $Cl^-$ was suggested not to occur after this transition. Thus, the absence of $Cl^-$ could specifically affect the $S_2$ state created. In this hypothesis, the function of $Cl^-$ was to bind to the system when a positive charge arrived on the water oxidation complex, stabilising the system, and to lead the $H^+$s away from the system by simultaneous unbinding from the system. The reversible binding of $Cl^-$ (high exchange rate) and the low binding energy ($\sim$ 3Kcal) supported this picture.

Coleman and Govindjee proposed a detailed picture as to how $Cl^-$ activates the base catalysis, *i.e.*, how it activates the

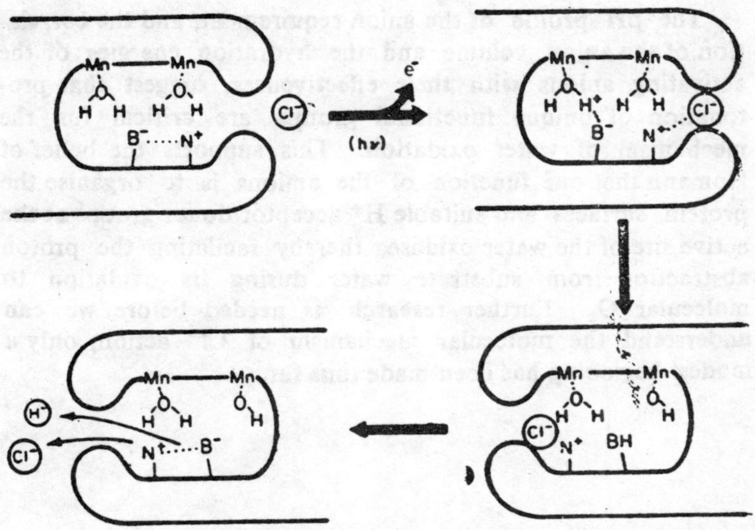


Fig. 5.7. A working model for $Cl^-$ activation of $O_2$ evolution. Here binding of the anion to a basic amino acid ($N^+$) shifts the pKa of a reactive group ($B^-$; an acidic amino acid) on the lumenal side of $D_1$ and $D_2$ polypeptides (also see Fig 5.2). This makes $B^-$ more reactive in 'pulling' water protons from water molecules bound at the active site. As soon as BH is made. $H^+$ (bound at $B^-$) and $Cl^-$ (on $N^+$) leave the binding sites restoring the system to its original state. In this hypothesis. $Cl^-$ is reversibly bound and unbound.

$H^+$ removal from water (see Fig. 5.7). Here, a proton accepting group $B^-$ functions to extract $H^+$s from $H_2O$, whereas Mn functions to extract an electron from $H_2O$, and $Cl^-$ functions to bind transiently (and thus, reversibly) to a positively charged group $N^+$ that raises the pKa of $B^-$. The $N^+$ may represent an amino acid histidine (see Fig. 5.3), and $B^-$ may represent a proton accepting group (Fig. 5.3). Thus, as $Cl^-$ binds to $N^+$, the pKa of the nearby $B^-$ changes, and this increases its affinity to catalytic $H_2O$ protons at the Mn-site making it easier to 'pull' $H^+$s out of water, as 'Mn' pulls electrons from $H_2O$. Both the $H^+$s and $Cl^-$s leave the catalytic site at the same time. $H^+$s may then be conducted through the 33 kDa $Cl^-$ -binding extrinsic polypeptide to the outside, *i.e.*, into the lumen.

The pH profile of the anion requirement, and the correlation of the anion volume and the hydration energies of the activating anions with their effectiveness, suggest that protonation of unique functional groups are critical for the mechanism of water oxidation. This supports the belief of Homann that one function of the amions is to organise the protein surfaces and suitable $H^+$ acceptor/donor groups at the active site of the water oxidase, thereby facilating the proton abstraction from substrate water during its oxidation to molecular $O_2$. Further research is needed before we can understand the molecular mechanism of $Cl^-$ action, only a modest beginning has been made thus far.

# 6

# Net Gas Exchange

Photosynthesis involves the coordinated operation of a series of biochemical and biophysical reactions starting with the absorption of photons and ending with the incorporation of inorganic carbon into stable organic compounds. All algae and the cyanobacteria, as well as terrestrial vascular plants, evolve oxygen during the light reactions that produce reductant ($NADP^+ \rightarrow NADPH$) and adenosine triphosphate (ATP). These products of the light reactions are subsequently consumed in $CO_2$ fixation by the photosynthetic carbon reduction cycle (PCRC). Competing with photosynthetic carbon reduction for NADPH and ATP are other processes, including photorespiration, nitrate reduction, mediated ion and gas transport between algal cell and environment, and various maintenance and synthetic processes. Although photosynthetic physiology can be conceptually isolated from the remainder of algal metabolism and in vitro separation of light and dark reactions of isolated chloroplasts has been accomplished in practice photosynthesis occurs simultaneously with, and cannot be separated from, the remainder of algal growth and maintenance processes. This complexity can lead to some confusion in the operational definition of the rate of photosynthesis.

The rate of photosynthesis can be defined as the rate of evolution of $O_2$ by the light reactions of photosynthesis or the rate of consumption of $CO_2$ and/or $HCO_3^-$ by the dark reactions. Measurement of these rates is complicated by simultaneous $O_2$ consumption and $CO_2$ evolution, and, in practice, only net gas exchange is readily quantified. The use of tracers such as $^{18}O$ and $^{14}C$ allows investigation of rates of partial reactions, but may be subject to some uncertainty associated with isotope disequilibrium. This chapter considers methods for measuring net gas exchange.

Net gas exchange is the oldest and most widely employed procedure for measuring plant photosynthesis. It remains the standard against which other measurements are often compared. Changes in concentration of either dissolved oxygen ($O_2$) or total inorganic carbon ($TCO_2$) in an aqueous sample or $CO_2$ in a gas sample are used to determine net-gas-exchange rates. Limits on the magnitude of measurable rates are set by the concentration of $O_2$ and $TCO_2$ in water and the accuracy of methods for determining $O_2$ and $TCO_2$ concentrations. Our discussion begins with a consideration of these factors.

### Factors Influencing the Distribution of $O_2$ and $CO_2$

The solubility of dissolved $O_2$ in water is determined by temperature, salt concentration, and atmospheric pressure. Over the range of conditions typical of most natural waters, the concentration of $O_2$ in equilibrium with the atmosphere varies by a factor of approximately three (Table 6.1), decreasing from 457 $mmol.m^{-3}$ in distilled water at 0°C, to 163 $mmol.m^{-3}$ at 40 parts per thousand (‰) salinity and 40°C (Benson and Krause, 1984). Wide variations from the thermodynamically defined equilibrium $O_2$ concentrations can result from biological and chemical oxygen-consuming or producing reactions. These variations are often expressed in terms of percent saturation or apparent oxygen utilization (AOU).

$$\%\ \text{Saturation} = [O_2]/[O_2]_{eq} \qquad (6.1)$$

$$\text{AOU} = [O_2] - [O_2]_{eq} \qquad (6.2)$$

Where $[O_2]$ is the dissolved oxygen concentration $[O_2]_{eq}$ is the thermodynamically defined equilibrium dissolved oxygen concentration at the ambient temperature and salinity.

**Table 6.1. Values for the equalibrium dissolved oxygen concentration mmol.m$^{-3}$ at 1 atm pressure acid various temperatures and salinities. For conversion from mmol.m$^{-3}$ to mg.1$^{-1}$ multiply by 3.19988 | × | 10$^{-2}$.**

| *Temperature* (°C) | *Salinity* (‰) | | | | |
|---|---|---|---|---|---|
| | *0* | *10* | *20* | *30* | *40* |
| 0 | 457 | 426 | 397 | 370 | 345 |
| 10 | 353 | 331 | 310 | 291 | 273 |
| 29 | 284 | 268 | 253 | 238 | 224 |
| 30 | 236 | 224 | 212 | 200 | 190 |
| 40 | 200 | 190 | 181 | 172 | 163 |

**Values calculated from :**

$$\ln (O_2)_{eq} = -133.90205 + 1.576.701 \times 10^5/T - 6.642308 \times 10^1/T^2 + 1.243800 \times 10^{10}/T^3 - 8.621949 \times 10^{11}/T^4 - S(0.017674 - 10754/T + 2140.7/T^2)$$

**where T is the absolute temperature (in°K) and S is the solinity (in %). For more details and corrections to account for variations in atmospheric pressure from 1 atm.**

In upper ocean waters characterized by low algal abundance (<1 μg chlorophyll *a* dm$^{-1}$) oxygen percent saturation ranges approximately from 90% to 110% with pronounced seasonal variations and smaller diel variations. In eutrophic waters, the range in oxygen percent saturation can be much more pronounced (< 10 to > 200%) over a diel cycle as a consequence of a higher biomass and greater rates of biological activity. The diel and seasonal ranges of dissolved oxygen concentration and percent saturation depend on the balance between biological activity, which is often limited by nutrient supply, advective and diffusive exchange between water masses, and exchange through the air-water interface. By correcting for exchange processes, it is possible to estimate rates of community photosynthesis and respiration either from diel measurements or from changes in oxygen concentration over longer time periods.

The concentration of dissolved $CO_2$ at air equilibrium in natural waters is much lower than that of $O_2$, ranging from approximately 25 mmol.m$^{-3}$ at 0°C in freshwater, to 6 mmol-$m^{-3}$ at 40°C in seawater of 40‰ salinity. However, air equilibrium is not always attained and $CO_2$ can be significantly enriched or depleted by the effects of biological activity on aqueous inorganic carbon equilibria. Ionized forms of inorganic carbon can be present in appreciable and often variable amounts depending on the pH and ionic content of the solution. Total inorganic carbon ($TCO_2$) is much more variable than dissolved $CO_2$ ranging from approximately 10 mmol.$m^{-3}$ in acidic freshwater at 20°C, increasing to 2.3 mol.$m^{-3}$ in seawater with a salinity of 35‰, and exceeding 100 mol.$m^{-3}$ in bicarbonate-rich waters at high pH.

Dissolved inorganic carbon exists in a number of readily interconvertable forms. The most important reactions involved in this system are

$$CO_2 + H_2O \leftrightarrow H_2CO_3 \quad (6.3)$$
$$H_2CO_3 \leftrightarrow HCO_3 + H^+ \quad (6.4)$$
$$HCO^-_3 \leftrightarrow CO^{2-}_3 + H^+ \quad (6.5)$$

where the forms of inorganic carbon are aqueous carbon dioxide gas dissolved in water ($CO_2$), carbonic acid ($H_2CO_3$), bicarbonate ion ($HCO^-_3$) and carbonate ion ($CO_3^{2-}$). The total inorganic carbon concentration ($TCO_2$) is the sum of all of these forms :

$$TCO_2 = [CO_2] + [H_2CO_3] + [HCO^-_3] + [CO_3^{2-}] \quad (6.6)$$

Because of the difficulty in separately measuring the equilibria for Eqs. 6.3 and 6.4 these equations are often combined.

$$CO_2 + H_2O \leftrightarrow HCO^-_3 + H^+ \quad (6.7)$$

Of the different forms of inorganic carbon present in water, only $CO_2$ and $TCO_2$ are amenable to direct experimental observation. The concentrations of the other forms must be calculated from thermodynamic considerations. The equilibrium inorganic carbon speciation is influenced by coexisting equilibria involving the ionization of water ($H_2O \leftrightarrow H^+ + OH^-$), ionization of weak acids such as boric acid ($H_3BO_3 + H_2O \leftrightarrow B(OH)^-_4 + H^+$), hydrogen sulfide ($H_2S \leftrightarrow HS^- + H^+$), or ammonium ($NH_3+H_2O \leftrightarrow NH^+_4+OH$), and the formation of ion pairs with various cations (Na, K, Ca, Mg) or the hydrolysis of aluminum ions ($Al^3+2H_2O \leftrightarrow Al(OH)^{2+}+2H^+$). The most important physical and chemical properties used for describing the equilibrium concentrations of various inorganic carbon species include temperature, pH, and carbonate alkalinity.

Alkalinity (equivalents $dm^{-1}$) is defined as the sum of the normalities (*i.e.*, the product of charge and molarity) of the anions of weak acids. In most waters, the dominant weak acids are those of the inorganic carbon system, and total alkalinity (TA) is defined as

$$TA=[HCO_3]+2[CO_3^{2-}]+[OH^-]-[H^+]+C_{ra} \qquad (6.8)$$

where $C_{ra}$ is the surplus alkalinity due to anions of weak acids other than carbonic acids. Carbonate alkalinity is the contribution of the first two terms on the left-hand side of Eq. 6.8. The contributions of [$OH^+$] and [$H^+$] will be significant under acidic or basic conditions, but can be ignored in most natural waters at a normal pH of 6-8. In sea waters, the contribution of borate to total alkalinity cannot be neglected, but a correction based on salinity can be applied. Organic acids may comprise a variable, but substantial, portion of total alkalinity in freshwaters and have been suggested to comprise a small but measurable component of alkalinity in seawater. The contributions of $H_2S$ and $NH^+_4$ to total alkalinity are usually unimportant under aerobic conditions where photosynthetic gas exchange is being measured, but may be important in suboxic or anoxic waters.

### Oxygen Determinations

The concentration of dissolved $O_2$ can be measured by a number of techniques. Many early studies employed manometry, which is still advocated by some researchers. However, most recent investigations have been undertaken with the Winkler method or with polarographic oxygen electrodes. Alternative methods, which were used in a limited number of investigations, but will not be considered further here, include gas chromatography for the simultaneous determination of

$O_2$ and other gases, microgasometry for the determination of $O_2$ in waterlogged soils and other fluids containing high concentrations of organic matter or solids and mass spectrometry.

### Winkler Method

The Winkler method (1888) is the oldest method used to quantitatively determine dissolved $O_2$. It is based on the quantitative oxidation of Mn(II) to Mn (III) by $O_2$ in an alkaline solution followed by the oxidation of $I^-$ to $I_2$ in acid solution and titration of $I_2$ with thiosulfate. This *fixation* of $O_2$, which comprises the initial steps of the Winkler technique, is summarized by the following reactions.

$$Mn^{2+}+2OH^- \rightarrow Mn(OH)_2 \tag{6.9}$$

$$2Mn(OH)_2+0.5O_2+H_2O \rightarrow 2Mn(OH)_3 \tag{6.10}$$

$$2Mn(OH)_3+6H^++2I^- \rightarrow 2Mn^{2+}+I_2+6H_2O \tag{6.11}$$

$$I_2+I^- \leftrightarrow I_3 \tag{6.12}$$

Once these reactions have been completed, a sample can be allowed to stand for a period of up to one day before completing the analysis. In the Winkler method, the $O_2$ content of a sample is determined without recourse to an $O_2$ standard. Instead, the technique relies on the quantitative stoichiometric conversion of $O_2$ to $I_2$ following by measurement of the amount of $I_2$ produced by titration with thiosulfate ($S_2O^2_3$) to the starch endpoint, although potentiometric or amperometric endpoints can also be employed. The strength of the thiosulfate titrant is normally determined by titrating a standard potassium iodate solution.

$$I_3+2S_2O^2_3 \rightarrow 3I^+ +S_4O^2_6 \tag{6.13}$$

Spectrophotometric measurements of $I_2$ at 287.5 nm or $I^-_3$ at 352 nm although less accurate than carefully conducted iodometric titrations, provide a quick method for determining $O_2$ content to two significant figures.

Modifications to the original Winkler method by Montgomery *et al.* (1964) and Carpenter (1965b) have resulted in significant improvements over the original formulation, and the accuracy of the modified Winkler method has been experimentally demonstrated. Descriptions of the modified Winkle technique, together with practical advice on its application are given by Hitchman (1978) and Grasshoff *et al.* (1983). Currently accepted protocols described by the American Public Health Association (1985) are not as precise as the protocols described by Montgomery *et al.* (1964) and Carpenter (1965*b*). Recent improvements in methodology involve automation of the titration using photometric or amperometric endpoints. A high accuracy (standard deviation of 0.3 $mmol.m^{-3}$) can be obtained with the Winkler technique provided that precise control is exercised over sample and reagent volumes.

Samples for $O_2$ analysis are usually collected in 50-300 $cm^3$ glass bottles with tapered ground-glass stoppers. Entrainment of atmospheric $O_2$ can be a major source of error in the technique and must be avoided during sample collection. Bottles are usually filled from a reservoir, or water bottle sampler, by inserting a tube to the bottom of the $O_2$ bottle and allowing liquid to overflow for at least three bottle volumes before addition of the Winkler reagents (American Public Health Association, 1985). Alternatively, smaller samples can be collected in 5.0-50 $cm^3$ glass syringes with tight-fitting ground-glass plungers, provided that care is taken to avoid atmospheric contamination and cavitation.

Loss of iodine can be a significant source of error in the Winkler technique, and a number of precautions can be taken to obtain the highest accuracy. These include protecting samples from bright light to prevent the photooxidative consumption of $I_2$ after acidification, and titrating the entire sample in the original battle (rather than removing a subsample for titration), to avoid volatilization of $I_2$. Other problems can be encountered with the Winkler technique in waters containing high concentrations of organic matter, nitrite, ferrous iron, or sulfites, and modifications to the recommended procedures have been developed to overcome these potential problems.

**Oxygen Electrodes**

Oxygen electrodes are most often used in investigations with laboratory cultures of microalgae or thalli of macroalgae in which the ratio of plant biomass to volume of suspending medium is high. Polarographic $O_2$ electrodes were introduced to photosynthesis research in the late 1930s, when a hanging-drop mercury electrode was used in studies of oxygen evolution and measurements of the photon yield of photosynthesis in *Chlorella*. Grasshoff (1981) describes the application of hanging-drop mercury electrodes for measuring $O_2$ with specific applications to oceanographic research, although the potentially toxic effects of mercury may limit the usefulness of this method for continuous measurement of oxygen exchange.

Fixed platinum electrodes were used to obtain qualitative measurements of photosynthesis in vascular plant leaves and algal thallus tissue by Blinks and Skow (1938) and Haxo and Blinks (1950), although reproducibility in these early studies was poor. Olson *et al.* (1949) measured the transient current associated with an alternating potential applied to the electrode to obtain continuous measurements of $O_2$ concentration with a response time of seconds. The method was used to measure the photon yield of $O_2$ evolution by *Chlorella pyrenoidosa*. The rapid response of bare platinum electrodes make them especially suitable for specialized measurements, such as the determination of photosynthetic action spectra, oxygen flash yields and measurements of S states but these electrodes may be subject to considerable error.

The most commonly employed oxygen electrode for routine use is the membrane-covered polarographic detector. This type of electrode, known as the Clark electrode, was designed to overcome problems associated with drift in calibration due to changes in the chemical composition of the sample and poisoning of the electrode from surface-active components that may contaminate bare electrodes (see above). The Clark type electrode consists of a platinum cathode, silver anode, and KCl salt solution, which are separated from the sample by a gas-permeable membrane. According to Grasshoff (1981) the reaction of the cathode ($O_2 + 4H^+ + 4e \rightarrow 2(H_2O)$ is coupled to the reaction at the anode ($4Ag \rightarrow 4Ag^+ + 4e$) by the electrolyte KCl ($Ag^+ + Cl^- \rightarrow AgCl$).

Membrane-convered electrodes are produced by a number of manufacturers and are routinely used in many laboratories. They are less subject to drift or contamination than bare platinum electrodes and allow much more rapid measurement of $O_2$ concentration than the Winkler method. These electrodes, typically, exhibit a 90% response time of 30 sec and a precision of 10 mmol.m$^{-3}$. The response time is considerably slower than that obtained with a bare platinum electrode, and the precision is 10 to 100 times less than can be obtained from the Winkler titration. Use of the membrane-covered electrode requires that the sample be stirred to break down boundary layers in the aqueous phase so that diffusion through the membrane becomes the step that limits $O_2$ supply to the electrode surface.

The output from membrane-covered electrodes is linearly related to the partial pressure of $O_2$ ($pO_2$) in the solution. To obtained absolute $O_2$ concentration requires additional information on temperature, salinity, and atmospheric pressure so that $pO_2$ can be converted to a concentration. Temperature must be carefully controlled, because electrode properties and response, particularly the diffusion of $O_2$ through the membrane, are highly temperature-dependent. Routine calibration of commercial electrodes at two points, usually oxygen-free water (a solution of 2% sodium sulfite or a supply of oxygen-free nitrogen gas) and air-saturated water, is recommended. Electrodes can also be calibrated in water-saturated air, however, care must be taken to ensure that the electrode surface is free of contaminants and that the air is completely saturated with water vapor.

Polarographic $O_2$ *micro*electrodes are becoming a standard component of some specialized areas of photosynthesis research. The construction and use of these electrodes is described by Revsbech and Jorgensen (1986). Electrodes with a tip diameter of only a few micrometers are useful for photosynthesis measurements that require high spatial resolution. These electrodes have been used to measure photosynthesis within benthic algal mats, by symbiotic algae associated with planktonic foraminifera, or by algae within macroscopic aggregates. They may also prove to be useful in investigation of oxygen fluxes and concentration within the diffusive boundary layer around macroalgae. Despite the expectation that diffusion through a boundary layer at the electrode surface should be negligible for microelectrodes, Gust *et al.* (1987) report that the response of microelectrodes in still water is only 10% to 50% of the maximum response in stirred water, which may lead to artifacts when measuring $O_2$ concentrations in gradients of water velocity such as occur at solid surfaces in flowing waters.

A computerized, pulsed membrane-covered electrode with a precision of 0.8 $mmol.m^{-3}$ approaches the sensitivity of the Winkler technique (Langdon, 1984). This approach relies on measuring the transient current associated with application of a pulsed potential across the electrochemical cell. The electrode does not require stirring and is thus suitable for use with fragile microalgae such as the dinoflagellate *Gonyaulax tamarensis* which may be damaged by conventional technique. An instrument based on the pulsed electrode can be deployed for periods of 3-4 months in the sea to provide continuous monitoring of oxygen in situ.

Direct comparisons of measurements of $O_2$ obtained with the Winkler technique and membrane-covered polarographic electrodes generally indicate good agreement within the limits set by the precision of the analytic techniques. When sufficient care is taken, the Winkler technique provides higher precision than the $O_2$ electrodes (Table 6.2) ; but, because it is destructive, the Winkler technique allows only one $O_2$ determination per sample. Therefore, at least two samples are required for each rate determination. The Winkler method is used primarily in field studies when a large number of samples are incubated for an extended period of time (usually 1-24 hr). The high precision and accuracy of the technique, together with its ability to rapidly fix the measurable oxygen concentration in a large number of samples for subsequent analysis makes this technique suitable for routine use in the field. Conventional oxygen electrodes are usually unsuitable for studies of phytoplankton photosynthesis, especially in oligotrophic waters.

**Table 6.2. Typical reported precision of various techniques for measurements of dissolved gas concentrations.**

| *Technique* | *Precision* | *Reference* |
|---|---|---|
| $O_2$ electrode-steady mode | 3.6 mmol.m$^{-3}$ | Hitchman, 1978 |
| $O_2$ electrode-pulsed mode | 0.8 mmol.m$^{-3}$ | Langdon, 1984, Bender *et al*. 1987 |
| Winkler titration-visual detection | 0.7 mmol.m$^{-3}$ | Strickland, 1960 Talling, 1976 |
| Winkler titration-automated detection | 0.3 mmol.m$^{-3}$ | Williams and Jenkinson, 1982, Culberson and Huang, 1987 |
| Total $CO_2$ from pH and aklalinity | 5 mmol.m$^{-3}$ | Smith and Key, 1975 |
| Total $CO_2$ from potentiometric titration | 2 mmol.m$^{-3}$ | Bradshaw and Brewer, 1988a, b |
| Total $CO_2$ from acidification and IRGA* | < 3-6 mmol.m$^{-3}$ | Johnson *et al*. 1981 Roberts and Smith, 1988 |
| Total $CO_2$ from coulometric titration | 1 mmol.m$^{-3}$ | Bender *et al*. 1987 Johnson *et al*. 1987 |

*Infrared gas analysis

However, Maccio and Langdon's (1988) pulsed electrode has been used to monitor diel changes ($O_2$) in open ocean waters.

**Inorganic Carbon Determinations**

Total inorganic carbon can be measured directly by acidfying a sample and determining the amount of liberated $CO_2$ by infrared gas analysis, by gas chromatography, or coulometric titration. These techniques have an accuracy of approximately 1-5 mmol.m$^{-3}$ $TCO_2$ in ocean waters where the $TCO_2$ concentration is typically 2 mol.m$^{-3}$. Thus, on a molar basis, the best $TCO_2$ measurements are approximately a factor of three less sensitive than the best $O_2$ determinations. Direct determination of net $TCO_2$ fluxes have been undertaken in a limited number of investigations.

Alternatively, $TCO_2$ can be calculated from observations of pH and alkalinity based on a thermodynamic treatment of the inorganic carbon dioxide system. Under the assumption that the inorganic carbon system is in equilibrium, total inorganic carbon is given by

$$[TCO_2] = [CA]\,\frac{1+K_2^1/H^+ + H^+/K_{L1}^1}{1+2K_2^1/H^+} \tag{6.14}$$

where $[TCO_2]$ is the total inorganic carbon concentration, [CA] is the carbonate alkalinity, $[H^+]$ is the hydrogen ion activity (*i.e.*, $10^{-pH}$), $K_{L1}^1$ is the first apparent dissociation constant of carbonic acid and $K_2^1$ is the second apparent dissociation constant of carbonic acid in water.

Photosynthetic $CO_2$ fixation produces an increase in the pH of the suspending medium without altering the alkalinity. Thus, changes in pH together with an initial value of the carbonate alkalinity, can be used to continuously assess changes of net inorganic carbon concentration using Eq. 6.14. The concentration of $TCO_2$ in seawater can be measured to an accuracy of $\pm 5$ mmol.m$^{-3}$ when pH is measured with an accuracy of $\pm 0.005$ units and alkalinity is measured to $\pm 5$ mmol.m$^{-3}$. This is the practical limit on measurement of $TCO_2$ in discrete samples. However, measurement of a much smaller change in $TCO_2$ concentration of only $\pm 0.5$ mmol.m$^{-3}$ is possible assuming that changes in pH of 0.0005 units can be detected by continuously monitoring pH in a single sample. Vary accurate determinations of alkalinity ($\pm 1$ mmol.m$^{-3}$) and $TCO_2$ ($\pm 1$ mmol.m$^{-3}$) in sea water by potentiometric titration requires stepwise addition of acid and application of elaborate algorithms to calculate the equivalence points on the titration curve for $H^+$, $HCO^-_3$, $HSO^-_4$, HF, $B(OH)^-_4$, $H_2PO^-_4$, and $SiO(OH)^-_3$.

$TCO_2$ concentrations can also be calculated from measurements of pH and aqueous $CO_2$ [$CO_2$(aq)] under the assumption that $CO_2$ in the gas and aqueous phases are in equilibrium :

$$[TCO_2] = K_H[CO_2(g)]\left(1+\frac{K_{L1}^1}{H^+}+\frac{K_{L1}^1K_2^1}{(H^+)^2}\right) \tag{6.15}$$

where $CO_2(g)$ is the concentration of $CO_2$ in the gas phase that is in equilibrium with the aqueous phase, and $K_H$ is Henry's law constant.

Inaccuracies in measuring pH due to differences in the ionic strength of the sample and buffers used to calibrate the pH meter may bias calculated rates of $TCO_2$ exchange. Axelsson (1988) showed that calculated rates of net photosynthesis of marine macroalgae were 20% greater when the pH meter was standardized against the National Bureau of Standards (NBS) pH scale than when the meter was calibrated against the Hansson (1973) seawater pH scale. Direct determination of the $TCO_2$ changes associated with the observed changes in pH were intermediate between those obtained from the indirect calculations using the NBS and Hansson scales. Difficulties in measuring pH may be exacerbated in waters with low ionic strength, particularly in the presence of high concentrations of dissolved organic carbon. In fact, it has been suggested that pH in some low alkalinity freshwaters containing high concentrations of dissolved organic carbon should be calculated from the more accurate measurements of $TCO_2$ and the partial pressure of $CO_2$ to overcome biases in pH measurements.

Changes in pH and alkalinity brought about by metabolic processes other than inorganic carbon exchange can lead to inaccuracies in estimates of $TCO_2$ exchange if constant alkalinity is assumed. The consumption or production of $H^+$ or $OH^-$ accompanies many metabolic processes as indicated in Table 6.3. These include assimilation of inorganic nitrogen, calcium carbonate precipitation, and organic acid excretion. The potential effects of these processes on estimates of $TCO_2$ exchange are now considered.

The typical stoichiometry of inorganic carbon and nitrogen assimilation by algae and the associated oxygen and proton fluxes is given by Eqs. 6.16-6.18.

$$106CO_2+138H_2O+16NO_3^- \Rightarrow (CH_2O)_{106}(NH_3)_{16}+16OH^-+138O_2 \quad (6.16)$$

$$106CO_2+106H_2O+16NH_4^+ \Rightarrow (CH_2O)_{106}(NH_3)_{16}+16H^++106O_2 \quad (6.17)$$

$$106CO_2+114H_2O+8CO(NH_2)_2 \Rightarrow (CH_2O)_{106}(NH_3)_{16}+8CO_2+106O_2 \quad (6.18)$$

**Table 6.3.** Summary of major processes leading to extra-cellular pH changes.

| *Metabolic Process* | *Direction of External pH Change* |
|---|---|
| Photosynthetic $CO_2$ fixation | Increase |
| Nitrogen assimilation : $NH_4^+$ entering cell | Decrease |
| Nitrogen assimilation : $NO_3$ entering cell | Increase |
| Excess cation influx, *e g.* $K^+$ carboxylate accumulation | Decrease |
| Excess anion flux, *e.g.*, $Cl^-/HCQ^-$ exchange | Increase |
| Organic acid assimilation—undissociated form entering | Increase |
| Excretion of organic acids in undissociated form | Decrease |
| Calcium carbonate precipitation | Increase |

Adapted from Smith and Raven (1976)

Note the one-to-one relationships between $NO_3^-$ and $OH^-$ in Eq. 6.16 and between $NH_4^+$ and $H^+$ in Eq. 6.17. Goldman and Brewer (1980) verified these stoichiometries for nitrogen assimilation and $H^+$ or $OH^-$ exchange in experiments with the marine chlorophyte *Dunaliella tertiolecta*. They demonstrated a one-to-one relationship between nitrogen assimilation and alkalinity changes when $NH^+_3$, $NO_2^-$, or $NO_3^-$, were the nitrogen sources. Assimilation of the cation ammonium decreased alkalinity, whereas assimilation of the nitrate or nitrite anions increased alkalinity. In contrast, assimilation of urea does not produce a net acid or base exchange, and alkalinity remained constant when phytoplankton used urea as a nitrogen source.

Warm ocean waters and many hard-water lakes are super-saturated with respect to calcium carbonate, and $CaCO_3$ precipitation can occur by biological or chemical processes that is,

$$Ca^{2+} + 2HCO_3^- \Rightarrow CaCO_3(s) + H_2CO_3 \qquad (6.19)$$

where (*s*) refers to the solid phase.Calcium carbonate precipitation reduces alkalinity, with a stoichiometry of 0.5 $mol.m^{-3}$

$CaCO_3$ formed producing a 1.0 mol.m$^{-3}$ reduction in total alkalinity. A wide variety of algae can precipitate $CaCO_3$ including the coraline algae, which are important contributors to reef ecosystems and the coccolithophorids, which are widespread in the phytoplankton as well as the microalgae associated with foraminifera and corals. However, calcium carbonate precipitation only occurs in waters that are nearly saturated or supersaturated with calcite or aragonite.

Calcification rates covaried with photosynthesis rates in four *Halimeda* species, with calcium carbonate precipitation proceeding at approximately 64% of the rate of organic carbon production. The rate of calcification was found to equal the photosynthesis rate in *Corallina officinalis* and the coccolithophorid *Emiliania huxleyi*, but lower ratios of $< 0.25$ were obtained for *Halimeda* spp. Both photosynthetic oxygen evolution and calcification rates could be described as exponential or hyperbolic tangent functions of irradiance in the reef-building corals *Manicura areolata*, *Acropora cervicornis*, and *Acropora formosa*. From the limited number of samples examined, both photosynthesis and calcification rates appeared to saturate at similar irradiances.

Organic acids are important early products of photosynthesis and photorespiration in algae, but the importance of these compounds in altering pH during photosynthesis experiments is expected to be minimal under conditions in which low-molecular-weight organic acids usually make up $<10\%$ of the exuded dissolved organic carbon (DOC). As DOC release is typically $< 10\%$ of gross photosynthetic carbon fixation this would not seem to be a problem. However, the presence of organic acids in fresh and seawaters may affect the determination of carbonate alkalinity and thus calculated $TCO_2$ concentrations. The error is not large in ocean waters where a careful comparison of indirect and direct estimates of $TCO_2$ concentration demonstrated that indirect estimates of $TCO_2$ calculated from high-precision measurements of alkalinity exceeded direct measurements of $TCO_2$ by approximately 1% (21 $\mu$mol.kg$^{-1}$). This difference between the two methods was attributed to an unknown protolyte, probably the anion of a weak organic acid with humic and fulvic acids suggested as likely candidates.

### Measuring $CO_2$ Exchange in Air

Some of the difficulties in measuring carbon exchange by algae can be overcome if plants are suspended in a moist atmosphere that allows rapid and direct measurement of $CO_2$ fluxes by infrared gas analysis (IRGA). Bidwell and McLachlan (1985) found that $CO_2$ assimilation rates were stable for periods of up to 8 hours, and that both photosynthesis and respiration procceded at similar rates whether plants were suspended in air or water. Addition of carbonic anhydrase to the plant surface was necessary to insure rapid intercover-sion of $CO_2$ and $HCO_3^-$ since $HCO_3^-$ was actively assimilated. A similar technique can be employed with microalgae, in which cells at a density of 2 mg chlorophyll-dm$^{-2}$ are supported in a thin film of water on a nylon screen. Bidwell (1977) reported steady rates of net photosynthetic or respiratory gas exchange of *Chlorella pyrenoidosa* for periods of up to 15 hr provided that incoming air was saturated with water and found similar rates in suspended cultures containing carbonic anhydrase and artificial leaves. More recently, Vonshak *et al.* (1989) have measured $O_2$ exchange rates of cells of the cyanobacterium *Spirulina platensis* suspended on glass-fiber filters. However, the extensive handling required to prepare an artificial leaf may make this technique inappropriate for many fragile species lacking a rigid cell wall.

Maintaining a high tissue water content is essential to the maintenance of photosynthesis rates of macroalgae in air. Photosynthesis by *Ulva* was not affected between the onset of desiccation and up to 20% water loss, but net photosynthesis was completely inhibited when plant water loss equalled 50-70%. However, even a 10% water loss resulted in approxi-mately a fourfold reduction in carbon fixation rates with *Enteromorpha prolifera*. In general, algae from the lower intertidial zone show reduced rates of photosynthesis in air when compared with submerged rates, whereas middle and upper intertidal algae appear to photosynthesize more rapidly in air than in water.

**Measuring Net Gas Exchange**

The simplest system for measuring net gas exchange consists of a closed container in which $O_2$ or $TCO_2$ is measured at the start and end of an incubation. The first measurements of photosynthesis in aquatic environments were obtained with such a method by using the Winkler technique to follow changes of dissolved oxygen concentration in illuminated and darkened bottles. This method consists of dispensing subsamples containing a plant population into clear and opaque glass bottles, measuring the initial dissolved oxygen concentration in one set of replicate bottles and measuring the final concentrations in sets of light and dark bottles. Oxygen evolution in the light is referred to as net productivity, and oxygen consumption in the dark provides a measure of respiration. It is commonly assumed that respiration in the dark bottle occurs at the same rate as respiration in the light bottle, and the difference between oxygen concentrations in the light and dark bottles at the end of the experiment is assumed to equal total photosynthetic oxygen evolution, termed gross productivity. The assumption that the rates of $O_2$ consumption in dark and light bottles are equal is a major limitation of the technique for estimating "true" gross photosynthesis because of the possibility of light-dependent $O_2$-consuming processes and modifications of mitochondrial respiration by light. The light-dark bottle technique, however, has been widely and repeatedly used in aquatic ecosystems and it remains a standard technique in limnological and oceanographic research on algal productivity.

The light-dark bottle oxygen technique was used in an extensive series of oceanic investigations by Riley which provided the first estimate of gross photosynthesis rates in the open ocean of 860 ± 560 mg $C.m^{-2}.d^{-1}$ (range of 165-1,550mg $C.m^{-2}.d^{-1}$). In some regions, incubations of up to 7d duration were required to obtain measurable differences in oxygen concentration between light and dark bottles. Riley's experiments in oligotrophic waters were later criticized by Steemann-Nielsen (1952) because of potential artifacts associated with the long incubation times. Steemann-Nielsen found a much lower rate of primary production with his new $^{14}C$ technique. Improvements in methodology now allow measurements of $O_2$ production and consumption rates in incubations that last less than a day even in the most oligotrophic open ocean waters. Rates of gross photosynthesis in the oligotrophic (<0.1 $\mu$g chlorophyll *a*.$dm^{-3}$) subtropical North Pacific gyre averaged 1.3 $\mu$mol $O_2.dm^{-3}.d^{-1}$ in the upper mixed layer. Assuming that this rate applies to the upper 50 m of the water column, and that the photosynthetic quotient is unity, an estimate of the areal rate of 780 mg $C.m^{-2}.d^{-1}$ can be obtained. Although subject to considerable uncertainty, Riley's (1941) earlier conclusions appear to be consistent with this extrapolation of the results of Williams *et al.* (1983).

Providing sufficient biomass to ensure measurable changes in $O_2$ or $TCO_2$ concentration does not pose a problem in measuring net gas exchange of macroalgae, and most investigators prefer to use $O_2$ electrodes rather than the Winkler titration to measure changes in $O_2$ concentration which generally exceed 10% of the initial value. In fact, care must be taken to ensure that optimum ratios of thallus weight to bottle volume are employed, because high ratios may lead to shading within the incubation bottle and/or rapid depletion of nutrients, including inorganic carbon, with affects on subsequent rates of photosynthesis during a prolonged incubation. Littler (1979) recommends a maximum dry weight to volume ratio of 0.03 $g.dm^{-3}$ for experiments of up to 6 hr duration, with the further provision that whenever feasible "the proportions of the container be commensurate with the size and metabolic rate of an entire representative thallus." The duration of the incubation also influences the choice of container volume and sample biomass, with shorter incubations requiring correspondingly higher biomass. Water movement is greatly reduced when algae are enclosed in bottles, and it is necessary to provide mechanical mixing to avoid rate limitations due to diffusion effects. Stirring appears to have more of an effect on photosynthesis than respiration (Table 6.4).

**Table 6.4.** Effect of water movement on rates of photosynthetic and respiratory gas exchange in macroalgae.

| *Macroalgae* | *Ratio of Rates in Mixed versus Stagnant Containers* *Photosynthesis* | *Respiration* |
|---|---|---|
| *Ulva lobata* | 3.6 | 1.3 |
| *Colpomenia sinuosa* | 1.1 | 0.8 |
| *Pudina* sp. | 2.6 | — |
| *Cladophora* sp. | 1.1 | 1 3 |
| *Oedogonium kurzii* | — | 1.7 |

Adapted from Littler (1979).

Oxygen and pH electrodes can be used to continuously measure $O_2$ or $TCO_2$ concentrations in either closed or open systems. When employed with closed systems under nearly constant environmental conditions, the rate of net gas exchange can be directly measured from the rate of $O_2$ or pH change. A constant rate of net gas exchange is usually established within minutes following a change of environmental conditions. Small chambers of approximately 1-10 $cm^3$ are often employed to measure net gas exchange of microalgae in the laboratory. Commercially available chambers are often illuminated from the side through a curved surface making determinations of irradiance within the experimental chamber difficult. Construction of a chamber that allows illumination through an optically flat surface to provide a well-defined light field is described by Dubinsky *et al.* (1987). A leaf disc electrode developed for use with terrestrial vascular plants can be employed to measure $O_2$ exchange by discs cut from a macroalgal thallus; however, wound respiration may be a problem with such measurements. Larger chambers can be used to measure rates of gas exchange of entire plants.

The change of $O_2$ and pH brought about by algal photosynthesis in closed chambers can influence the subsequent gas exchange rates. In some circumstances the large excursions of $O_2$ and pH that are possible in closed systems provide the data desired by an investigator. For example, the upper limit of pH at which net gas exchange ceases can be used to infer the chemical forms of inorganic carbon used for photosynthesis and the effect of $pO_2$ on net gas exchange provides information on the rate of photorespiratory metabolism in algae. However, in many cases, large excursions of $O_2$ or pH are an undesirable feature of a closed system and must be compensated for by periodic renewal of $TCO_2$ or removal of $O_2$. As a practical limit, many investigators attempt to control $pO_2$ to within 80% to 120% saturation in order to limit $O_2$-sensitive changes in gas exchange rates when determining the light or temperature dependence of photosynthesis. Exchange of $O_2$ between gas and liquid phases is relatively rapid, and intermittent aeration can be used to maintain $pO_2$ near air saturation. However, equilibration of pH and $TCO_2$ by aeration is much slower. Harris and Lott (1973) describe a system for measuring photosynthesis of freshwater phytoplankton where pH and $O_2$ are controlled by titrating with $CO_2$ and bubbling with $N_2$.

Open systems have been used to provide improved control of the environmental conditions within an experimental chamber. In systems that are open to fluid flow but closed to the atmosphere, measurements of the change of $O_2$ or $TCO_2$ between the entrance and exit ports of the chamber can be used to obtain gas exchange rates when the volume of the chamber and the flow rate through the chamber are known. Chambers that are open to the atmosphere but closed to fluid flow are also used in algal photosynthesis research. An IRGA can be used to continuously monitor $TCO_2$ exchange provided that aeration is vigorous enough to overcome rate limitation by $CO_2$ equilibration between gas and liquid phases. Measuring changes in $O_2$ of a gas stream used to flush a liquid sample is impractical because of the high concentration of $O_2$ in the atmosphere. However, differences in $pO_2$ between aqueous and gas phases can be used to estimate net gas exchange, provided that the kinetics of gas transfer through the airwater interface have been adequately characterized.

**Sampling and Subsampling**

The organisms under consideration here range over seven orders of magnitude in size from the < 1 μm diam chrococcoid cyanobacteria, chloroxybacteria, and some eucaryotic microalgae, to the > 10 m length of the giant kelp *Macrocystis pyrifera*. Various problems arise in attempts to manipulate organisms over this size continuum. Gas exchange rates of individuals of the very smallest and very largest organisms are difficult to measure directly. In most instances, populations, of microalgae in samples of 1 $cm^3$ to 1 $dm^3$ volume, or subsamples of 1-100 $cm^2$ area cut from macroalgae, provide the material for physiological and biochemical studies. Information on individual microalgal cells are not common, although such measurements are possible. This section considers some of the subsampling problems associated with measuring net photosynthesis and respiration rates of macroalgae.

It is common to assess photosynthesis rates of macroalgae by measuring gas exchange rates of "representative" discs removed from the fronds. Sample size is often dictated by the size of the enclosing container. Littler (1979) has recommended the optimum ratio of sample weight to container volume. Considerable variation in photosynthetic and respiratory gas exchange rates of tissue discs is often found. For example, rates of photosynthesis of *Macrocystis pyrifera* measured by different investigators varied from 0.06 to 3.4 mg $C.g^{-1}$ dry $wt.h^{-1}$ and 0.002 to 0.216 g $C.m^{-2}.h^{-1}$. Much of this variability appears to be due to the type of tissue under investigation, its physiologic state (*i.e.*, light and nutrient history), physical injury, wounding, sloughing, and the presence or absence of epiphytes. To overcome the large variability observed in gas exchange rates of different subsamples, Dor and Levy (1987) recommend making sequential observations of photosynthesis and respiration *via* $O_2$ exchange on the same sample. Although this allows more precise estimates of relative rates of respiration and photosynthesis, it neglects the basic problem of within-plant and between-plant variability.

One way to overcome this subsampling problem is to enclose entire plants or fronds. Hatcher (1977) described a procedure for in situ incubation of macroalgae on the seafloor.

Algae were placed in a 76 $dm^3$ PVC container (30×100 cm). Provision was made for filtering seawater to exclude planktonic organisms, continuous mixing to minimize diffusive boundary layers at the plant surface, replacement of seawater during 24-hr incubations to avoid extreme changes in water chemistry, and reduction of $O_2$ by bubbling with $N_2$ gas to avoid super-saturation. Enclosures have also been used in experiments with freshwater algae. For example, Kairesalo *et al.* (1987) used 56 or 112 $dm^3$ (40×40×35 or 70 cm) acrylic plastic chambers to enclose intact *Nitella opaca* (Charophyceae).

Gerard (1986) measured net changes in $O_2$ concentration during 1-2 hr in situ incubations of a single blade or apex of *Macrocystis pyrifera* enclosed in 12-17 $dm^3$ of ambient seawater in a 21-$dm^3$ capacity polyethylene bag. In addition to overcoming problems associated with subsampling (*i.e.*, obtaining representative samples and avoiding potential metabolic shocks in damaged tissues), enclosing a sample in a plastic bag can maintain a more realistic turbulence regime, which may be important in controlling diffusive fluxes of dissolved gases and nutrients. Losses of $O_2$ through plastic may, however, introduce an error into measurements of photosynthesis and respiration, although these were found to be negligible provided that the gradient across the bag remained $< 50$ $mmol.m^{-3}$.

It is not always possible or desirable to enclose entire plants in chambers for gas exchange measurements, and it may be necessary to employ excised tissue for some rate measurements. Two factors must be considered in using tissue discs : the effects of wounding on gas exchange characteristics and intrablade variability. Wounding may increase rates of extracellular product formation and oxygen consumption. For example, rates of $O_2$ consumption were five times higher in freshly cut material than in intact *Laminaria* sp. 3.4 times higher in freshly cut than in intact *Macrocystis pyrifera* and 1.4 times higher in blade discs of *Laminaria saccharina* com ared to intact plants. Release of phenolics from damaged tissue has been suggested as a possible reason for enhanced $O_2$ consumption. Bidwell and McLachlan (1985) using "moist air suspensions" report that an $O_2$-dependent respiration that could be confused with photorespiration was actually an artifact due to wounding. Wounding appears to have less of an effect on net photosynthesis, with increases of < 10% reported for *Macrocystis pyrifera* immediately following tissue excision and decreases of approximately 25% in blade discs of *Laminaria saccharina*. Prolonged incubation of tissue discs generally leads to rates of photosynthesis and respiration that are lower than those observed in freshly excised material.

Within-blade variability of photosynthesis and respiration rates have been reported for *Laminaria* spp. and *Macrocystis* spp. Differences in rates of photosynthesis and respiration are perhaps best documented for the morphologically complex *Macrocystis pyrifera* (Table 6.5) The highest rates were found in apical scimitar and blades, which make up the bulk of the photosynthetic tissue, whereas lower rates were found in sporophylls, stipe, and holdfast. Within and between-blade variability in *Macrocystis pyrifera* was high, two-to fivefold differences in mass-specific photosynthesis rates along blades and up to fivefold differences between blades from mature plants. It is perhaps not surprising that gas exchange rates vary widely between different tissues, given the structural and morphological complexity of large seaweeds.

**Table 6.5.** Rates of photosynthesis and respiration in Macrocystis pyrifera.

| | *Photosynthesis* | | *Respiration* | |
|---|---|---|---|---|
| | *(mmol $O_2.g^{-1}$ dry wt.$h^{-1}$)* | | | |
| | *Mean* | *SD* | *Mean* | *SD* |
| Apical scimitar | 0.32 | ±0.11 | 0.10 | ±0.033 |
| Blade | 0.26 | ±0.04 | 0.056 | ±0.021 |
| Sporophylls | 0.12 | ±0.05 | 0.036 | ±0.018 |
| Stipe | 0.013 | ±0.005 | 0.011 | ±0.005 |
| Holdfast | >0.000 | | 0.015 | ±0 003 |

**Comparison of $O_2$ and $CO_2$ Exchange**

Many investigations of the photosynthetic or respiratory metabolism of algae employ either $O_2$ exchange or $TCO_2$

exchange but rarely both in combination, even though it has long been recognized that $O_2$ and $TCO_2$ fluxes may not be equivalent. The ratio of the rates of oxygen evolution to carbon dioxide assimilation during photosynthesis is defined as the photosynthetic quotient (designated PQ). Similarly, a respiratory quotient (designated RQ) can be defined as the ratio of oxygen consumption to carbon dioxide evolution during respiration. Photosynthetic and respiratory quotients of unity are commonly assumed when direct observations are lacking, based on the synthesis or remineralization of carbohydrate :

$$CO_2 + H_2O \leftrightarrow O_2 + CH_2O \tag{6.20}$$

Although a photosynthetic quotient of unity is often assumed, there are many situations in which large deviations ($\pm$ 50%) are to be expected.

Oxygen and inorganic carbon exchanges are the result of a number of biochemical reactions, some of which operate largely independently. Oxygen evolution provides a measure of the electrons available for biological redox reactions. In light, the evolution of an $O_2$ molecule is associated with the flow of four electrons from $H_2O$ to reduce $2NADP^+$ to 2NADPH during linear electron flow. Coupled to this electron flow is a variable amount of ATP synthesis of $<3$ ATP/$O_2$ $<$ 3-4 ATP. Inorganic carbon is only one of the electron acceptors that compete for the reductant produced as a consequence of linear electron flow. Other electron acceptors include nitrate, nitrite, sulphate, or molecular oxygen. Nitrite reduction to ammonium can account for approximately 40% of photoreduction at low irradiances, decreasing to approximately 10% at irradiances that saturate photosynthetic oxygen evolution. Changes in the photosynthetic quotient with irradiance may largely reflect a changing stoichiometry of nitrate and/or nitrite reduction to $CO_2$ fixation.

Photosynthetic quotients exceeding unity are expected for growth of micro and macroalgae based on stoichiometries of element exchanges. Reduction of nitrate to ammonium, which requires the transfer of eight electrons for each nitrogen atom reduced, accounts in part for photosynthetic quotients that exceed unity in algae. These additional electrons can be accounted for in a mass balance equation.

$$nCO_2+(n+1)\ H_2O+HNO_3 \Rightarrow (CH_2O)_n(NH_3)+(n+2)O_2 \qquad (6.21)$$

Given the assumptions that nitrate is the inorganic nitrogen source and carbon is reduced to the level of carbohydrate, mass balance equations for microalgal and macroalgal growth are as follows.

For microalgae:

$$106\ CO_2+122\ H_2O+16\ HNO_3+H_3PO_4 \rightarrow (CH_2O)_{106}(NH_3)_{16}(H_3PO_4)+138\ O_2 \qquad (6.22)$$

For macroalgae:

$$550\ CO_2+580\ H_2O+30\ HNO_3+H_3PO_4 \rightarrow (CH_2O)_{550}(NH_3)_{30}(H_3PO_4)+610\ O_2 \qquad (6.23)$$

Although considered typical of micro and macroalgal growth, the preceding equations do not apply in all situations. The Redfield stoichiometry was derived for the average elemental composition of organic matter in the oceans, and significant local variations above and below these proportion are found. The C : N and C : P ratios in phytoplankton can vary widely from the Redfield proportions depending on nutrient availability and irradiance with molar C : N ratios ranging 4-25, and C : P ratios ranging 75-625. The data for macroalgae span a range of C : N ratios of $< 10$ to 60 and C : P ratios of 140-3550. Both the irradiance and availability of inorganic nutrients affect elemental composition ratios. Low C : N and C : P ratios are found under nutrient-sufficient, but light-limited conditions, whereas high C : N and C : P ratios are indicative of light-saturated, but nitrogen-limited or phosphorus-limited conditions. The N : P ratio reflects the relative availability of these two important inorganic nutrients and may vary from 5-115 in microalgae or 5-180 in macroalgae.

Further increases in the PQ over those calculated on the basis of nitrate reduction (Eq. 6.21) can be attributed to the fact that organic carbon in lipids and proteins is more highly reduced than organic carbon in carbohydrate. Protein accounts for 10% to 50% of microalgal dry weight, and lipid accounts for 15% to 70% of dry weight. In contrast, macroalgae are comprised predominantly of carbohydrate (80%), with protein accounting for only 15% and lipid $< 5\%$ of ash-free, dry weight.

Calculations based on the biochemical composition of microalgae lead to estimates of the photosynthetic quotient for net cell synthesis of 1.1-1.2 when a reduced nitrogen source such as urea or ammonia is available and 1.4 when nitrate is the nitrogen source. Nitrogen-fixing cells should have a PQ of approximately 1.24. It is thus somewhat paradoxical that early observations of the PQ obtained from gas exchange measurements often yielded values close to unity, consistent with carbohydrate as the major product of photosynthesis. The low PQs observed by early investigators are not representative of balanced algal growth, and Myers and Cramer (1948) considered a PQ of unity to be a consequence of measuring photosynthesis at very high irradiances in cultures that were acclimated to low light. These are the conditions under which a large amount of carbohydrate synthesis and relatively small amounts of protein and lipid synthesis are expected. Exponentially growing phytoplankton cultures have PQs consistent with calculations based on the elemental composition and nitrogen source (Table 6.6). However, large variations may be observed in stationary-phase cultures or immediately following subculturing.

Photosynthetic quotients of $1.21 \pm 0.06$ Buesa (1980) and $1.17 \pm 0.02$ have been obtained for macroalgae from the Chlorophyceae, Phaeophyceae, and Rhodophyceae, with no significant difference in the values for different taxonomic groups. These values were obtained at air-equilibrium $O_2$ concentrations in natural seawater. Buesa (1980) measured $O_2$ and $CO_2$ exchange using a Warburg manometer and Axelsson (1988) measured pH and $O_2$ in a well-mixed flow through chamber. In both cases, diffusive boundary layers at the algal surface are expected to be minimized. Similar values have been obtained by other investigators.

**Table 6.6.** Stoichiometries of phytoplankton growth on different nitrogen sources.

| | | | |
|---|---|---|---|
| $0.89\ CO_2 + 0.61\ H_2O + 0.127\ NH_3$ | $\Rightarrow 0.89$ C-mol biomass $+ O_2$ | PQ = 1.12 | C : N = 7.01 |
| $0.82\ CO_2 + 0.62\ H_2O + 0.067\ N_2H_4CO$ | $\Rightarrow 0.89$ C-mol biomass $+ O_2$ | PQ = 1.12 | C : N = 6.64 |
| $0.71\ CO_2 + 0.59\ H_2O + 0.101\ HNO_3$ | $\Rightarrow 0.71$ C-mol biomass $+ O_2$ | PQ = 1.41 | C : N = 7.02 |
| $0.81\ CO_2 + 0.73\ H_2O + 0.058\ N_2$ | $\Rightarrow 0.81$ C-mol biomass $+ O_2$ | PQ = 1.23 | C : N = 6.98 |

Variations in PQ are not always explicable in terms of nitrogen metabolism, and other metabolic processes may also have to be considered. Photorespiratory metabolism can contribute to PQs significantly less than unity. Littler (1979) found that the PQ decreased from unity at low $O_2$ (< 1 $\mu$mol. mol$^{-1}$) to 0.5 at air-saturated $O_2$ (6.6 $\mu$mol.mol$^{-1}$) and to 0.25 at supersaturated $O_2$ (15.5 $\mu$mol.mol$^{-1}$) in *Ulva lobata* at 20°C and an irradiance of 140 $\mu$mol.m$^{-2}$ . s$^{-1}$. Burris (1981) demonstrated an inverse relationship between PQ and the partial pressure of $O_2$ in three marine microalgae at a temperature of 20°C and an irradiance of $1.3 \times 10^{17}$ photons.cm$^{-2}$. s$^{-1}$. In *Glenodinium* sp. and *Pavlova lutheri,* PQ decreased linearly from 1.2-1.8 at $pO_2$ of 0.1 atm (110 $\mu$mol $O_2$.dm$^{-3}$) to 0.5-0.7 at $pO_2$ of 0.3 atm (330 $\mu$mol $O_2$.dm$^{-3}$). Kaplan and Berry (1981) found that decreases in the PQ of *Chlamydomonas reinhardtii* were correlated with increases in the excretion of glycolate, which is one of the products of ribulose bisphosphate carboxylase-oxygenase (RUBISCO) oxygenase activity. In contrast, Bidwell and McLachlan (1985) found no evidence for photorespiration except in an atmosphere with simultaneously elevated $O_2$ and very low $CO_2$ ($50 \times 10^{-4}$ cm$^3$.dm$^{-3}$), although an $O_2$-sensitive wound respiration that could potentially be confused with photorespiration was observed. Variations in the PQ of aquatic macrophytes (with PQ often less than unity) may result from differences in the accumulation of oxygen in the lacunal gas-space.

**Changes in Biomass**

The autotrophic biomass or standing stock is the concentration of plant material per unit volume (g.m$^{-3}$) or per unit area (g.m$^{-2}$). Biomass is usually measured as wet weight, dry weight, ash-free dry weight, or organic carbon, although a number of other variables can also be employed. In studies of phytoplankton photosynthesis, chlorophyll concentration is the most readily and unambiguously measured biomass indicator. The increase in autotrophic biomass over a given time interval is the yield, which, like standing stock, has units of g.m$^{-2}$ or g.m$^{-3}$. Under natural conditions the yield is determined by the difference between net algal photosynthesis and losses to grazers, export, and excretion, where all variables are reported in the same units.

Yield = Net Photosynthesis—Excretion—Grazing—Export (6.24).

Problems in balancing Eq. 6.24 can arise if measurements of the different variables are obtained in different units. For example, net photosynthesis is often obtained from oxygen evolution, excretion from release of dissolved organic $^{14}C$, and yield from the accumulation of dry weight. A parameter related to the yield is the net production rate defined as the rate of increase of autotrophic biomass. Yield can be considered as the integrated value of net production rate over time. In axenic cultures, the export and grazing terms are by definition zero, and the increase in biomass provides an independent check on measurements of net gas exchange and excretion.

In field samples, export can be eliminated by enclosing a sample, but the problem of eliminating grazing remains. Using enclosures may also introduce other artifacts such as a reduction in turbulence, adsorption of nutrients on container surfaces with consequent changes in the chemical composition of the aqueous phase, and elimination of external nutrient sources. With macroalgae, grazing can be reduced by physically removing large macroscopic grazers or by choosing samples that are free from grazers. It is difficult to separate microalgae from protozoan grazers, but a protocol allowing phytoplankton growth to be uncoupled from microscopic grazers has recently been developed.

**Coupling of Growth and Photosynthesis**

Changes in biomass are also used to calculate the specific growth rate :

$$\mu = \ln [B_t/B_0]/\delta t = \ln [(B_0+\gamma)/B_0]/\delta t \qquad (6.25)$$

where $B_0$ is the initial biomass, $B_f$ is the final biomass. $Y = B_f - B_0$ is the yield, and $\delta t$ is the time interval between measurements of $B_0$ and $B_f$. Growth is said to be balanced when the specific growth rates (Eq. 6.25) obtained from all measured biomass variables are equal. Differences in the specific rates of synthesis of various biomass variables typifies the unbalanced growth that occurs following changes in environmental conditions, such as a shift in irradiance or exhaustion of nutrients.

Algal growth is not always directly coupled to the rate of photosynthesis. Uncoupling of growth and photosynthesis is evident on diurnal time scales, with net photosynthesis and carbohydrate accumulation limited to the illuminated part of the day, whereas net protein synthesis often continues in darkness. Uncoupling growth and photosynthesis is also evident when nutrient-limited algae are exposed to elevated nutrient concentrations. Storage carbohydrates and lipids are accumulated under energy-sufficient but nutrient-limited conditions and can be rapidly mobilized following nutrient additions. Ammonium assimilation into low-molecular-weight compounds occurs at a faster rate than macromolecular synthesis when nitrogen-limited algae are provided with increased ammonium concentrations. This occurs because the maximum cell-specific ammonium uptake rate is largely independent of growth rate, whereas the rate of macromolecular synthesis is nearly proportional to growth rate in ammonium-limited chemostat cultures. As a consequence, the pool of low-molecular-weight nitrogen compounds shows a transient increase following ammonium addition to nitrogen-deficient cells, and the ratio of short-term ammonium uptake to nitrogen incorporation into macromolecules provides a good indicator of nitrogen deficiency. This increase can be measured by adding $^{15}NH_4$ to a sample and following the accumulation of $^{15}N$ label in low-molecular-weight (hot ethanol extract) and macromolecular (hot ethanol insoluble) pools. Using this approach, Kanda *et al.* (1988) found both nitrogen-sufficient and nitrogen-deficient phytoplankton populations in oligotrophic waters of the western North Pacific near Japan.

An extreme case of the uncoupling of photosynthesis and growth has been observed in *Laminaria solidungula* from the high Arctic. Carbohydrate reserves (laminarin) are built up during the summer when light is available but inorganic nutrients are low, and growth occurs at the expense of stored carbohydrates over winter when nutrients are available but light levels are low. Less extreme cases of uncoupling of photosynthesis, nutrient assimilation and growth have been documented for other macroalgae. The most common pattern is similar to that observed in *L. solidungulu* and involves an accumulation of carbohydrate reserves during the summer when temperatures are high and nutrients are low and an accumulation of tissue nitrogen during the winter when growth rate is limited by temperature but nutrients are available.

## Community Metabolism

All of the methods that require enclosing the sample also at least partially isolate the enclosed organisms from the rest of their environment. This isolation may have several effects including (1) reducing vertical and horizontal mixing, (2) reducing turbulent mixing, particularly in studies of macrophytes and benthic microalgae, (3) removing large grazers and sources of regenerated nutrients, and (4) introducing contaminants that may alter gas exchange rates. These problems can be overcome if gas exchange rates can be estimated in unconfined conditions.

Photosynthetic and respiratory activity changes oxygen concentration and pH within a waterbody. Oxygen accumulates, $TCO_2$ is consumed, and pH rises when photosynthesis exceeds respiration, whereas $O_2$ is consumed, $TCO_2$ is released, and pH drops when respiration exceeds photosynthesis. These processes produce patterns of in situ $O_2$, $TCO_2$, and pH variations on time scales ranging from hours to weeks that can often be used to calculate biological gas exchange rates and net productivity of aquatic ecosystems. However, these patterns are strongly influenced by atmospheric exchange, which tends to reduce in situ variations in shallow, well-mixed waterbodies, whereas stratification in deeper water columns is reduced by vertical mixing.

The method for estimating community metabolic rates from diurnal changes in $O_2$ and/or $TCO_2$ in freshwater and marine environments is often associated with the work of Odum and coworkers. Although Odum is often credited with developing the method, the approach was first used to estimate phytoplankton productivity by workers in the Soviet Union. With increases in the accuracy of $O_2$ and $TCO_2$ determinations (Table 6.1), open-water techniques are being used more often in studies of open-ocean primary production.

Open-water methods for estimating primary production have been advocated by some investigators, because this approach eliminates any potential artifacts associated with enclosing planktonic or benthic organisms within bottles or bell jars. The deficiencies inherent in open systems include a lower level of precision, which is associated with heterogeneity in the water mass under investigation, a requirement to estimate the fluxes associated with physical exchange across the air-water interface, and the need to measure exchanges with other water masses. The spatial and temporal resolution of a sampling program must accommodate the spatial and temporal variability of $O_2$ or $TCO_2$ concentrations in the environment examined if accurate rates of community photosynthesis and respiration are to be obtained. In general, constraints on manpower and equipment limit the spatial resolution to one or a few locations in the horizontal, one or a few depths in the vertical, and the temporal resolution to several hours between samples. However, automation of data collection and computer-aided analysis can greatly extend the application of open-water techniques.

Homogeneity in both the physical environment and in the biological community is required if open-water techniques are to be applied successfully. In addition, biological oxygen-producing and consuming processes must exceed rates of physical exchange across the system boundaries if meaningful data are to be obtained. Otherwise, where physical processes dominate the $O_2$ budget, small errors in estimates of physical exchange can lead to very large errors in calculated biological fluxes.

The $O_2$ budget for a well-mixed, horizontally homogeneous water column can be expressed as

$$d[O_2]/dt = P - R + k([O_2] - [O_2]_{eq})/D \quad (6.26)$$

where $[O_2]$ is the dissolved oxygen concentration ($mmol.dm^{-3}$), P is the rate of photosynthetic $O_2$ production ($mmol.dm^{-3}.s^{-1}$), R is the rate of respiratory $O_2$ consumption ($mmol.dm^{-3}.s^{-1}$), $[O_2]_{eq}$ is the equilibrium dissolved oxygen concentration predicted from temperature and salinity, *k* is the atmospheric exchange coefficient ($m.s^{-1}$), and D is the depth of the mixed layer (*m*). The exchange coefficient *k* is strongly dependent on wind speed (V) with $k \approx V^2$. The importance of atmospheric exchange is greatest in shallow mixed layers and when $[O_2]$ deviates greatly from $[O_2]_{eq}$. Hartman aud Hammond (1985) recently reexamined the relationship between wind speed and the gas exchange coefficient and proposed the following empirical relationship, which fits available data to within 20% for wind speeds of 3-12 $m.s^{-1}$ :

Demonstrated light-enhanced $O_2$ consumption in highly colored lake waters that had been filtered to remove particulate matter (Whatman GF/A) and poisoned with mercuric chloride to prevent microbial respiration. They found that the rate of $O_2$ consumption was greater in waters with higher humic acid content, was faster at higher pH, and increased with increasing total Fe concentration. The rates of abiological oxygen consumption in the dark were only 20% of the light-enhanced rates measured at an irradiance of 500 $\mu mol.m^{-2}.s^{-1}$. Oxygen consumption in the light was accompanied by $CO_2$ evolution with a stoichiometry of $1/2O_2 : CO_2$, consistent with Fe-mediated photooxidation of aminopolycarboxylates.

$$RHNCH_2COOH + 0.5O_2 \Rightarrow RNH_2 + CO_2 + CH_2O \quad (6.27)$$

Iron-mediated, light-dependent $O_2$ consumption has been demonstrated for several organic compounds containing amino, carboxyl, and hydroxyl groups whereas some other compounds, such as tannic acid, promote $O_2$ consumption in both light and darkness in the pH range 2.11.

Production of superoxide ($O_2^-.$) by near ultraviolet light (<400 nm) and consumption of this $O_2.$ in the oxidation of organic matter provides another mechanism for photochemical oxygen consumption concluded that photochemical processes accounted for $O_2$ consumption in seawater filtered through double layers of Whatman GF/F filters or 0.2-$\mu m$ pore membrane filters and implicated singlet oxygen production as a factor in the photo-oxidation. Higher rates of $O_2$ consumption were observed in quartz bottles and blood transfusion bags than in glass bottles, presumably because of greater penetration of ultraviolet radiation in the former. Rates of photochemical $O_2$ consumption varied from about 0.2 $\mu mol.dm^{-3}.h^{-1}$ in the Sargasso Sea to 165 $\mu mol.dm^{3}.h^{-1}$ in the eutrophic Lake

**Table 6.7.** Comparison of photosynthetic oxygen production and photochemical oxygen consumption rates in selected natural waters.

| *Aquatic Ecosystem* | *Net Photosynthetic $O_2$ Production ($mmol\ O_2.m^{-2}.d^{-1}$)* | *Photochemical $O_2$ Consumption ($mmol\ O_2.m^{-2}.d^{-1}$)* |
|---|---|---|
| Caribbean Sea | 5 | 1 |
| North Sea | 10 | 4 |
| Ems-Dollart Estuary | 4 | 0.2 |
| Lake Tjeukemeer | 19 | 3 |

Tjeukermeer in the Netherlands (Table 6.7) and accounted for the consumption of up to 40% of net photosynthetic $O_2$ evolution.

Diurnal trends in a number of photochemical products have been demonstrated in natural seawater : these include aldehydes and ketones hydrogen peroxide, carbon monoxide and carbonyl sulfide, Kieber and Mopper (1987) recently demonstrated photochemical formation of glyoxylic and pyruvic acid at rates comparable to biological production and consumption rates in ocean waters ranging from the oligotrophic Sargasso Sea to Florida Bay. In contrast to conventional wisdom, which assumes that virtually all of the cycling of low-molecular-weight compounds in the sea is mediated by biological processes, recent studies suggest an important role for photochemical reactions in the natural cycles of these compounds.

# 7

# Exchange of Gases in Plants

In every living plant cell an astonishing number of diverse chemical activities are constantly going on. These, like the chemical reactions in the world of nonliving matter, involve energy transformations. Some activities—such as photosynthesis, assimilation, and the synthesis of complex foodstuffs and other materials from simpler substances—build up new compounds and hence are constructive. Other reactions, the most important of which are respiration and digestion, are destructive, since complex food materials are broken down into simpler compounds. It is principally with these destructive reactions that this chapter is concerned.

Constructive, or building-up, processes and destructive, or tearing-down, processes are interrelated and interdependent, and commonly go on at the same time in the living cell. Destructive processes are as essential to cell survival and growth as constructive processes. The term *metabolism* has long been applied to the sum total of the chemical changes associated with the behaviour of protoplasm. *Anabolism* is the term for the constructive, and *catabolism* for the destructive processes.

## RESPIRATION

In some respects the plant is like an automobile engine, which burns gasoline to release energy. Like the engine, the plant derives its energy from the combustion of organic materials. But it differs, among other ways, in that the burning process goes on at a much slower rate and the living cells in which combustion takes place carry on, simultaneously, many other activities.

The energy necessary for plant activities is made available through the process of *respiration*. In respiration, a sugar, commonly glucose, is transformed into simpler substances. This may be represented by the equation.

$$\underset{\text{glucose}}{C_6H_{12}O_6} + \underset{\text{oxygen}}{6O_2} \longrightarrow \underset{\text{carbon dioxide}}{6CO_2} + \underset{\text{water}}{6H_2O} + \underset{\text{energy}}{674 \text{ kcal.}}$$

Note that this reaction, from the energy standpoint, is the exact opposite of photosynthesis. In photosynthesis, energy is stored ; in respiration, energy is released. Respiration is classed as an oxidation process, in which oxygen reacts with the carbon and hydrogen of the sugar to form water and carbon dioxide, releasing energy. This is an inadequate description of respiration, which involves a succession of intricate steps, a number of enzymes, and the formation of many intermediate compounds. It indicates, however, the raw materials and the final products.

*How energy is used.* It is clear that energy, the ability to do work, is involved in lifting a weight or in performing other mechanical work. It is not so clear that energy is used in the accomplishment of the internal work which allows a plant to live and grow. The energy liberated in respiration appears partly in the form of heat, which is dissipated, and partly in what may be termed useful energy. It is the latter which is responsible for the activities of the plant body.

A small amount of useful energy expresses itself as motion—in the streaming of protoplasm, the movement of chromosomes, and in other minor ways. Much of the useful energy is expended in *assimilation*, the processes by which food is converted into cell walls and protoplasm. Energy is required in the utilization of carbohydrates to build cell walls and in the chemical reactions by which such foods as proteins and fats are synthesized from the product of photosynthesis. Another important way in which useful energy is employed is in the accumulation of solutes from the soil by the roots.

In addition, the energy of respiration is used to synthesize a number of miscellaneous substances in plants, none of which are foods. Some are of vital importance to the plant, but the significance of others is not yet known. These substances include the pigments, enzymes, vitamins essential oils, gums, resins, the organic acids (such as malic, citric, oxalic and tartaric), the tannins, the glucosides (such as digitalin), and the alkaloids (such as caffeine, nicotine, and strychnine).

The internal work of the plant manifests itself externally in growth. Leaves, flowers, seeds, and fruits are formed. The plant increases in height and diameter. Roots extend through the soil. The utilization of energy, then, does not result in conspicuous motion, as it does in most animals, and this is an essential difference between most plants and most animals. But beneath this difference is found basic similarity, for although most plants are sedentary and most animals mobile, energy is required to carry on the life processes within the cells of both. In the use of energy to carry on both destructive and constructive cellular processes, all life is essentially the same.

*Respiration and enzymes*. Organic catalytic agents, known as enzymes, possess the property of accelerating a chemical reaction without being altered by it. They may enter into temporary combination during the course of the reaction but are not used up during the process and appear unchanged at the end. Enzymes, unlike other catalytic agents, are produced only by living protoplasm. All the known enzymes are colloidal, and all are composed of protein, although some have nonprotein substances attached to them. A number of enzymes have been isolated in pure crystalline form. Extensive experiments, mostly on fungi and bacteria, have shown that the ability of an organism to elaborate, or construct, a particular enzyme may be dependent upon the presence of a particular gene. In addition to those mentioned above, other characteristics of enzymes are as follows :

(1) They are effective in low concentrations. The rate at which enzymatic reactions proceed, however, is directly proportional to the concentration of the enzyme, provided other factors are favourable.

(2) They are specific-that is, a given enzyme can act only upon a given substance or group of closely related substances. Since many chemical reactions are known to proceed within living cells, this means that each cell must produce a large number of enzymes. Hundreds of enzymes are, in fact, known, although it does not follow that all of them are produced within any specific cell or tissue.

(3) Enzymes are able to catalyze both synthetic, or building-up, reactions and the splitting of complex compounds into simpler substances. In some reactions the same enzyme is responsible for both these processes—that is, the reaction is reversible. The synthesis of every substance, however, is not necessarily brought about by the same enzyme which breaks it down ; in a number of cases a different enzyme is known to be involved in the synthetic process.

(4) Reactions catalyzed by enzymes are hastened by an increase in temperature until a certain temperature optimum has been attained. For many plant enzymes this is about 50°C. (122°F). Above this point the enzyme is rapidly inactivated and is destroyed within the range 60° to 80°C. (140° to 176°F.).

Most, perhaps all, chemical reactions that proceed within living cells are catalyzed by enzymes. The role of enzymes in digestion is familiar, but enzymes are concerned in many other processes, among them respiration. Enzymes make chemical changes possible which otherwise would proceed only at much higher temperatures or in the presence of chemical reagents. Sugar, outside the plant or animal body, for example, well remain indefinitely in the state familiar to us unless subjected to combustion, which breaks it down into carbon dioxide and water. Within the plant, however, the sugar is oxidized and energy released at ordinary temperatures, because of the respiratory enzymes produced by the living cell.

*Evidences of respiration.* Respiration is usually accompanied by several phenomena which indicate the nature and rate of the process. A study of these should not, however, obscure the most significant aspect of respiration, the release of energy. Some of the more obvious aspects of respiration can be demonstrated directly.

*Rise in Temperature.* Much of the energy which is released in respiration is in the form of heat. In warm-blooded animals, this heat is important, for it maintains the temperature of the body. In plants, it is commonly viewed as wasted energy, for it is quickly dispersed into the environment, and the temperature of the plant is, therefore, approximately that of the surrounding air. In this respect plants resemble cold-blooded animals, which possess no mechanism for conserving the heat of respiration. Actively growing tissues and organs, such as germinating seeds, young fruits, and opening buds, have a higher metabolism and so respire at a higher rate than mature ones. If the heat generated by such tissues is confined, the temperature rises.

This is readily demonstrated by placing a number of seeds, previously soaked in water, in a vacuum flask and inserting a thermometer. After germination has begun, the rise in temperature may amount to as much as 20 centigrade degrees, depending upon the kinds of seeds used. The water should be sterile and the seeds treated to kill bacteria and fungi, whose respiration would otherwise contribute to the rise in temperature.

When the outer layers of a mass of plant materials act as insulators, a temperature rise due to respiration can also be noted. For example, the interior of a crate of celery or green beans at room temperature may register 5.5 centigrade degrees (10 Fahrenheit degrees) higher than the outside air. Another example is afforded by the observation that the interior of a heap of grass cuttings reached a temperature of 38°C. (100°F.) within 18 hours of mowing when the temperature of the air was 15.5°C. (60°F.).

*Release of Carbon Dioxide.* The rate at which carbon dioxide is released by living plant tissue is a generally accepted index of the rate of respiration. It can be measured by several methods. Qualitatively, the emission of carbon dioxide can be demonstrated by partly filling a large flask with germinating peas. The flask is stoppered and allowed to stand overnight. If a burning splint is then lowered into the flask, the flame will be extinguished. This, of course, indicates the absence of oxygen, rather than the presence of carbon dioxide, but it may be inferred that the oxygen has been used by the seeds and replaced by carbon dioxide.

A more direct approach makes use of the fact that calcium hydroxide solution (limewater) reacts with carbon dioxide to form insoluble calcium carbonate. This precipitate turns the colourless solution cloudy. Two stoppered flasks, one containing dry peas, the other an equal number of germinating peas, are set up and allowed to stand for several hours. The air in the flasks is then displaced by adding water through a funnel. The displaced air is conducted by a tube to a solution of calcium hydroxide. The air from the dry seeds produces no change in the limewater. That from germinating seeds, on the other hand, brings about the rapid formation of calcium carbonate, as evidenced by a milky precipitate (Fig. 7.1).

*Loss in Weight.* That respiration results in a loss in weight is quite readily demonstrated. An experiment to show this in a green plant must, of course, be carried on in the dark, for in the light the plant will gain in weight as a result of photosynthesis. The loss in weight due to respiration is easily established with germinating seeds. A number of seeds are weighed and then germinated in the dark. After the seedlings are well developed and the stored food in the seeds is largely exhausted, the dry weight is obtained by drying the plant material in an oven. The dry weight of the seedlings and seed remnants will be much less than the weight of the seeds with which the experiment started. The difference in weight represents the water and carbon dioxide lost in respiration.

Since most seeds contain a small but measurable quantity of water to begin with, more accurate results can be obtained

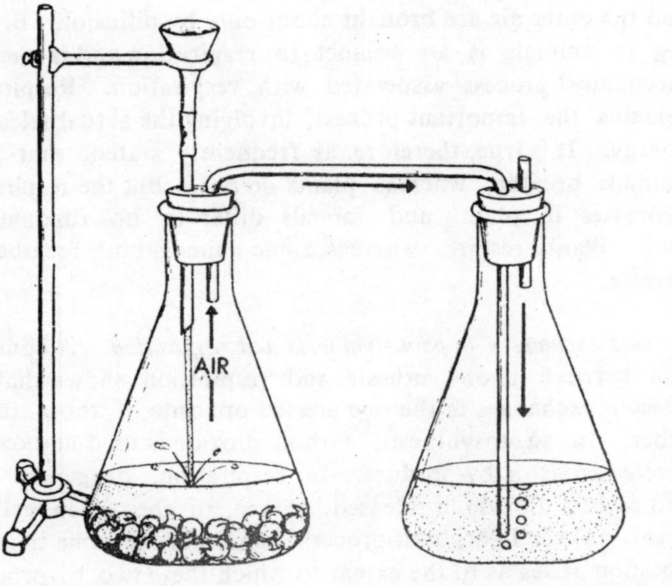


Fig. 7 1. Demonstration of release of carbon dioxide by germinating peas. The air from the flask at the left passes through limewater in the flask at the right. A milky precipitate, calcium carbonate, forms.

by starting with two lots of seeds of equal weight, determining the dry weight of one lot, and germinating the other. The dry weight of the seeds and seedlings of the second lot is then compared with the dry weight of the seeds of the first lot.

*Respiration and breathing*. Respiration must occur in every living cell if life is to be maintained. Not only plants but also all animals respire. In addition, some animals also breathe, and the distinction between respiration and breathing must be clearly understood—particularly because in common usage respiration is used incorrectly as synonymous with breathing. Higher animals have a breathing mechanism, including lungs, by which normal air passes into the body and air containing a higher content of water vapour and carbon dioxide moves out. Plants, however, possess no such mechanism, and gaseous exchanges between the internal tissues and the outer air are brought about only by diffusion. Breathing in animals is an adjunct to respiration and is merely a mechanical process associated with respiration. Respiration remains the important process, involving the actual release of energy. It is true, therefore, as frequently stated, that some animals breathe, whereas plants do not. But the respiratory processes in plants and animals differ in no fundamental way. Plants respire, whereas some animals both breathe and respire.

*Gas exchanges in photosynthesis and respiration.* A comparison between photosynthesis and respiration shows that the gaseous exchanges in the one are the opposite of those in the other. In photosynthesis, carbon dioxide is used and oxygen is released as a by-product. In respiration, oxygen is used and carbon dioxide in released. Since, in the green cells of plants in the light, both processes go on at the same time, the question arises as to the extent to which these two by-products are utilized.

The evidence is clear that the carbon dioxide released in respiration is used in photosynthesis and that the oxygen released in photosynthesis is used in respiration. Under light of low intensity, these two processes may exactly balance each other, and carbon dioxide and oxygen will move neither out of nor into the leaf. The light intensity at which this balance occurs in a leaf is called the *compensation point.*

The light intensity at the compensation point is quite low—about 1 percent of full sunlight, but it varies widely in different kinds of plants. Temperature and other factors which affect either respiration or photosynthesis also affect the compensation point. At the compensation point there is neither a gain nor a loss in the dry weight of the leaves during the day. Under such conditions, however, a plant could not long survive, for during the day the non-green parts of the plant respire and use food, and at night respiration alone occurs. The plant as a whole, then, can survive at the compensation point only as long as stored food is available.

During the day, however, photosynthesis usually proceeds at a rate at least ten times that of respiration. All the carbon doxide released in respiration is therefore used in photosynthesis, and additional carbon dioxide diffuses into the leaf from the air. Respiration uses some of the oxygen of photosynthesis ; the excess diffuses from the leaf into the air. These gaseous exchanges are shown diagrammatically in Fig. 7.2. At night, of course, the plant utilizes only oxygen and liberates carbon dioxide. So rapid is the process of photosynthesis, however, that, considering day and night together, the plant liberates far more oxygen than it uses, and it uses more carbon dioxide than it releases to the air.

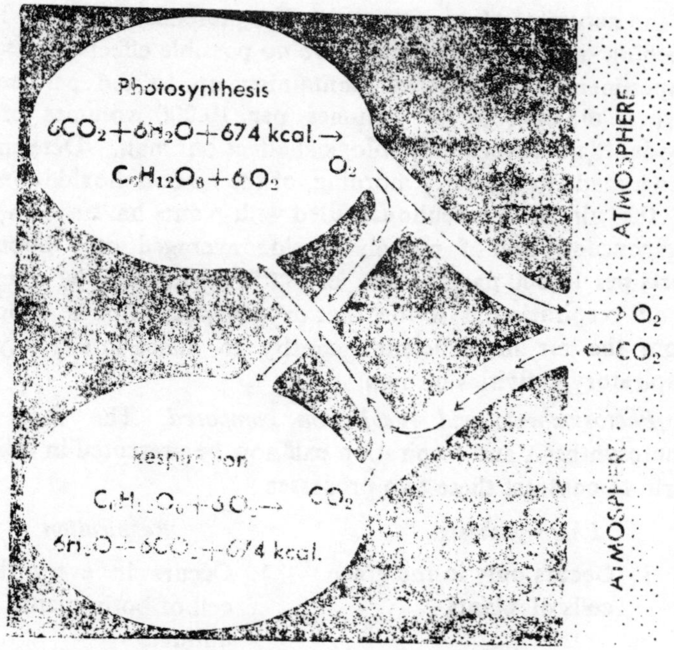


Fig 7.2. Diagram showing gas exchanges between the atmosphere and a green cell of a leaf in the sunlight.

Since photosynthesis exceeds respiration, the green plant manufactures more food than it respires. Because photosynthesis proceeds at a more rapid rate than respiration throughout the life of the plant, food is available for immediate assimilation and for accumulation within storage cells. Such food makes it possible for growth to occur, since growth depends upon recently formed excess food or upon stored-up food reserves.

*Effects of respiration on air of rooms.* The amount of carbon dioxide liberated in plant respiration is sometimes overestimated by those who have not studied the process; hence the belief that it is unhealthy to permit plants to remain in bedrooms during the night. In some hospitals potted plants and cut flowers are removed from rooms in which patients sleep. This procedure may be the result of misunderstanding of the nature of plant activities. The presence of plants in a sleeping room at night can have no possible effects on health. Experiments show that air containing up to 0.6 per cent of carbon dioxide, or 60 volumes per 10,000 volumes of air, produces no adverse physiological effect on man. Determinations, made in the early morning, of the carbon dioxide content of the air of a greenhouse filled with plants have shown that the concentration of carbon dioxide averaged only about 5 parts per 10,000 parts of air, or 0.05 per cent. The rate of plant respiration is so low that oxygen would not be removed from the air in sufficient quantity to affect in any way the respiratory activities of man.

*Photosynthesis and respiration compared.* The facts on photosynthesis and respiration can now be presented in tubular form to contrast these two processes :

| *Photosynthesis* | *Respiration* |
|---|---|
| 1. Occurs only in the green cells of plants | 1. Occurs in every living cell of both plants and animals |
| 2. Takes place only in the presence of light | 2. Takes place at all times during the life of the cell, both in the light and in the dark |
| 3. Uses water and carbon dioxide | 3. Uses food and oxygen |

| | |
|---|---|
| 4. Releases oxygen as a by-product | 4. Releases water and carbon dioxide as end-products |
| 5. Solar (radiant) energy is converted into chemical energy, which is stored | 5. Energy is released by oxidation. Chemical energy is converted into heat and useful energy |
| 6. Results in an increase in weight | 6. Results in a decrease in weight |
| 7. Food is built up | 7. Food is broken down. |

*Factors affecting the respiration rate*. The rate of respiration in plant tissues is affected by temperature, moisture, injuries, the age and kind of tissues concerned, the concentration of carbon dioxide and oxygen, the amount of available food, and other factors. Although all of these are of interest from the standpoint of understanding plant activities under both natural and experimental conditions, they become a matter of immediate and practical concern in connection with the transportation and storage of cereal grains, fruits, and vegetables.

The production of food in our civilization, with its great concentration of population in cities, is not only a problem of agriculture but also one of storage and distribution. The problem is complicated because the activities of plant tissues do not cease with the harvest. Citrus, apple, and other fruits, potatoes, and vegetables of all kinds continue to live and respire—sometimes for months—before they are used. Unless properly handled and stored, these foods will soon deteriorate, either through the effect of their own activities or by the action of bacteria and fungi. The handling of such products, involving what has been termed "after-harvest" botany concerns both botanists and engineers and leads to many problems of ventilation and storage.

Immediately after harvesting, or during subsequent storage, the enzymes of fruits and vegetables act on various materials within their tissues and gradually bring about changes in colour, texture, and chemical composition. The most important of these changes are caused by the enzymes of respiration, and an important objective of proper storage is to slow down the rate of this process. As temperature increases, up to a point, so does the rate of respiration ; hence the wide use of refrigeration as a method of food preservation.

Much of the effectiveness of low storage temperatures is due to the slowing up of the reaction of respiratory and other enzymes of higher plant tissues and of bacteria and molds. The relation between temperature and respiration is similar to that between temperature and other chemical reactions, for within certain limits the rate of respiration is doubled or tripled with each rise of 10 centigrade degrees (18 Fahrenheit degrees). The effect of a rise in temperature of strawberries, which have a high respiratory rate, is shown below.

The obvious effects of the impairment of food materials are rot and decay, but less apparent results of respiration may be seen from the table below.

Milligrams of carbon dioxide evolved per kilogram of fresh weight per hour

| 32°F. | 40°F. | 50°F. | 60°F. | 70°F. | 80°F. |
|---|---|---|---|---|---|
| 14.4 mg. | 24.6 mg. | 49.2 mg. | 76.9 mg. | 123.9 mg. | 186.7 mg. |

Calculations of carbon and glucose lost per 100 grams dry weight of vegetables held at 30°C. for 26 hours after harvest.

| *Vegetables* | *Dry matter, per cent* | *Weight of $CO_2$ evolved between 2d and 26th hour, grams* | *Weight of carbon lost between 2d and 26th hour, grams* | *Calculated weight of glucose lost between 2d and 26th hour, grams* |
|---|---|---|---|---|
| asparagus | 8.3 | 20.066 | 5.473 | 13.682 |
| lettuce | 6.2 | 9.424 | 2.570 | 6.425 |
| green beans | 7.7 | 9.273 | 2.529 | 6.322 |
| green onions | 9.3 | 6.627 | 1.807 | 4.518 |
| carrot | 13.9 | 4.659 | 1.271 | 3.178 |
| tomato | 6.1 | 3.920 | 1.069 | 2.672 |

The loss in dry weight, as indicated in the table on page 160, is primarily the result of the respiration of glucose. But the loss of weight in stored fruit and vegetables also involves the loss of water, since in time plant products lose weight by evaporation. The humidity of the air in storage rooms is therefore directly related to the keeping qualities of the products held in them.

In addition to these losses, other changes occur after harvesting. For example, in asparagus, some sugar is converted into cell-wall substance ; the sugar of green peas is, in part, transformed into starch ; and in sweet corn the sugar which has accumulated as a temporary reserve in the seeds is also converted into starch. Because of such after-harvest changes, vegetables picked fresh from the garden are sweeter and more tender than those that are picked three to five days before they appear on the grocery counter. The conversion of sugar to starch in sweet corn, for example, is noticeable in a few hours and is nearly complete in a day.

Poor storage conditions, such as lack of proper ventilation and high temperatures, give rise to physiological diseases, such as blackheart in potatoes, scald and browning in apples, and other conditions which reduce the food value and marketability of the products of farm and orchard.

*Aerobic and anaerobic respiration.* The type of respiration thus far discussed takes place under a sufficient supply of oxygen and is termed *aerobic* respiration. Under certain conditions, respiration may also occur in the absence of oxygen, a kind of respiration called *anaerobic.* Like aerobic respiration, anaerobic respiration proceeds under the influence of enzymes and yields energy. But it does not progress to the usual end products, carbon dioxide and water. A number of compounds may be formed, including alcohol, carbon dioxide, and organic acids. The yield of energy is far less than in aerobic respiration. Consequently, it is a rather inefficient process. A common equation for anaerobic respiration follows, although this reaction is known to proceed in several stages :

$$\underset{\text{sugar}}{C_6H_{12}O_6} \xrightarrow[\text{enzymes}]{} \underset{\text{alcohol}}{2\,C_2H_5OH} + \underset{\text{carbon dioxide}}{2\,CO_2} + \underset{\text{energy}}{22\text{ kcal.}}$$

Anaerobic respiration is inefficient not only in the production of energy but also in the utilization of sugar, for much more sugar is required to yield the same amount of energy as in aerobic respiration. The energy released in anaerobic respiration results from atomic rearrangement in the molecules of the substances respired. Although the amounts and kinds of end-products of anaerobic respiration differ from those of aerobic respiration, the earlier stages in the aerobic process appear to be identical with those which occur during respiration in the absence of oxygen.

Anaerobic respiration is more characteristic of the metabolism of the bacteria and yeasts than of the higher plants. The process is more commonly known as fermentation, and the products of this respiration are important industrially. Nevertheless, anaerobic respiration is significant in green plants as well. Most plants that respire aerobically can also respire anaerobically, at least for a time. The length of time that plant tissues can live under anaerobic conditions is usually short, owing, in part, to the accumulation of toxic products. The anaerobic type of respiration may normally occur in the early stages of the germination of seeds, especially those of aquatic and semi-aquatic plants. It may also occur for brief periods in the roots of plants growing in poorly aerated soils.

Roots and underground stems require oxygen as well as the more visible plant parts. In ordinary well-drained soil, this oxygen is available. In water-logged soils, the oxygen content may be reduced to a small fraction of the normal supply. Plants differ widely in their ability to survive in soils that lack oxygen. Swamp and aquatic plants are much more successful in this respect than upland plants, which may die quickly when the soil is flooded. It has been shown, in fact, that the roots and rhizomes of a number of kinds of semi-submerged plants (those with roots or rhizomes in mud and their leaves in the air) are able to carry on anaerobic respiration for a considerable period of time without apparent injury. In one experiment alcohol was distilled from freshly dug rhizomes of a water lily in sufficient amounts to be detected by a delicate colour test. Such submerged organs can also respire aerobically, and probably carry on both types of respiration simultaneously in different cells or tissues.

## PHYSIOLOGICAL CHEMISTRY OF RESPIRATION

Respiration is one of the most fundamental phenomenon of life and indeed life itself is insaparable from respiration in some forms. Respiration involves the oxidative break-down of certain organic substrates present in the cells. Respiration, therefore, is not the concern of special organs or tissues, it is an integral part of the life of every cell. Cellular respiration is the aerobic oxidation of carbohydrates in the cell with the help of atmospheric oxygen, in the course of which energy is liberated and carbon dioxide is given out. In glycolysis and fermentation, glucose is oxidized in the absence of oxygen, *i.e.*, *anaerobically*. Organisms that live in the presence of oxygen are aerobes, those that live in its absence are anaerobes.

### Cellular Respiration

The first series of reactions in oxidative break-down of glucose is called the *Embden-Meyerhof (E-M) pathway*. This was described by two German biochemists *Embden* and *Mayerhof*. The term *glycolysis* is sometimes used as a synonym for the E-M pathway, but biochemists usually apply the term glycolysis to the E-M pathway plus a form of fermentation, which together lead from glycolysis to lactic acid. The E-M pathway is a several-step process that occurs in the cytoplasmic matrix of cell. It can proceed in either the presence or absence of oxygen but makes available only a small portion of the potential energy that could be expected from glucose.

### I. The E-M Pathway

From six-carbon glucose molecule the E-M pathway starts and the formation of 2-three-carbon molecules of pyruvic acid is the final step of this pathway. Following steps take place in the entire pathway.

1. *Phosphorylation.* In the first step glucose reacts with adenosine triphosphate (ATP) to form glucose-6-phosphate in the presence of *hexokinase* and $Mg^{++}$ ions.

$$\text{Glucose} + \text{ATP} \xrightarrow[\text{Mg}^{++}]{\text{hexokinase}} \text{Glucose-6-phosphate} + \text{ADP}.$$

This reaction is exergonic in nature so that it is irreversible.

2. *Isomerization.* Glucose-6-$PO_4$, in the presence of the enzyme *phospho-hexose-isomerase*, is isomerised into fructose-6-$PO_4$.

$$\text{Glucose-6-phosphate} \xrightarrow{\text{Phosphohexose-isomerase}} \text{Fructose-6-phosphate.}$$

3. *Phosphorylation.* Fructose-6-$PO_4$ reacts with one molecule of ATP in the presence of $Mg^{++}$ ions and *phosphofructokinase* to form fructose-1, 6- diphosphate. This reaction is also irreversible due to its strongly exergonic nature.

$$\text{Fructose-6-phosphate} + \text{ATP} \xrightarrow[\text{Mg}^{++}]{\text{Phosphofructokinase}} \text{Fructose-1-6-disphosphate.}$$

4. *Cleavage.* Fructose-1, 6-diphosphate is cleaved under the influence of an enzyme *aldolase* with the formation of dihydroxy acetone phosphate or phosphodihydroxy acetone (DHAP) and 3-phosphoglyceraldhyde or glyceraldehyde-3-phosphate (PGAL). Both the products are triose phosphates. The two trioses can be reversibly changed into one another by enzyme *Triosephosphate isomerase.*

Fructose-1-6-diphosphate —Aldolase→ →Dihydroxyacetone phosphate (DHAP)
Triose phosphate isomerase
→3-phosphoglyceraldehyde (PGAL)

5. ***Phosphorylation and oxidative dehydrogenation.*** This is the first reaction in the energy-yielding phase of the E-M pathway. 3-phosphoglyceraldehyde (PGAL) reacts with phosphoric acid ($H_3PO_4$) to form 1-3-diphosphoglyceraldehyde in the presence of enzyme *glyceraldehyde-3-phosphate dehydrogenase.*

3-phosphoglyceraldehyde + $H_3PO_4$ ⟶ 1, 3-diphosphoglyceraldehyde.
(2 molecules) (2 mole.) (2 mole.)

Coenzyme nicotinamide adenine dinucleotide (NAD) reacts with 1-3-diphosphoglyceraldehyde removing 2 electrons and 2 protons (the equivalent of 2 H atoms) forming 1, 3-diphosphoglyceric acid.

1, 3-diphosphoglyceraldehyde + $2NAD^+$ —Glyceraldehyde-3-phosphate dehydrogenase⟶
(2 mole).
1, 3-diphosphoglyceric acid + 2NADH + H
(2 mole.)

6. *ATP generation.* The high-energy phosphate group in 1-3-diphosphoglyceric acid is transferred to ADP resulting in the formation of 3-phosphoglyceric acid (3 PGA) and ATP. This reaction takes place in the presence of enzyme *phosphoglycerate kinase* and $Mg^{++}$ ions.

1, 3-diphosphoglyceric acid + 2ADP —phosphoglycerate kinase, $Mg^{++}$⟶
(2 mol)
3-phosphoglyceric acid.
+ 2ATP (2 mole.)

7. *Isomerization.* 3-phosphoglyceric acid (3-PGA) is isomerized into 2-phosphoglyceric acid (2-PGA) with the help of enzyme *phosphoglyceromutase.* This enzyme catalyse the transfer of a phosphate group from one carbon to another carbon of the same compound.

3-phosphoglyceric acid —Phosphoglyceromutase⟶ 2-phosphoglyceric acid
(2 mole.)

8. *Dehyderation of 2-PGA.* The 2-phosphoglyceric acid molecule is dehydrated by the action of enzyme enolase to yield phosphoenol pyruic acid (PEPA). Presence of $Mg^{++}$ ions is also necessary.

2-phosphoglyceric acid —Enolase, $Mg^{++}$⟶ 2-phosphoenol pyruvic acic + $2H_2O$
(2 mole.) (2 mole.)

9. *ATP generation.* In the final step of E-M pathway the high energy phosphate group in 2-phosphoenol-pyruvic acid (PEPA) is transferred to ADP in the presence of *pyruvate kinase* and $Mg^{++}$ ions with the formation of ATP and pyruvic acid.

$$\text{2-phosphoenolpyruvic acid} + 2ADP \; (2 \text{ mole.}) \xrightarrow[Mg^{++}]{\text{Pyruvate kinase}} \text{Pyruvic acid} \; (2 \text{ mole.}) + 2ATP$$

**Energy Yield of the E-M Reactions**

How much energy of ATP has the cell gained after all these reactions ? That depends upon whether the reactions have taken place under anaerobic or aerobic conditions. In general, the steps of energy production in E-M pathway are (1) during conversion of 1, 3-diphosphoglyceric acid into 3-phosphoglyceric acid and (2) during conversion of 2-phosphoenol pyruvic acid into pyruvic acid. If ADP is present in these steps, energy of energyrich compounds is captured with the formation of ATP. One molecule of ATP at each of the above two steps in synthesized per molecule of 1, 3-diphosphoglyceric acid or 2-phosphoenolpyruvic acid utalized for forward reaction. Thus, in all, 4 molecules of ATP are formed at these two steps per molecules of glucose used up under anaerobic conditions.

Besides these two steps, these exists one more site of energy production which comes into play under aerobic conditions. This step takes place during conversion of 3-phosphoglyceraldehyde into 1, 3-diphosphoglyceric acid when one molecule of reduced NAD, is oxidized with the formation of 1, 3-diphosphoglyceric acid.

In order to evaluate the net gain of energy the energy expenditure has also got to be taken into account and must be substracted from total energy production. Under anaerobic conditions 2ATP molecules are used up during reactions and there is a net gain of only 2ATP molecules.

**II. Oxidative decarboxylation**

Pyruvic acid, which enters the mitochondrion does not itself enter Krebs' cycle reactions. Entry into the Krebs' cycle requires preparation of a 2-carbon fragment called *acetyl group*. This takes place in several steps. In the first step there is a decarboxylation (a carbon dioxide molecule is released), that shortens the carbon chain by one carbon. A molecule of NAD is also reduced. The remaining 2-carbon fragment *i.e.*, acetyl group is attached to a carrier molecule, *coenzyme A*. This complex, made up of two carbon fragment and coenzyme A is called *acetyl CoA*. The acetyl CoA is in an active form ready to be fed into Krebs' cycle.

$$\underset{\text{(Pyruvic acid)}}{CH_3COCOOH} + \underset{\text{(CoA)}}{CoASH} + NAD \xrightarrow[\text{Thymine pyrophosphate, Lipoic acid amide}]{\text{Pyruvic dehydrogenase multienzyme complex}} \underset{\text{(acetyl-S-CoA).}}{CH_3COS\ CoA} + NADH + H^+ + CO_2$$

**III. Krebs' cycle or Tricarboxylic acid cycle (TCA)**

The Krebs' cycle completes the oxidation of carbohydrates and supplies electrons to be passed through electron transport system. Enzymes catalyzing Krebs' cycle reactions are located in the mitochondrial matrix.

The cyclical nature of the reactions by which further oxidations beyond pyruvic acid occur was established after a series of experimental analyses by *Albert Szent-Gyorgyi*, *Hans Krebs* and others. *Krebs'* was awarded Noble Prize in 1953 for this work.

The Krebs' cycle is a *cycle* because it is a series of reaction events, each catalyzed by a specific enzyme, that ends up back at its original point. The reactions involve a series of organic acid with different number of carbon atoms.

(*i*) *Formation of citric acid*. Acetyle Co-A formed from pyruvic acid condenses with oxaloacetic acid forming citric acid. This reaction is catalyzed by enzyme *citrate synthatase* or condensing enzyme.

$$CH_3CO\text{-}S\text{-}CoA + CoA + \text{Oxalic acid} + H_2O \xrightarrow[\text{CoASH}]{\text{Citrate synthetase}} \text{Citric acid} + CoASH$$

(*ii*) *Dehyderation*. Citric acid is converted into *cis*-aconitic acid by the enzyme *aconitase* with release of one molecule of water.

$$\text{Citric acid} \xrightarrow{\text{Aconitase}} \textit{Cis}\text{-aconitic acid} + H_2O$$

(*iii*) *Hyderation I.* By hydration of *Cis*-aconitic acid, isocitric acid is formed. This reaction is catalyzed by enzyme *aconitase*.

$$\textit{Cis}\text{-aconitic acid} + H_2O \xrightarrow{\text{Aconitase}} \text{Isocitric acid.}$$

(*iv*) *Dehyderogenation I.* Isocitric acid is dehydrogenated to form oxalo succinic acid in the presence of *isocitric dehydrogenase*. One pair of hydrogen aloms removed is accepted by $NAD^+$ or $NADP^+$ which forms $NADH + H^+$ or $NADPH + H^+$ (Mitochondria have been found to contain two types of isocitric dehydrogenases, one is NAD specific and other is NADP specific).

$$\text{Isocitric acid} + NAD^+ \text{ or NADP} \xrightarrow{\text{Isocitric dehydrogenase}} \text{Oxalosuccinic acid} + NADH + H^+ \text{ or } NADPH + H^+$$

(*v*) *Decarboxylation I.* In the presence of enzyme *carboxylase*, the oxalosuccinic acid undergoes decarboxylation to form α-Ketoglutaric acid. One $CO_2$ is removed. Thus α-Ketoglutaric acid is 5-carbon molecule.

$$\text{Oxalosuccinic acid} \xrightarrow{\text{carboxylase}} \alpha\text{-Ketoglutaric acid} + CO_2$$

(*vi*) *Dehydration II and Carboxylation II.* α-Ketoglutaric acid undergoes simultaneous decarboxylation and dehyderation and joins with coenzyme A to form succinyl Co-A, a 4-carbon atom derivative of coenzyme A. This reaction is catalyzed by enzyme *α-Ketoglutaric dehydrogenase* in the presence of thiamine pyrophosphate (TPP) and lopoic acid. The hydrogen atoms released are accepted by $NAD^+$ to form $NADH + H^+$

$$\alpha\text{-Ketoglutaric acid} + \text{Co—ASH} + NHD^+ \xrightarrow[\text{dehydrogenase}]{\alpha\text{-Ketoglutaric}} \text{Succinyl CoA} + CO_2 + NADH + H^+$$

(*vii*) *Phosphorylation.* Succinyl CoA on hydrolysis yields to the formation of succinic acid and CoA. Succinyl CoA possesses a high-energy thiol group. In the presence of guanosine diphosphate (GDP) and inorganic phosphate the energy of thiol group is transferred to GDP with the formation of guanosine triphosphate (GTP). This reaction is catalyzed by enzyme *succinic thiokinase*.

$$\text{Succinyl}-\text{CoA}+\text{GDP}+\text{H}_3\text{PO}_4 \xrightarrow{\text{Succinic thiokinase}} \text{Succinic acid}+\text{GTP}+\text{CoASH}.$$

(*viii*) *Dehydrogenation III.* Succinic acid undergoes dehydrogenation under the catalytic action of *succinic dehydrogenase* to form fumaric acid. This dehydrogenase contains flavin adenine dinucleotide ($FAD^+$) as its prosthetic group which accepts the pair of hydrogen atoms removed from succinic acid.

$$\text{Succinic acid}+\text{FAD}^+ \xrightarrow{\text{Succinic dehydrogenase}} \text{Fumaric acid}+\text{FADH}+\text{H}^+$$

(*ix*) *Hydration II.* Fumaric acid is hydrated to form malic acid. This reaction is catalyzed by the enzyme *fumarase*.

$$\text{Fumaric acid}+\text{H}_2\text{O} \xrightarrow{\text{Fumarase}} \text{Malic acid.}$$

(*x*) *Dehydrogenation IV.* This is the final reaction. The Krebs' cycle. The malic acid is transferred to oxaloacetic acid in the presence of *malate dehydrogenase* $NAD^+$ accepts hydrogen atoms to form $NADH+H^+$.

$$\text{Malic acid}+\text{NAD}^+ \xrightarrow{\textit{malate dehydrogenase}} \text{Oxaloacetic acid}+\text{NADH}+\text{H}^+$$

It should be noted that two molecules of pyruvic acid are formed per molecule of glucose during E-M pathway. Therefore, two molecules of oxaloacetic acid per molecule of glucose are formed at the end of Krebs' cycle. Only one pyruvic acid molecule fate has been traced.

During transfer of hydrogen atoms from one enzyme carrier to another in electron transport system, large amount of energy is released (as shown in Table). The enzymes which participate in the process are called respiratory chain enzymes or electron transport system enzymes. All such enzymes are found in the inner membrane of mitochondria. These are arranged in the following sequence—succinic dehydrogenase which remains associated with coenzyme NAD, flavoprotein (FAD), non-haem ion protein or iron-sulphur protein, flavoprotein (FAD or FP), cytochrome *b*, ubiquinone or Co Q, cytochrome $c_1$, cytochrome *c*, cytochrome *a*, cytochrome $a_3$ and three coupling sites where phosphorylation coupled with oxidation leads to the production of ATP.

The respiratory chain system manufactures the ATP molecules during oxidation of reduced coenzymes in the following steps :

1. *Dehyderogenase* enzymes accept hydrogen from different substrates of Krebs' cycle. The hydrogen atom becomes ionized into protons ($2H^+$) and electrons (2e).

Protons or hydrogen ions reduce the oxidized NAD.

$$NAD^+ + 2H^+ \rightarrow NADH + H^+$$

The NADH molecule passes through M-side of inner mitochondrial membrane and enters the respiratory chain system.

2. The NADH is oxidized into $NAD^+$ by transferring its proton ($H^+$) to flavoprotein FAD. The FAD acts as a hydrogen carrier. From the flavoproteins each proton is discharged into perimitochondrial space through the C-side of inner mitochondrial membrane and electrons are passed onto non-haem iron protein.

3. In normal course, reduced FADH is reoxidized by ubiquinone (Co Q) which is the next member of ETS. The ubiquinone is reduced to dihydroubiquinone which is oxidized by ferric ($Fe^{++}$) cytochrome-c ($Fe^{+++}$) which is reduced to ferrocytochrome. At this level of oxidation, hydrogen atoms undergo ionization to form $2H^+$ and $2e^-$. The protons cannot be further carried by ETS, however, the electrons can be carried easily by the cytochromes. One molecule of each cytochrome is capable of carrying only one electron at a time. Thus, for

each molecule of hydrogen which releases two electrons, two molecules of each cytochrome are required. At the level of cytochrome-*b*, two protons per molecule of hydrogen are released in the mitochondrial system. Ferrocytochrome *b* is oxidized in the presence of ferric cytochrome $c_1$. From ferro-cytochrome $c_1$ these electrons are taken by cytochrome *a* nad than to $a_3$ (Both cyto-*a* and $a_3$ together called cytochrome oxidase).

4. These electrons from cytochrome oxidase react with one atom of oxygen making it reactive.

5. This activated oxygen react with $2H^+$ (protons) forming molecule of water.

$$O + 2e \rightarrow O^- \xrightarrow{2H^+} H_2O$$

With the release of 2 electrons and 2 H atoms from every molecule of $NADH + H^+$ and their reaction with oxygen, 3ATP molecules are synthesized. Energy released at these positions reach and stored in ATP molecules. The process is known as oxidative phosphorylation because it occurs in respiration in the presence of oxygen hence it is called oxidative phosphorylation.

Now we can summarize the energy return from the complete oxidation of a glucose molecule under aerobic conditions. Under aerobic conditions, the NADH produced in the E-M pathway can transfer electrons to the ETS inside mitochondria, these by salvaging more usable energy than is possible under anaerobic conditions.

However, reduced NAD cannot enter a mitochondrion directly, so it cannot pass its electrons directly to an ETS in the inner mitochondrial membrane. Instead, the electrons are passed to a shuttal carrier that takes them into the mitochondrion and donates them to an electron transport system.

These are several such electron shuttles for example malate-asperate shuttle, glycerol phosphate shuttle etc., but the common one is *glycerol-phosphate shuttle*, which picks up electrons from reduced NAD, carries them into mitochondrion and donates them not to NAD, but to ubiquinone. In this case the electrons do not start at the 'top' of the ETS, so the possibility for one phosphorylation is by passed. If mitochondria use this particular mechanism, a reduced NAD in the cytoplasm ends up being worth only 2ATP molecules. In this case complete oxidation of glucose will result in a gain of 36 ATP molecules.

In *malate-asperate shuttle*, the electrons are transferred from NADH in cytoplasm to malate which traverses the inner mitochondrial membrane and reoxidized to form $NADH+H^+$ thus resulting in the formation of oxaloacetate. Oxaloacetate does not readily cross the inner mitochondrial membrane 20 a transmission reaction is needed to form asperate which does traverse this barrier. 3ATP molecules are generated for each pair of electrons. In this case complete oxidation of glucose will result in a gain of 38 ATP molecules.

Table 7.1 Showing ATP Formation During Oxidation of Glucose

| *Embden-Meyerhof Pathway* | *ATP calculations* | *Sum* |
|---|---|---|
| ATP expanded to phosphorylate sugar molecule | -2 | |
| Substrate-level phosphorylation | 4 | |
| ATP formed as a result of electron transport from two NADS reduced during E-M pathway reactions (2 or 3 ATP each) | 4 or 6 | |
| Net gain from E-M pathway under aerobic conditions | | 6 or 8 |
| Two pyruvic acid to two acetyl CoA | | |
| Two NAD reduced (3 ATP each) | 6 | |
| | | 6 |
| Kerbs' cycle (two turns) | | |
| 6 NAD reduced (3ATP each) | 18 | |
| 2 FAD reduced (2ATP each) | 4 | |
| Two substrate-level phosphorylation (1ATP each) | 2 | |
| Total from Krebs' cycle (2 turns) | | 24 |
| Total ATP formation per molecule of glucose | | 36 or 38 |

**Table 7.2. Free-Energy Change at Different Steps in the Electron Transport Chain.**

| *Steps of Transport* | *Difference in Redox-potential* ($\Delta E°$), *(Volt)* | *Free energy change (Kcals)* |
|---|---|---|
| NAD $\xrightarrow[2e^-]{2H^+}$ FAD | +0.22 | −10.1 |
| FAD $\xrightarrow{2e^-}$ cyt.b | +0.15 | −6.9 |
| 2Cyt.b $\xrightarrow{2e^-}$ 2 cyt.c | +0.19 | −8.77 |
| 2Cyt.c $\xrightarrow{2e^-}$ 2cyt.a | +0.05 | −2.30 |
| 2Cyt.a $\xrightarrow{2e^-}$ Oxygen | +0.52 | −24.00 |
| Total | 1.13 Volts | 52.07 Kcals |

**Oxidative Phosphrylation Mechanism**

Oxidative phosphorylation was discovered in 1939 but the mechanism of coupling between ETS and ATP synthesis is not yet known. However, three principal hypothesis have been proposed for the coupling of oxidation and phosphorylation. These are :

1. Chemical Coupling hypothesis.
2. Cheiniosmotic Coupling hypothesis.
3. Conformational hypothesis.

1. *Chemical Coupling hypothesis.* According to this hypothesis the ET is coupled to ATP synthesis by a sequence of consecutive reactions in which a high-energy covalent intermediate is formed by electron transport and subsequently is cleaved and donates its energy to ADP to form ATP. For example, when an oxidoreduction reaction occurs between $A^{red}$, and $B^{oxi}$, the factor I is incorporated into the formation of an energy rich $A^{oxi} \sim I$ complex (here ~ represents a linkage having an energy-rich nature).

$$A^{red} + I + B^{oxi} \rightleftharpoons A^{oxi} \sim I + B^{red}$$

In the following reaction an enzyme (E) replaces $A^{oxi}$ from the above complex forming an energy rich E~I complex.

Later on inorganic phosphate ($P_i$) reacts with E~I complex, forming phosphoenzyme complex (E~P). The phosphoenzyme complex reacts with ADP to form ATP liberating enzyme free.

$$A^{oxi} \sim I + E \rightleftharpoons A^{oxi} + E \sim I$$
$$E \sim I + P_i \rightleftharpoons E \sim P + I$$
$$E \sim P + ADP \rightleftharpoons ATP + E$$

An extensive search has been made for the nature of E~I and E~P, but none have been found in mitochondria. The chemical coupling hypothesis is only of historical importance today.

2. *Chemiosmotic Coupling hypothesis.* This hypothesis was proposed by *Peter Mitchell* in 1961. He proposed that electron transport and ATP synthesis are coupled by a proton gradient, rather, than by a covalent high-energy intermediate or an activated protein. This theory suggests that a transmembrane electrochemical gradient is critically important to the phosphorylation process.

Chemiosmotic coupling hypothesis suggests that during electron transport, a gradient of $H^+$ ions is produced across the inner mitochondrial membrane. The mobile electron carrier that gains $H^+$ ion is reduced one side of the membrane and loses proton as it is oxidized near the other side. These reactions lead to the transport of a pair of $H^+$ across the inner mitochondrial membrane from the matrix to the space between the mitochondrial membrane. Thus there is a $H^+$ ion gradient across the membrane. There is a strong tendency for these $H^+$ ions to move back toward the matrix because of the excess of +ve charges in the space between the two membranes.

*Mitchell* further suggests that there are special channels through which the $H^+$ ions can move in response to the forces acting on them (electrical as well as concentration differences). In some way, the movement of $H^+$ ions back through the membrane is coupled in ATP synthatase complex to the phosphorylation of ADP to form ATP.

*Oxidation-reduction loop. Mitchell* has proposed that reducing equivalents are transferred as H atoms by some of the electron carriers (ubiquinone) and as electrons by others (Fe-S center and cytochromes). According to him hydrogen carrying and electron carrying proteins alternate in the respiratory chain to form three functional loops called *oxidation-reduction loops* O/y loops).

A single loop consists of a hydrogen carrier and an electron carrier. In each loop two $H^+$ are carried outward through the inner membrane and delivers two $H^+$ to the C side, the corresponding pair of electron is then carried back from outer to the inner surface of the membrane. Thus, each pair of reducing equivalents passing through such loop carries $2H^+$ from the matrix to outside. Each loop is though to provide the osmotic energy to form one molecule of ATP.

Michell's hypothesis is supported by a wealth of evidences.

3. *Conformation Coupling hypothesis*. According to this proposed by *Boyer* (1965), the electron transport and phosphorylating systems are in molecular contact and that information is transmitted by short-range interactions which may, in part, be electrostatic in nature. It is postulated that upon acceptance of the two electrons, each electron transfer protein undergoes a conformational change which is than transmitted to the ATPase by short range interactions. However, the mode of conformational changes that take place in the inner mitochondrial membrane is not yet clearly understood.

According to some recent views the conformational and chemiosmotic mechanisms of electron transport linkage to phosphorylation do not represent mutually exclusive alternatives but may indeed be two aspects of the same mechanism.

## PRACTICAL WORK

**To demonstrate that heat is generated in respiration**

Two thermos flasks are taken. In one of the flasks about 100 dry, and in other, about 100 water-soaked gram seeds are taken. Both the flasks are fitted with air-tight corks through which are inserted thermometers (Fig. 7.3.) The bulbs of the thermometers are kept embedded within the mass of seeds. The initial temperature is noted in either case.

After few hours it is observed that while the thermometer does not show any rise in temperature in the case of dry seeds. there is a distinct rise in temperature in the case of germinating seeds. This rise in temperature is obviously due to the liberation of heat energy by the respiring seeds.

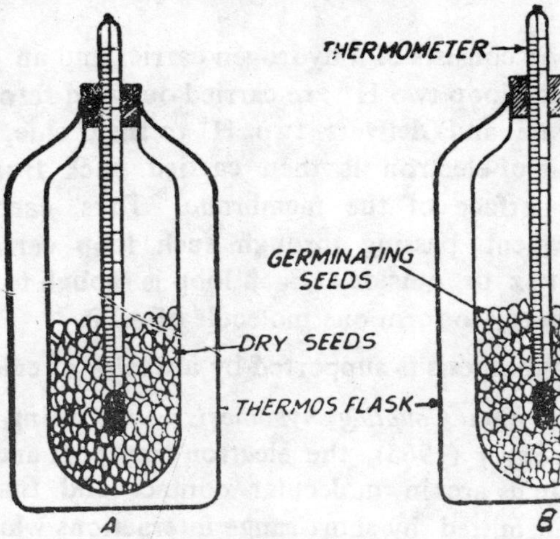


Fig. 7.3. Apparatus to show that heat is liberated in respiration.

**To demonstrate that roots also respire**

Two bottles with air tight rubber corks are taken. Both the bottles are filled with water which is made slightly alkaline. Few drops of phenolphthalein are added to the water in either case. Water gives a pink colour. In one of the bottles the root system of a plant is introduced through a hole in the cork. The other bottle is fitted with the cork and is allowed to remain as such to act as a control. After half an hour or so the pink colour of the water containing the roots disappears whereas there is no such change in the control.

Carbon dioxide liberated by the respiring roots is converted into carbonic acid. The resulting acidity neutralizes the alkalinity of water and, therefore, the pink colour disappears.

**To measure the liberation of $O_2$**

The instruments used are known as respirometers. Various types of respirometers are described as follow :

### (*i*) *Simple Respirometer*

The set-up consists of a flask and a delivery tube dipping in saline water as shown in Fig. 7.4. The level of water in the delivery tube is marked in the beginning of the experiment. The arrangements are air tight.

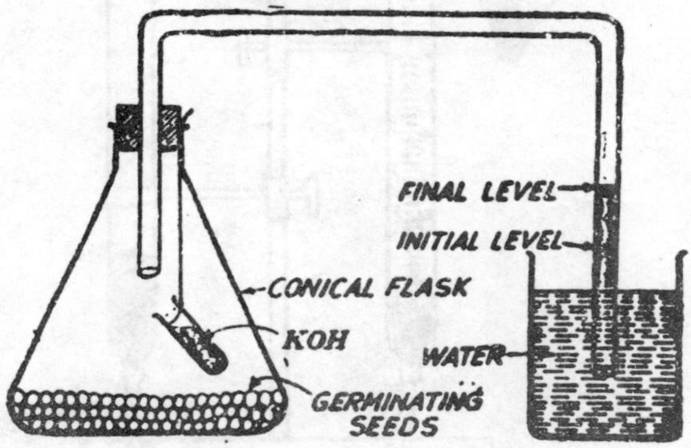


Fig. 7.4. Simple respirometer.

Oxygen is taken in during the process of respiration and $CO_2$ given off is absorbed by KOH in the flask. Due to utilisation of oxygen in respiration the vacuum created is filled up by upward movement of water in the delivery tube. The rise of water shows the amount of $O_2$ taken in.

### (*ii*) *Ganong's Respirometer*

The apparatus consists of a bulb with a graduated tube. The stopper of the bulb and its neck are provided with holes which can be brought opposite to each other. The total volume of the apparatus is about 102 ml. The graduated tube is connected with a levelling tube by means of a rubber tubing. Respiratory material is placed in the bulb. KOH solution is poured into the levelling tube and the levelling tube is so adjusted that KOH is at 100 ml. mark in the graduated tube, while the hole in the stopper and neck of bulb are opposite to each other Fig. 7.5. The stopper in the neck of the bulb is turned so that the two holes get apart. The apparatus is allowed to stand for a definite length of time.

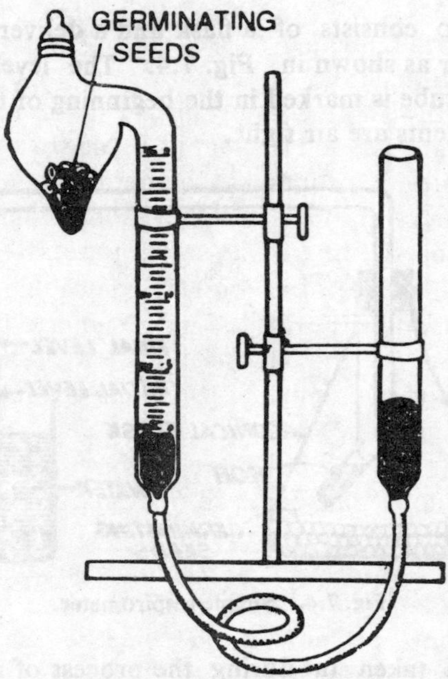


Fig. 7.5. Ganong's Respirometer.

**$CO_2$ produced in respiration is absorbed by KOH and the KOH solution begins to rise in the graduated tube.**

**After some definite time period the levelling tube is raised to bring the solution in two tubes to the same level again. Say, now it is at 80 ml. mark in the graduated tube. This indicates that 20 ml. of oxygen has been absorbed during respiration. The amount of $CO_2$ given out is roughly equal to the oxygen absorbed.**

**To measure the Respiratory Quotients by means of a pair of Respiroscopes**

A pair of respiroscopes is fixed on a vertical wooden stand as shown in the Fig. 7.6. In the upper wider portions of the two tubes are taken equal quantities of the given material. In one of the tubes, say B, a small glass tube containing pellets of KOH is hung from the top. The lower ends of both the tubes are dipped below the surface of water present in the beakers.

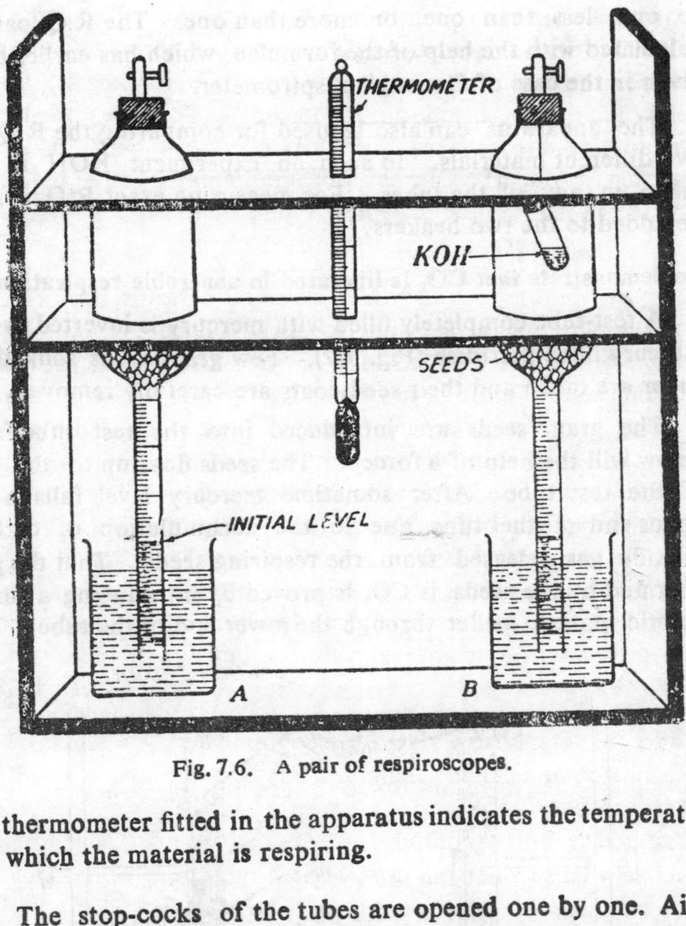


Fig. 7.6. A pair of respiroscopes.

A thermometer fitted in the apparatus indicates the temperature at which the material is respiring.

The stop-cocks of the tubes are opened one by one. Air is sucked out with the help of a rubber tubing till water rises to a predetermined mark in the graduated tube. The initial levels are kept the same in both the tubes.

The stop-cocks are closed immediately. The tubes are made air-tight. After half an hour the levels in the two tubes are noted. In tube B there is always a rise in the level because $CO_2$ liberated in respiration is absorbed by KOH. The rise in the level, therefore, indicates the amount of $CO_2$ released in respiration. In tube A, the level may remain constant or may rise or fall depending upon the fact whether the R.Q. is equal to one, less than one, or more than one. The R.Q. can be calculated with the help of the formulae which has earlier been given in the case of Ganong's respirometer.

The apparatus can also be used for comparing the R.Q. of two different materials. In such an experiment KOH is not taken in any of the tubes. For measuring exact R.Q. it may be added to the two beakers.

**To demonstrate that $CO_2$ is liberated in anaerobic respiration**

A test-tube completely filled with mercury is inverted over a mercury filled petridish (Fig. 7.7). Few gram seeds soaked in water are taken and their seed-coats are carefully removed.

The gram seeds are introduced into the test tube from below will the help of a forcep. The seeds float up to the top of the test tube. After sometime mercury level falls in the upper end of the tube due to the accumulation of carbon dioxide gas released from the respiring seeds. That the gas, liberated by the seeds, is $CO_2$ is proved by introducing a small caustic potash pellet through the lower end of the tube. The

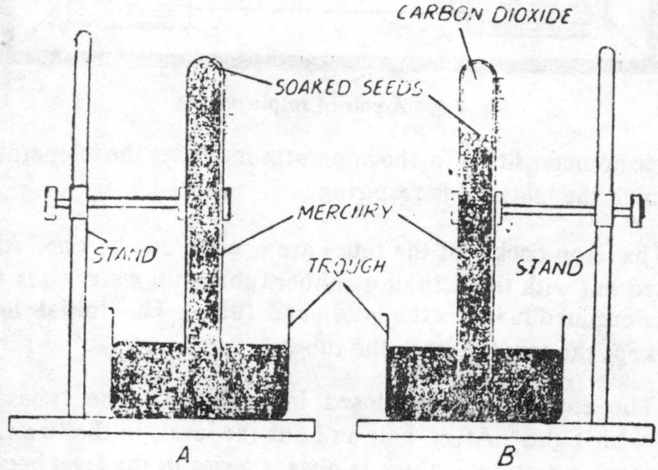


Fig. 7.7. Evolution of $CO_2$ in anaerobic respiration.

gas is absorbed by KOH. The mercury level, therefore, rises once again.

**To prove that $CO_2$ is liberated in aerobic respiration**

The apparatus is fitted as shown in the Fig. 7.8. In the case of germinating seeds, the bell jar is replaced by a U-tube within which the seeds are kept. The connections are made air-tight. When the aspirator is on, air enters first into the

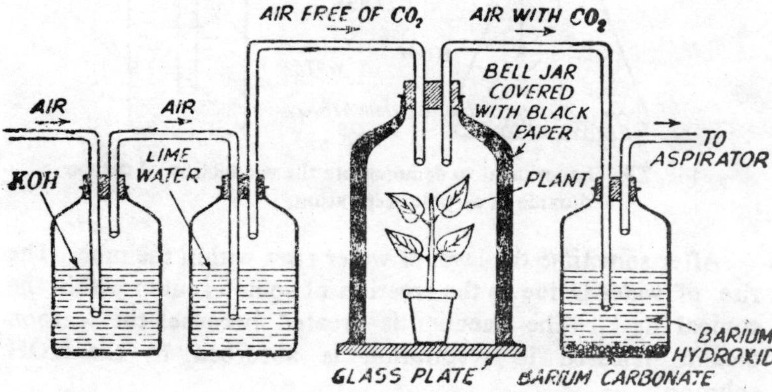


Fig. 7.8. Apparatus for the experiment on aerobic respiration.

KOH solution where $CO_2$ of the air is absorbed. Air almost free from $CO_2$ then enters into the second bottle which contains lime water. Whatever $CO_2$ is still left in the air is absorbed by the lime water. Air completely free of $CO_2$ then enters into the bell-jar where the plant respires with the help of atmospheric oxygen. Air, enriched with the $CO_2$ released in the aerobic respiration of the plant, enters the bottle containing $Ba(OH)_2$ solution. Carbon dioxide reacts with baryta water to give a white precipitate of barium carbonate.

**To prove that $CO_2$ is evolved in aerobic respiration by conical flask method**

A good number of moist germinating seeds is taken in a conical flask. The flask is fitted with a twice bent tube fitted with the cork. A glass tube containing KOH solution is hung down from the top with the help of a thread. The other end of the bent tube is kept dipping in water present in a beaker Fig. 7.9. The conical flask is made air-tight. The level of water in the bent tube is noted.

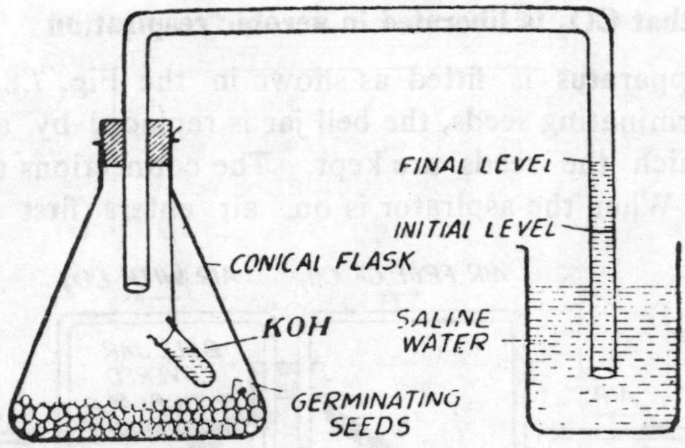


Fig. 7.9. **Experiment to demonstrate the production of carbon dioxide in aerobic respiration.**

**After sometime the level of water rises within the tube. The rise of water is due to the creation of some vacuum within the conical flask. The vacuum is created because the carbon dioxide released in respiration is absorbed by the KOH solution.**

**To measure the rate of respiration**

**There are several quantitative ways of measuring the rates of respiration of the living materials. They are all cumbersome methods since quantitative estimation of the amount of $CO_2$ released in respiration can be found out only by weighing the precipitate of $BaCO_3$ or by titrating NaOH which had absorbed the $CO_2$.**

**A simple manometric method for determining the amount of $CO_2$ is by means of *Warburg* manometer (Fig. 7.10). $CO_2$ released in respiration of a living material is absorbed by KOH kept within the conical flask.**

**The changes in the gas pressure due to the respiration is measured by observing a rise or fall of *Brodie's solution* in the graduated manometer tube of the apparatus.**

## FACTORS AFFECTING RESPIRATION

**The various factors affecting respiration may be divided into two groups, external factors and internal factors.**

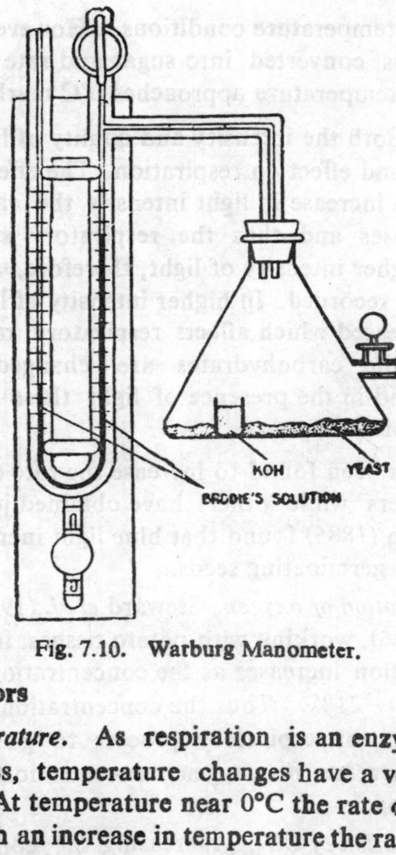


Fig. 7.10. Warburg Manometer.

**External Factors**

1. *Temperature.* As respiration is an enzymatically controlled process, temperature changes have a very pronounced effect on it. At temperature near 0°C the rate of respiration is very low. With an increase in temperature the rate of respiration increases up to 35°C. In most of the tissues maximum rate of respiration is attained somewhere between 35°C and 45°C. Beyond 45°C the rate of respiration decreases as the enzymes involved are inactivated.

While studying the effect of temperature on respiration it is essential to consider the length of time an organ or plant is exposed to any particular degree of temperature. In general the higher the temperature the shorter is the time period before the rate of respiration falls.

Temperature factor is important in the cold storage of plant and food. When we decrease the temperature the rate of respiration falls and it will take very long for the oxidation of food under low temperature conditions. However, in potatoes at 0°C starch is converted into sugar and rate of respiration increases as the temperature approaches 0°C mark.

2. *Light*. Both the intensity and quality of light are found to have a profound effect on respiration. The effect is however, indirect, with an increase in light intensity the rate of photosynthesis increases and thus the respiratory substrate is increased. At higher intensity of light, therefore, a higher rate of respiration is recorded. In higher intensity of light, temperature is also increased which affects respiratory rate indirectly. In succulents the carbohydrates are changed to organic acids at night and in the presence of light these organic acids are broken to form $CO_2$.

Red light has been found to increase the rate of respiration by some workers while others have obtained just the reverse results. Mangin (1885) found that blue light increased the rate of respiration in germinating seeds.

3. *Concentration of oxygen*. Steward *et al.* (1936), Hoagland and Broyer (1936), working with potato tissues, found that the rate of respiration increases as the concentration of oxygen is increased from 0—21%. Thus the concentration of $O_2$ present in air is optimum for respiratory process to go on. Beyond 21% further increase in oxygen concentration decreases the rate of respiration.

4. *Concentration of $CO_2$*. Increasing the concentration of $CO_2$ has a depressing effect on respiration. Kidd (1915) working with germinating mustard seeds got similar results. Heath (1950) has shown that increased concentration of $CO_2$ causes closure of stomata which increase the internal concentration of $CO_2$. This has a limiting effect on respiration.

5. *Minerals*. Many of the inorganic ions form prosthetic groups of respiratory enzymes. Absence of these ions decreases the activity of the concerned enzymes. Some of the minerals actually enter into the structure of enzymes. Increasing phosphorus and nitrogen supply to plant increases the rate of respiration. Potassium on the other hand increases the rate of respiration at lower concentration. Higher concentration of $K^+$ impairs respiration.

**Internal Factors**

6. *Respiratory Substrate.* As respiratory substrates are the materials which are oxidised in respiration the rate of respiration is affected by quality and concentration of substrate. Increase in sugar concentration increases the rate of respiration.

7. *Water Content.* With an increase in the hydration of tissues the rate of respiration increases. Water content of seeds is low. When water is given to a seed the food is changed into soluble form and is oxidised at a fast rate ultimately resulting into germination. It is for this reason that seeds are stored in dry places.

8. *Mechanical Injury.* It is generally believed that wounding of plant organs stimulates respiration. Hopkin (1927) found in case of potatoes that sugar concentration was increased around a cut. Perhaps increase in respiration due to injury is because of increased availability of respiratory substrate.

9. *Protoplasmic Factor.* Young growing organs like apices of shoot and root, germinating seeds, etc., respire more than the differentiated older tissues. The rate of respiration is dependent upon the quality of protoplasm in a cell.

Distinction between photorespiring and non-photorespiring plants

| *Character* | *Photorespiring plants* | *Non-Photorespiring plants* |
|---|---|---|
| 1. Geographical distribution | Generally temperate | Generally tropical |
| 2. Photosynthetic efficiency | Low | High |
| 3. $CO_2$ evolution in light | High | Absent |
| 4. Leaf-bundle sheath | Absent | Present |
| 5. Leaf chloroplasts | Mostly similar | Mostly dimorphic |
| 6. Carbon pathway | Calvin cycle ($C_3$ cycle) | $C_4$ cycle |
| 7. Reassimilation of respiratory $CO_2$ | Mostly none | Probably occurs |

## SOME IMPORTANT RELATED TOPICS

### Distinction between fermentation and anaerobic respiration

| *Fermentation* | *Anaerobic respiration* |
| --- | --- |
| 1. *Occurrence* : Yeast, bacteria and certain other micro-organisms can ferment carbohydrates to produce $CO_2$, alcohol and other products almost in the same manner as in anaerobic respiration. | 1. Tissues of higher plants deprived of $O_2$ are generally capable of surviving at least for a few hours with the help of anaerobic respiration. In fact, anaerobic respiration takes place in certain tissues even under normal conditions when $O_2$ is available. |
| 2. *Substrate and product* : In fermentation, almost all kinds of carbohydrates (mono, di or polysaccharides) may be converted into $CO_2$, ethyl alcohol, butanol, acetone etc. Hence, terminal reaction of fermentation is slightly different from anaerobic respiration. | 2. In plant tissues, maintained under anerobic conditions, pyruvic acid produced from glucose through glycolytic pathway is converted to $CO_2$ and ethyl alcohol. Glucose → Pyruvate → $CO_2$ + ethyl alcohol |
| 3. *Catalytic enzymes* : A group of enzymes, earlier known as zymase complex, is involved in the process. The enzymes may act intracellularly or extracellularly depending on the circumstances. | 3. The catalytic enzymes are located in the cytoplasm of cells. Two enzymes, pyruvic acid decarboxylase and alcohol dehydrogenase are involved in the reaction. |

Distinction between photorespiration and respiration

| *Photorespiration* | *Respiration* |
|---|---|
| 1. *Occurrence* : In most green plants photorespiration, which is induced by light, occurs. | 1. Respiration does not need the presence of light. |
| 2. *Site* : Takes place only in photosynthetic cells in the presence of light. Chloroplasts, peroxysome, mitochondria and cytoplasm are involved in the process. | 2. Takes place in all living cells. Only cytoplasm and mitochondria are involved in the process. |
| 3. *Light dependence* : Involves light dependent uptake of oxygen and evolution of $CO_2$. | 3. $O_2$ uptake and $CO_2$ evolution are independent of light. |
| 4. *$O_2$ sensitivity* : Increases with rise in $O_2$ concentration from 0% to 100%. | 4. Saturates at a very low $O_2$ concentration (about 2%) |
| 5. *Substrate* : Glycollate is the substrate. | 5. Usually simple carbohydrates, or acetyl CoA. |
| 6. *ATP synthesis* : Absent | 6. Highly efficient for ATP synthesis. |
| 7. *Biological significance* : Not clearly known, may be a unavoidable wasteful process. | 7. The only oxidative process that meets energy requirements of the body. |

# 8

# Phosphorylation in Higher Plants

In the economy of life processes, phosphorylation occupies a place comparable to that of nucleic acids in protein synthesis and heredity transmission, of proteins as specific catalysts, and of oxidation-reduction systems in energy transfer.

Phosphorylation may be defined as the biochemical process by which phosphate or phosphoryl radicals are transported to an acceptor by a transfer reaction. As a consequence of phosphorylation the reactivity of a compound increases. This is due, in many cases, to the capacity for substituting the phosphoryl group with other groups or radicals (synthetic reactions); or, in other cases, to a higher capacity for binding to active constituents, for instance enzymatic proteins, or for forming complexes with metal ions.

As a result of phosphorylation, activation energy barriers are lowered and otherwise unfavorable thermodynamic conditions are overcome. The number of reactions chemically feasible in biological systems thus is enormously increased.

Our knowledge of phosphorylation in higher plants has been built up mainly through studies aimed at elucidating mechanisms already well known and thoroughly investigated in animal cells and in microorganisms. All available evidence indicates a fundamental identity for the process in all living organisms.

The main purpose of the present chapter is to present a general view of phosphorylation in higher plants. Particular attention will be paid to the physiological aspects and implications of this process. The importance of phosphorylating reactions for the control of such functions as respiration, growth, and solute uptake *in vivo* will thus be treated in some detail. The evidence concerning the regulation of phosphate metabolism in intact cells will also be briefly discussed.

The central point of oxidative phosphorylation at the mitochondrial level will be considered extensively in this *Review* by Crane, and is thus only superficially treated in the present chapter. Photosynthetic phosphorylation has also been reviewed and is not considered in this chapter.

## METHODS FOR THE STUDY OF PHOSPHATE METABOLISM

While many of the classical procedures for the determination and fractionation of phosphorylated compounds are still valid and widely used, a few recently developed procedures seem worth mentioning here, either because of their specificity and convenience, or because they permit us to overcome some technical difficulties characteristic of plant materials.

A convenient adaptation of the Fiske and Subba Row method to the rapid determination of very small amounts of phosphate in cell-free systems has been elaborated by Taussky and Shorr. Rapid and quantitative separation of organic from inorganic phosphate has been obtained with the isobutanol-benzene method by different scientists, it may be useful to observe that the presence of reducing agents such as ascorbate interferes with the procedure. This inconvenience is eliminated by the omission of benzene as in the original Berenblum and Chain method. The integration of this method with the use of radioactive phosphate is very convenient for the detection of low rates of phosphorylation in cell-free systems. Inorganic phosphate, hexose phosphates, and adenosine phosphates can be efficiently determined in cell-free systems through a procedure combining paper chromatography and the use of radioactive phosphate.

Good results in the separation by column chromatography of phosphate intermediates in extracts from whole cells are reported by Ducet and Mencl, who used a modification of the Khim and Conn method and by Bartlett. Paper chromatography has been widely used for the determination and identification of $^{32}P$- and $^{14}C$-labeled intermediates in higher plant tissues.

All of these methods, though giving excellent result with cell-free, partially purified systems, are often unsatisfactory for the quantitative determination of phosphate compounds in extracts from higher plant tissues, in which various substances may interfere with the clear-cut separation of the compounds investigated. Thus, the classical method of fraction with barium and alcohol fails, in many cases, to give reproducible and quantitative results. On the other hand, the presence in plants of phosphate-containing materials still unknown or unsufficiently characterized may cause significant errors. A convenient approach to this technical problem is the combination of a preliminary, partial fractionation, with the enzymatic measurement of the various phosphate metabolites by means of specific enzymes. This type of procedure has been successfully employed for ATP determination by Marre and Forti who combined nucleotide adsorption on Norit A charcoal with the use of rabbit muscle ATPase.

## LEVEL OF PHOSPHORYLATED INTERMEDIATES IN HIGHER PLANTS

The presence in higher plants of most of the phosphate compounds that occur in animal tissues and in lower organisms is well documented. However, relatively few data are available concerning the average concentration of the main phosphorylated constituents and their changes in relation to various physiological situations. A survey of recent reports in this field indicates that hexose phosphate and adenosine phosphate, together with inorganic phosphate, are probably the quantitatively more important and widespread components of the acid soluble fraction. In Table 8.1 some data on hexose phosphate, ADP, and ATP concentrations in higher plant tissues are

Table 8.1. Average level of acid-Insoluble Phosphates in some Higher Plant Tissues.

| Material | Total acid-soluble phosphate | Inorganic phosphate | Glucose-1-phosphate | Glucose-6-phosphate | Fructose-6-phosphate | Fructose-1.6-diphosphate | ATP | ADP |
|---|---|---|---|---|---|---|---|---|
| | *μM/gm fresh weight** | | | | | | | |
| Pollinated or parthenocarpic tomato ovaries (2 to 6 days from pollination) (33) | 15-19 | 10-12 | 0.6-0.9 | 2.1-2.9 | 0.8-0.9 | 1.9-2.2 | | |
| Developing pea seeds (12 to 21 days from anthesis) (34) | 17 | 9 | | 1-6 | | 0.15-0.3 | 0.3-0.8 | 0.2-0.5 |
| Mature pea seeds (imbibed overnight, aerobiosis) (35) | | 2 | | 1-9 | | 0.4-1 | 0.5-1 | 0.4-2 |
| Etiolated pea internode segments (with $5 \times 10^{-5}$M IAA) (29, 36) | | 6 | | | 0.5 | | 0.2-0.3 | 0.3-0.4 |

*Data roughly approximated from calculation of values reported in the original papers.

reported. It appears that hexosemonophosphates predominate among sugar testers their concentration ranging between 1 and 6 $\mu$M/gm fresh weight, and the equilibrium between glucose-6-phosphate and fructose-6-phosphate being near enough to that found *in vitro* for the phosphohexoseisomerase reaction. Glucose-1-phosphate and fructose-1,6-diphosphate concentrations are significantly lower, ranging between 0.3 and 2 $\mu$M/gm fresh weight. A consideration of the equilibrium constants of the reactions involved in the interconversion of the hexose esters suggests that the activity of phosphofuctokinase—besides, obviously, the availability of ATP—may be a factor of major importance in the control of this important point of carbohydrate metabolism.

According to the few data available, ATP and ADP concentrations vary around 0.2 to 1 $\mu$M/gm fresh weight for each of these compounds. A qualitative evaluation by Cherry and Hageman of other nucleotide polyphosphates in pea tissues shows that guanosine triphosphate and uridine triphosphate (UTP) are present at a much lower concentration. The same report also indicates the presence in this material of small amounts of a guanosine-adenine dinucleotide, which could possibly be related to Albaum's report of a polymerized form of ATP in plant tissues.

The scarcity of investigations in this field makes it difficult to draw any general picture correlating the changes in concentration of the above mentioned major phosphorylated metabolites with known physiological activities. In tomato ovaries as in pea seeds, the initial stages of development and the initiation of protein and starch synthesis correspond to a significant increase of hexosephosphate and to a decrease of inorganic phosphate, while the ATP/ADP ratio, in the single case where it has been determined, does not show significant changes. In tomato and pea fruits the intense starch synthesis is accompanied by a progressive, strong increase of phosphorylase activity ; on the other hand, the glucose-1-phosphate/inorganic phosphate ratio is such that the reaction catalyzed by this enzyme should follow the phosphorolytic direction.

One has to consider the difficulty of interpreting these data without a knowledge of the real concentration of reagents at the level of reaction sites. In this connection, Rowan and Turner suggested that the lack of correlation between starch synthesis and the glucose-1-phosphate/phosphate ratio might be due to a large difference between the apparent ratio and the one prevailing at the cell site where starch is synthesized. It is interesting to quote, in this regard, that Wu and Racker suggest that in isolated ascites tumor cells the active concentration of inorganic phosphate available for the triose phosphate dehydrogenase reaction is probably considerably lower than that appearing from the analysis of the cell extracts. An analogous suggestion of the presence of separate "stores" of ATP and ADP in intact ascites tumor cells has been advanced recently by Chance and Hess.

The few available data on ATP, ADP, and "high energy" phosphate concentrations are of obvious interest. Among other important phosphorylated compounds, the presence in higher plants of almost all phosphorylated coenzymes and cofactors has been qualitatively demonstrated.

Quantitative data are available for pyridine nucleotides; Anderson and Vennesland found in green leaves of several species a concentration of DPN ranging between 0.01 and 0.02 μmoles/gm fresh weight and an approximately equivalent amount of TPN. Quite recently, Krogmann reported values of 4 μmoles/gm chlorophyll for DPN and TPN (not distinguished) in chloroplasts. According to Bevilacqua the concentration of total pyridine nucleotide coenzymes during the first week of germination varies from about 2 to $5 \times 10^{-5}$ M, for wheat embryo, and from 1 to $3 \times 10^{-5}$ M, for pea embryos. Enzymatic determination of TPN in isolated, etiolated pea internodes gave values of $1.1 \times 10^{-5}$ and $0.9 \times 10^{-5}$ M, for reduced and oxidized TPN respectively. The ratio between the reduced and the oxidized form was consistently shifted towards reduction by the presence of indoleacetic acid in the medium.

## PHOSPHATE TURNOVER

In higher plants, as well as in all other organisms, phosphate turnover comparises three distinct phases. In a first phase, inorganic phosphate is taken up and combined with organic molecules or radicals, forming amide, ester, or anhydride bonds (phosphorylation *sensu strictu*). From the primarily phosphorylated compound thus formed, the phosphoryl group may be transported to other molecules (transphosphorylation). Finally, phosphate—or pyrophosphate—is split from the phosphorylated intermediates either through hydrolytic cleavage, or through its substitution by an organic radical.

The phosphate bonds of biologically interesting compounds are usually classified, since the classical work of Lipmann, as high- and low-energy bonds on the basis of the free energy change of hydrolysis. However, the transition between the two categories is rather continuous. For this, and for other general considerations the proposal by Klotz to indicate the "energy wealth" of phosphate bonds as the Phosphoryl transfer potential (Ptp) seems to be accepted, and this term will be adopted in the present chapter in many instances. The value of the Ptp, expressed in kcal. corresponds to the amount of biochemical work that can be obtained from a given phosphate bond.

Table 8.2 reports some data on the free energy change of hydrolysis (equivalent to the Ptp) of several important phosphorylated metabolites. Among these, the correction should be noted for ATP (terminal bond) from the higher early value of 10 to 12 kcal. to one of about 7 kcal.

In a study of the electronic structure of the phosphate bond Pullman and Pullman have recently calculated the theoretical contribution of different factors to the Ptp of several energy-rich phosphorylated compounds. According to these calculations, opposing resonance and electrostatic repulsion between the positively charged phosphate atoms for ATP and ADP, opposing resonance and free energy of ionization for acyl phosphates, and opposing resonance and enol-keto tautomerism for phospho*enol* pyruvate, appear to be the main factors in the

**Table 8.2. Free energy changes of Hydrolysis of some phosphorylated metabolites.**

| *Metabolite* | *Temperature* (°C.) | *pH* | *Free energy change in kcal. at specified temperature and pH* |
|---|---|---|---|
| ATP (→ ADP + inorganic phosphate) | 30 | 7 | — 7.0 |
| ATP (→ AMP + pyrophosphate) | 37 | 7.5 | — 8.6 |
| ADP (→ AMP + inorganic phosphate) | 25 | 7.0 | — 6.4 |
| Pyrophosphate | 30 | 7.0 | — 6.5 |
| Glucose-1-phosphate | 30 | 7.0 | — 5.0 |
| Galactose-1-phosphate | 25 | 7.0 | — 5.0 |
| Glucose-6-phosphate | 25 | 7.0 | — 3.3 |
| Fructose-6-phosphate | 30 | 7.0 | — 3.8 |
| Fructose-1-phosphate | 38 | 5.8 | — 3.5 |
| Glycerol-phosphate | 25 | 7 | — 2.1 |
| Glyceraldehyde-1,3-diphosphate | 25 | 6.9 | —11.8 |
| Phosphoenolpyruvate | 30 | 7.4 | —12.7 |
| Acetylphosphate | 29 | 7.0 | —10.1 |

very high free energy change of hydrolysis of these phosphate compounds.

## PRIMARY PHOSPHORYLATION

The mechanism of phosphate incorporation into organic compounds appears, at least in the cases most extensively investigated, substantially similar to that found in animal cells. The main source of energy for this endergonic process is the redox potential energy that is set free in oxidative metabolism, even if some minor contribution may come from the bond energy (group transfer potential) of compounds attacked by phosphate in phosphorolytic reactions, as in the case of phosphorolysis of glucosidic linkages.

A problem still open is whether phosphorolytic cleavage of starch or other polysaccharides significantly contributes to phosphorylation in higher plants, by saving, as in animal tissues, a large part of the energy of the glucosidic 1-4 bond and thus sparing one of the two molecules of ATP required to raise the free hexoses to the level of fructose diphosphate.

Starch phosphorylase is a widespread enzyme in plant tissues : moreover, as stated in an above section, the apparent glucose-1-phosphate/inorganic phosphate ratio in plant tissues is much in favor of starch breakdown. However, all available evidence seems to indicate that maximal phosphorylase activity in a tissue accompanies the stages of synthesis, while starch breakdown is concomitant with an increase in amylase activity. It seems possible that two different pathway are involved in starch breakdown : one, hydrolytic, used for rapid, massive production of soluble sugars to be transported to other cells ; and another, quantitatively minor, leading to the phosphorolytic synthesis of hexosemonophosphates for respiration or anabolic purposes.

The occurrence of a phosphorylating step at the level of glyceraldehyde-3-phosphate oxidation is well established, and the respective enzymes (glycer-aldehyde-3-phosphate dehydrogenase and phosphoglyceric acid kinase) have been partially purified and characterized. While the glyceraldehyde-3-phosphate dehydrogenase of non-photosynthetic tissues is DPN dependent and very similar to the animal and yeast enzyme, two additional dehydrogenases acting on the same substrate are present in green tissues. One of these is TPN specific and phosphate or arsenate dependent, and the second is also TPN specific but is active even in the absence of arsenate or phosphate, and thus apparently is uncoupled from the normal phosphorylating reaction. It has been suggested that TPN-linked triosephosphate dehydrogenase may be important for photosynthetic metabolism.

A large body of evidence indicates that in higher plants, as in animal tissues, the bulk of phosphorylation occurs at the mitochondrial level during the stepwise transfer of electrons from respiratory substrates to oxygen. The mechanism of oxidative phosphorylation being discussed in another chapter of this volume, only a few general points of this process will be touched here. Phosphorylation coupled to oxidation of the tricarboxylic acid cycle substrates has been demonstrated to occur in the mitochondria of a large number of species.

Isotope studies show that the Krebs cycle is usually the major route of oxidative carbohydrate breakdown. An electron transfer chain involving pyridine nucleotide coenzymes, flavoproteins, cytochromes-*b* and -*c*, and cytochrome oxidase as the terminal oxidase appears to be the most important system involved in mitochondrial oxidative phosphorylation, and, even in whole plant cells, modern work tends so assign a predominant role to cytochrome oxidase in mediating overall oxygen uptake. Uncoupling agents such as 2,4-DNP affect mitochondria of higher plants in the same way as they do those of animals, by completely uncoupling oxidations at the electron transport chain level from phosphorylation, while they do not affect phosphorylation at the substrate level.

On the other hand, the possibilily is still to be considered that other respiratory pathways besides the Krebs cycle might bring on, in higher plants, phosphorylation coupled to oxidations. This possibility appears of particular interest in the case of the pentose phosphate pathway, which recent researches indicate as one of the main respiratory routes, at least under particular physiological conditions. During the oxidation of glucose-6-phosphate to pentose, two TPN molecules are reduced. A problem arises then of the system mediating the oxidation of this TPNH, and its possible role in phosphorylation. This point is still largely obscure; however, some data are available that can be useful at least to start its discussion.

Investigations *in vitro* show the presence, in higher plant tissues, of at least five distinct systems which could mediate electron transfer from TPNH to oxygen :

(*a*) TPNH + pyruvate + $CO_2$ (malic enzyme) → malate → mitochondrial oxidations

(*b*) TPNH + *a*-ketoglutarate + $CO_2$ → citrate

(*c*) TPNH → glutathione → ascorbat/dehydroascorbate → ascorbate oxidase

(*d*) TPNH → glutathione → ascorbate → quinones → polyphenoloxidase

(*e*) TPNH → ascorbate/monodehydroascorbate → ascorbate oxidase or quinones-polyphenoloxidase, or cytochromes.

The first of these pathways appears as a mixed system, in which the pentose phosphate pathway would provide the primary electron donors and the initial electron carrier; then electrons from TPNH, pyruvate from glycolysis, and $CO_2$ would combine to form malate, a substrate for the mitochondrial oxidations of the Krebs cycle. In the step from malate to oxalacetate, three phosphorylating steps would occur, obviously, even if undirectly, at the expense of the oxidation-reduction energy primarily stored in TPNH by pentose phosphate pathway oxidations. Evidences in favor of the operation of this pathway under physiological conditions are: first, the presence of an active TPNH specific malic enzyme in the soluble fraction of higher plants and second, the demonstrated capacity of plant tissues to incorporate labeled $CO_2$ rapidly into malate. One must also consider that a large part of the carbon fragment entering the Krebs cycle is used on the way towards oxalacetate for anabolic purposes, so oxalacetate has to be continuously regenerated in order to maintain a constant efficiency of the cycle. The synthesis of malate from TPNH, $CO_2$ and pyruvate could be an active system for the maintenance of a steady oxalacetate level in mitochondria. The quantitative importance of such a system should probably be greater when large amounts of organic acids are removed from the Krebs cycle, for instance, during active synthesis of proteins. On the other hand the presence of other $CO_2$-fixing, dicarboxylic acid-producing mechanisms must be remembered. In fact, the phosphaenol pyruvate carboxylase system appears as perhaps the most active mechanism implicated in the $CO_2$ incorporation into malate in higher plants.

For the pathway involving glutathione and ascorbic acid, no considerable difficulty is met so far as hydrogen transfer from TPNH to ascorbate is concerned, as the enzymes involved have been shown to be of wide occurrence in higher plants and the occurrence of the reaction between glutathione and dehydroascorbic acid has been demonstrated *in vivo*. The doubtful step in this sequence concerns rather the efficiency of both ascorbic acid oxidase and polyphenoloxidase as terminal oxidases in whole cells. Eichenberger and Thimann found that in pea stem tissue only a few per cent of total respiration can be mediated by ascorbic oxidase although this enzyme shows very high activity in extracts from this material. In aged potato slices, where the operation of the pentose phosphate pathway appears very active both from enzymatic and from isotope measurements and which are a rich source of polyphenoloxidase, quantitative importance of the latter enzyme as a terminal oxidase seems contradicted by the very high affinity for oxygen and the considerable cyanide resistance of respiration.

No clear-cut evidence supports or denies the interesting possibility of a cytochrome-mediated oxidation of ascorbate. Many scientists reported *in vitro* oxidation of ascorbate by mitochondrial preparations, but the involvement of cytochrome oxidase could be demonstrated only in the presence of exogenous cytochrome-*c*, which non-enzymatically oxidizes ascorbate. Significant phosphorylation coupled to ascorbate oxidation—still in the presence of added cytochrome-*c*—was shown by Fritz and Naylor. As cytochrome-*c* seems strictly localized in mitochondria, to accept the hypothesis of a cytochrome-oxidasemediated, phosphorylating ascorbate oxidation one should postulate the presence of an unknown intermediate electron carrier in plant cells, to substitute for cytochrome-*c* in connecting ascorbate with the endomitochondrial cytochrome chain.

A different approach to the problem of an involvement of the pentose phosphate pathway in phosphorylation comes from the finding that some fraction of the total amount of the pentose phosphate pathway enzymes is bound to particulate elements of plant cell cytoplasm. Forti *et al.* showed that the glucose-6-phosphate-oxidizing particles are different from true mitochondria in as much as they sediment at lower centrifugal forces, do not contain cytochrome oxidase and succinic dehydrogenase, and are endowed with considerable TPN glutathione reductase, dehydroascorbic acid reductase, and ascorbic acid oxdiase activity. The particle-bound ascorbate oxidase differs from that found in the supernatant in being strongly activated by ADP, and also, at a lower level, by 2,4-DNP. This suggested the involvement of a phosphorylating step; but neither net phosphate uptake nor $^{32}$P-ATP exchange could be detected. It is true that the presence of a very active ATPase in this particulate fraction is a serious obstacle in these studies.

No direct evidence exists in favour of direct oxidation of extra-mitochondrial TPNH through the phosphorylating cytochrome chain, such as could be mediated by transhydrogenases, even though the presence of these enzymes in higher plants has been reported.

## THE POOL OF PHOSPHATE TRANSFER POTENTIAL

Since the first unequivocal demonstration by Albaum of the presence of adenosinepolyphosphates in higher plants, evidence has accumulated that these compounds represent the main Ptp reservoir in these organisms. No extensive research has been carried on concerning the existence of other auxiliary pools such as the creatine-phosphate and arginine-phosphate of animals. Inorganic polyphosphates and metaphosphates—which contain energy-rich phosphate bonds—are abundant in lower plants; but their presence in higher plants has been investigated to only a slight extent. The point is interesting, and further research in this field is desirable.

Besides ATP, other nucleoside polyphosphates playing an important role in phosphate metabolism (such as UTP and its sugar derivatives, and guanosine triphosphate) have been identified in higher plants. Among the enzymes catalyzing the reversible transport of phosphate between the various nucleotides, adenylic acid kinase has been partially purified and studied by various authors and the presence of uridinemonophosphokinase and of uridinediphosphokinase in pea extracts has been reported.

## UTILIZATION OF PHOSPHATE TRANSFER POTENTIAL

*General.* Utilization of the Ptp stored in the ATP system follows a substantially common pattern in all organisms, and involves activation of substrates by kinases or phosphate transferases, catalysis of synthetic reactions by phosphorylases or pyrophosphorylases, and finally regeneration of inorganic phosphate by phosphatases.

Brilliant work with $^{18}O$ and $^{32}P$ has demonstrated that in the kinasic and in the phosphatasic reactions a phosphoryl, rather than a phosphate group, is transported to the acceptor ; while in phosphorylasic and pyrophosphorylasic reactions the P-O bond of phosphate remains intact. The former type of reaction may thus be considered as a nucleophilic attack on the terminal phosphorus atom of ATP by the acceptor, and the latter reaction as a nucleophilic displacement of the unphosphorylated moiety of the attacked molecule by phosphate or pyrophosphate. As far as phosphatases are concerned, several interesting investigations have followed the important observation by Axelrod that some of these enzymes may efficiently transfer phosphate between organic compounds. Of particular interest in this regard seems the occurrence in higher plants of nucleoside transphosphorylases, catalyzing the reaction :

$$\text{nucleotide } (a) + \text{nucleoside } (b) \rightarrow \text{nucleoside } (a) + \text{nucleotide } (b).$$

Recent evidence suggests that, at least under conditions of low water availability at the level of the reaction site, enzymes of the phosphatase group could be involved in transphosphorylations utilizing compounds endowed with a relatively low Ptp as phosphate donors.

*Interconversion of carbohydrates and related compounds.* The scheme of carbohydrate interconversion presented in Fig. 8.1 is based on the demonstration *in vitro* of the presence in higher plants of all of the enzymes involved, and is substantially supported by the evidence obtained through isotope studies *in vivo*. The scheme permits us to distinguish between three distinct, major pathways of sugar metabolism, communicating between each other at the hexose phosphate level and each characterized by the hexose phosphate from which it originates. The glucose-1-phosphate sequence involves the phosphorolytic mechanism of glucose-1-phosphate ⇌ starch interconversion, and the UTP ⇌ UDP sugar chain of reactions, leading to glucocides, xylose and xylosides, and pectins. The hexose-6-

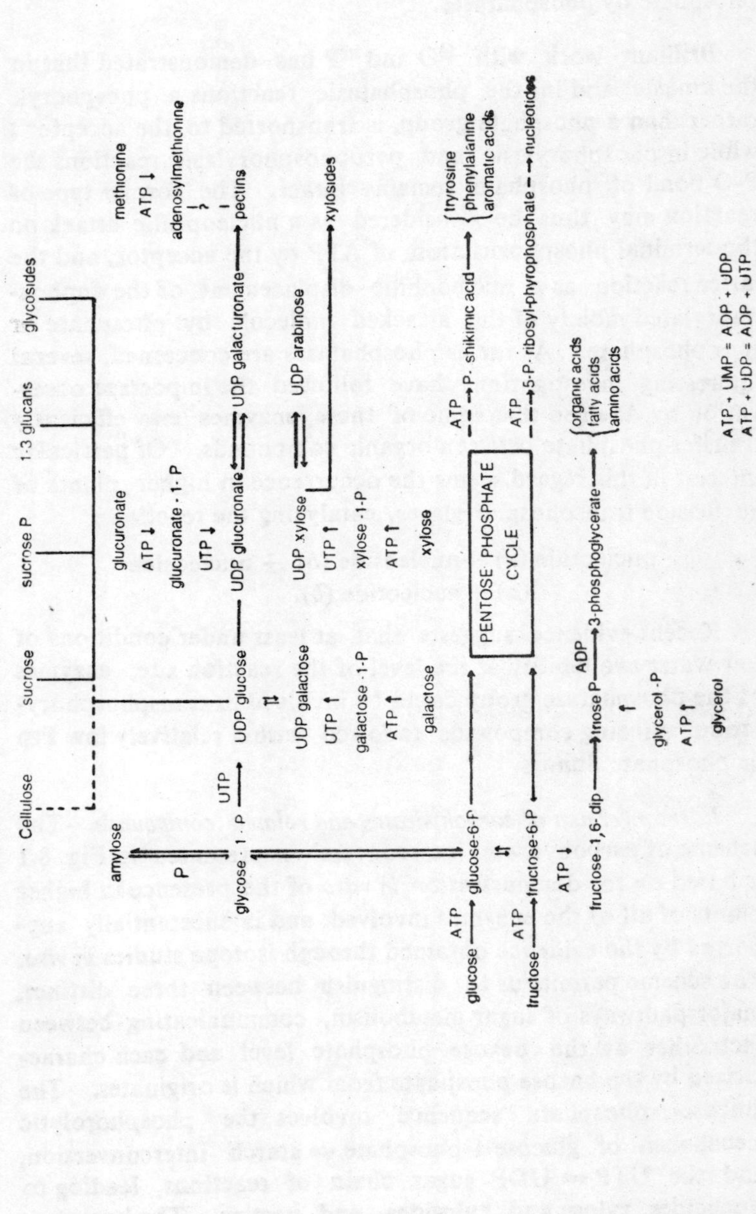


**Fig. 8.1. Role of adenosine and uridine polyphosphates in carbohydrate interconversion in higher plants.**

phosphate sequence involves the whole pentose phosphate pathway and is thought to lead to synthesis of nucleotidic pentose, and of various aromatic compounds including shikimic acid, the probable precursor of phenylalanine and tyrosine. The fructose diphosphate sequence involves the initial steps of glycolysis and of the mechanism of carbohydrate ⇌ organic acid interconversion.

From the point of view of distinguishing possible key points of control of metabolism by phosphate turnover, attention is called to the following points. The concentration of ATP and the activity of the various kinases must be of the utmost importance in determining the relative efficiency of the different pathways, a number of distinct enzymes in competition for ATP being involved in the scheme. The same consideration is valid for the UTP level and the relative activities of UMP and UDP kinases, and of the various UDP-sugar pyrophosphorylases. Pyrophosphorylasic syntheses are generally considered irreversible, as a consequence of hydrolitic cleavage of pyrophosphate by the widespread pyrophosphatases as in the example :

(*a*) UTP+glucose-1-phosphate ⇌ UDPG+pyrophosphate
(*b*) pyrophosphate+$H_2O$ → 2 inorganic phosphate

where the reversible Reaction (*a*) is driven from left to right by the strongly exergonic Reaction (*b*) ($-\Delta F.=7$kcal). The evidence *in vivo* supports this view, since $1^{-14}$ C-labeled galactose and galacturonic acid led to intact cells do not significantly contribute to respiratory $CO_2$, while they are rapidly incorporated into pectins and cell wall xylosides.

In pectin synthesis another important ATP-requiring step is probably involved, mediating the methylation of the galacturonic acid units. It is known that in yeast and in animals transmethylation requires the activation by ATP of the methyl donor methionine :

ATP+methionine → $P_i$+pyrophosphate+S—adenosylmethionine

adenosylmethionine being the general methylating agent. In higher plants methyl groups of methionine have been shown to be employed for the methylation of the uronic units of pectins in a transmethylation reaction. The recent finding by Mudd, that barley extracts catalyze both the synthesis of S—adenosylmethionine and the methyl transfer from this compound to suitable acceptors suggests that such a mechanism could also be involved in pectin synthesis.

*Glycolysis.* The chain of reactions leading from triose phosphate to pyruvate includes three reactions in which phosphorylation or transphosphorylation are involved. The coupling of glyceraldehyde-phosphate oxidation and DPN reduction to ATP synthesis from ADP makes this reaction easily reversible. The relation of this fact to the control of metabolism by phosphate-turnover will be discussed in a further section.

Phosphorylation of ADP by phosphoenolpyruvate and pyruvate kinase is a largely exergonic reaction ; however, recent studies by Krimsky on a cell-free reconstructed system suggest that it could be reversed, even *in vivo*, under conditions of high ATP/ADP and pyruvate/phosphoenolpyruvate ratios. For the use of pyruvate in the reversal of glycolysis, Krebs has postulated an ingenious ATP- and TPNH-involving mechanism :

(*a*) pyruvate+$CO_2$+TPNH → malate (malic enzyme)

(*b*) malate+$DPN^+$ → oxalacetate+DPNH

(*c*) oxalacetate+ATP → $CO_2$+phosphoenol pyruvate+ADP.

All the enzymes of this system have been shown in higher plants and recent studies consistently support the idea that Reaction (*c*) is the main route of reversal of glycolysis in sugar synthesis from fatty acids during fatty seed germination. Finally, phosphorylation may play an indirect role in the interconversion between the triose phosphates as Stumpf demonstrated that higher plant extracts may catalyze the sequence :

(*a*) dioxyacetone-phosphate+DPNH → glycerolphosphate +$DPN^+$

(*b*) glycerolphosphate+1/2 $O_2$ (mitochondria) → glyceraldehyde-3-phosphate+$H_2O$

as mitochondrial oxidation of glycerolphosphate is coupled to phosphorylation.

*Amino acid metabolism.* Recent work on animal materials and microorganisms shows that peptide bond synthesis involves as a first step the activation of amino acids by an ATP-dependent pyrophosphorylasic reaction : (*a*) ATP+amino acid → amino acyl—adenylate+pyrophosphate, followed by RNA requiring polymerization of the activated amino acyl units. Studies of this mechanism are largely based on the demonstration *in vitro* of an exchange between radioactive pyrophosphate and ATP, such as should be expected when Reaction (*a*) is occurring. Amino acid-dependent ATP-pyrophosphate exchange accompanied by amino acid incorporation into proteins has been demonstrated in cell-free preparations from a number of plant tissues.

ATP is also required for glutamine synthesis from ammonia and glutamic acid and for activation of shikimic acid by shikimic kinase during the biosynthesis of phenylalanine and tyrosine.

*Fat metabolism.* Several reports show that the synthesis of fatty acids from acetate and the activation of fatty acids is also an ATP and CoA requiring process in higher plants. Stumpf recently reported palmitate and oleate synthesis by a clear extract from avocado particles, in the presence of ATP, CoA and TPNH. The mechanism of the acetate activating reaction seems to be, as in animals and microorganisms, pyrophosphorylasic The presence in peanut mitochondria of an enzyme catalyzing phosphorylation of α, β-diglycerides and α-monoglycerides has been reported by Bradbeer and Stumpf.

ATP is necessary in lipid breakdown for the initial fatty acid activation, and also for glycerol phosphorylation. The enzyme catalyzing this reaction, glycerol kinase, has been recently demonstrated by Stumpf in pea cotyledons. An ATP requirement for phospholipid synthesis in mitochondrial preparations has been recently reported.

Researches by Shneour and Zabin on lycopene synthesis in cell-free preparations from tomato fruits indicate that mevalonic acid and ATP are involved in carotenoid synthesis.

*Sulfur metabolism.* It has been shown (98 to 100) that activation of sulfate in yeast involves two steps :

(*a*) ATP+sulfate ⇌ adenosinephosphosulfate (APS)+ pyrophosphate

(*b*) APS+ATP → 3-phosphoadenosine-5-phosphosulfate+ ADP

Reaction (*a*) is endergonic, but is driven from left to right by both pyrophosphate hydrolysis and by the large free energy change of Reaction (*b*). The enzyme catalyzing Reaction (*a*) has been named ATP sulforylase. Selenate can substitute for sulfate, adenosine phosphoselenate being formed. In this connection the recent finding of selenate-containing amino acids in *Astragalus* seed proteins is of interest. The enzyme ATP sulphorylase has been found in higher plants (Squire and Bandurski, personal communication). Reduction of active sulfate 3-phosphoadenosine-5-phosphosulfate to sulfite has been demonstrated in yeast extracts. According to recent studies, TPNH or lipoic acid mav function as electron donors in this reaction.

## PHOSPHATE TRANSFER POTENTIAL UTILIZATION IN PHYSIOLOGICAL PROCESSES

While the dependence of the above discussed mechanisms on Ptp energy has been demonstrated at the molecular level, the involvement of phosphorylation in a number of physiological processes is strongly suggested by observations *in vivo*.

Active uptake of solutes, growth by cell extension, and control of the rate of respiration, for which more data are available, will be briefly discussed here.

*Solute uptake.* The relationship between respiration and ion absorption, and the problem of whether active ion uptake depends directly on electron transfer, or on Ptp generated by oxidative reactions, has been discussed by several Scientists. A few additional remarks on the possible role of phosphorylation in this process may be made.

The main indication in favour of the requirement of Ptp for ion uptake is its inhibition by uncoupling agents such as

2, 4-DNP. Robertson suggested that the DNP effect could depend on the fact that electron transfer is effective for active ion uptake only when it proceeds coupled to phosphorylation, even if ATP synthesis is not the determining factor. This interpretation, however, is difficult to test experimentally and remains purely hypothetical.

Recent evidence indicates that ion uptake is inhibited not only by uncoupling agents that do not change the pathway of electron transfer, but also by cytochrome oxidase inhibitors, such as cyanide, CO, etc., even in the cases in which overall oxygen uptake is not affected. Hackett and co-workers report inhibition of inorganic phosphate uptake by KCN in aged potato disks, respiration of which is largely KCN insensitive, while Shealtiel and Ducet find that CO inhibits Rb absorption by barley roots by 75 per cent, without affecting oxygen uptake. Active amino acid uptake by carrot slices has been shown to be strongly inhibited by DNP as well as by KCN.

On the other hand, not only active ion uptake, but also the uptake of non-ionized compounds, such as sugars, is inhibited both by pure uncouplers and by inhibitors of cytochrome oxidase. Pennell and Weatherley found that KCN, axide, and DNP, as well as anaerobiosis, inhibit the penetration of sugars intc leaf disks ; sugar uptake was the same in air and in 2 per cent oxygen although $O_2$ uptake was inhibited by 35 per cent. Similarly, the rapid uptake of labeled glucose by aged potato disks is strongly inhibited by catechol at concentrations inhibiting respiration by 25 per cent, while a decrease of oxygen to 4 per cent, though decreasing respiration to the same or larger extent, does not affect sugar uptake at all.

In animal cells and in microorganisms, sugar uptake against a concentration gradient has been shown to be completely inhibited by uncoupling agents.

The above evidence, together with the difficulty of establishing a strict quantitative correlation between solute uptake and solute-dependent oxygen uptake, strongly suggests that some common basic Ptp energy-utilizing mechanism is required for the active uptake of any kind of solute. This of course does not contradict the evidence in favour of the existence of specific carriers or mechanisms for the uptake of the different molecular species, as the very synthesis or functioning of these carriers or mechanisms could depend on Ptp. In this connection, Laties' results on chloride uptake by potato slices at different temperatures may be quoted ; Laties found that in aging disks the capacity for chloride uptake rapidly increases, the increase being reversibly inhibited either by low temperature or by 2, 4-DNP. As both conditions obviously induce a lowering of Ptp in the cells, these data would be easily interpreted by the assumption of a Ptp energy-dependent synthesis or activation of a carrier. It may be pertinent to mention here that carrier synthesis has been demonstrated in bacteria as a substrate-inducible phenomenon in the case of galactose permease.

*Growth by cell extension*. Growth by cell division and protoplasm synthesis is an obviously Ptp-dependent process in several essential aspects, such as protein and nucleic acid synthesis. The requirement for Ptp energy in growth by cell extention is less clearly understood. A number of facts indicate that treatments inhibiting phosphorylation also inhibit cell extension. Carbon monoxide and cyanide inhibit water uptake even when, as in aged potato disks, they do not affect oxygen uptake. The simplest interpretation of these results is that the level of Ptp conditions this type of growth. Analysis of the mechanism of cell extension suggests that the metabolism of some component of the cell wall is essentially involved, as cell wall plasticity increases during cell extension. Recent important work shows that the metabolism of a pectin fraction is strongly enhanced during hormone-induced stimulation of cell extension. The dependence of cell extension on methylation of pectin is also indicated by its rapid inhibition by ethionine—which blocks methyl transfer from methionine to pection—and by the complete reversal of the inhibition by methionine. If one accepts the idea of a central role of pectin anabolism in cell extension, the necessary dependence of this type of growth on Ptp appears as a logical corollary. As discussed in a preceding section, ATP is required in two, or quite probably three, steps of pectin biosynthesis : hexose (or uronic acid) phosphorylation, UTP synthesis for UDP-sugar formation, and activation of methionine. Similarly, the synthesis of other cell wall components, as xylosides [and possibly cellulose utilizing the UDP-sugar pathway, is obviously Ptp energy-dependent. It is thus understandable that no growth by cell extension can be observed when the Ptp level in the cells falls below a threshold value. This seems to provide the more likely explanation for the inhibitory effect on growth by all respiratory inhibitors and uncouplers. On the other hand, it seems clear that such a basic involvement of Ptp energy in growth does not imply that hormonal regulation of the same process is necessarily mediated by an effect on phosphorylative systems.

## THE CONTROL OF RESPIRATORY METABOLISM BY THE PHOSPHORYLATING SYSTEM

As essential phosphorylative steps are involved in practically all metabolic pathways, it appears obvious that the efficiency of enzymes and the level of substrates involved in phosphorylation reactions closely control almost any important process in the cell. To restrict the field to a single, basic point, only the relationship between glycolytic and respiratory activity and ATP, ADP, and inorganic phosphate levels will be dealt with here. The problem has already been discussed by Laties, Axelrod and Beevers and Hackett.

The suggestion by Lardy that the level of ADP and inorganic phosphate could control the rate of oxidative reactions *in vivo* has found support in a number of experiments using widely varied materials. The possibility that this mechanism may operate is based on the assumption that in the whole cell—as is usual in isolated mitochondria—oxidations must be tightly coupled to phosphorylations ; thus the rate of $O_2$ uptake should depend on the availability of ADP and inorganic phosphate which, on the other hand, depends on the rate of transphosphorylations utilizing ATP as phosphate donor.

The physiological importance of such a mechanism would be that of equating at any particular moment the activity of the oxidative machinery to the requirement for Ptp (that is, for ATP high energy bonds) for biological work.

Highly interesting experimental analyses of the problem have recently been made by Gatt and Racker, by Wu and Racker and by Wu. Gatt and Racker, using a reconstructed system containing mitochondrial and glycolytic enzymes, found that when both systems are operating a strong competition for both ADP and inorganic phosphate between glycolysis and mitochondrial ovidations appears. In the inhibition of respiration by glycolysis, the more important factor seems to the ADP, while inorganic phosphate is rate limiting only at very low concentrations (less than $3 \times 10^{-4}$ M). In contrast, glycolysis requires, for full operation, much higher concentrations of inorganic phosphate, as for concentrations lower than $5 \times 10^{-3}$ M a strong inhibition of both glucose phosphorylation and lactate synthesis by mitochondrial oxidations is observed. This suggests that even *in vivo* determinations of $O_2$ uptake could make evident only a state of competition for phosphate acceptors, while the detection of competition for inorganic phosphate would require the determination of glycolytic activity.

The main conclusions of this work *in vitro* have been tested *in vivo* by Wu and Racker, on isolated ascites tumer cells. In this material, glycolysis is increased either by anaerobiosis or by 2,4-DNP by about 80 per cent, either treatment inducing a marked rise of inorganic phosphate, while no significant changes of the ATP/ADP ratio are observed. The addition of inorganic phosphate to the medium under aerobic conditions increases both glycolysis and glucose uptake, while the addition of AMP was ineffective on these processes, although it increased the level of internal ATP, ADP, and AMP. The observation that the apparent internal concentration of inorganic phosphate was about $9 \times 10^{-3}$ M, while the affinity constant of 3-phosphoglyceraldehyde dehydrogenase is of about $1.5 \times 10^{-3}$ M, leads the authors to suggest that some cell compartmentalization factor makes inorganic phosphate incompletely available for this enzyme, and that this condition could be attenuated to some extent during anaerobiosis.

An important role of compartmentalization of phosphate intermediates in the cell is also suggested by the work by Chance and Hess who determined spectrophotometrically the state of oxidation of the electron carrier chain in short time experiments in intact, freshly prepared ascites tumor cells. Additions of glucose to the medium induced at first a rapid rise of respiration and the oxidation of mitochondrial carriers, immediately followed by a period of inhibition of $O_2$ uptake, concomitant with the return of cytochrome-*b* to a state of reduction equal to or slightly higher than the initial. The suggested interpretation is that in the initial phase ADP increases because of hexose phosphorylation, while in the following period it decreases because of the slow rate of diffusion of the ATP synthesized at the mitochondrial level (mitochondrial ATP store) to the site where it would be used for hexose phosphorylation, and thus regenerate ADP. These results clearly emphasize the dominant role of ADP in the control of oxidative metabolism. They do not seem, however, to contradict the evidence brought by Racker and co-workers in favour of the importance of inorganic phosphate level for the control of glycolysis.

A number of data indicate that also in higher plants ATP, ADP, and inorganic phosphate levels may control glycolytic and oxidative metabolism. Many reports show that plant mitochondria require ADP and inorganic phosphate, -or uncoupling by 2,4-DNP or other substituted phenols to show full oxidative efficiency. Hatch and Turner demonstrated that the operation of typical glycolysis in an acetone powder extract from pea seeds requires inorganic phosphate and adenine nucleotides.

*In view*, the knowledge of the role of phosphate-acceptor level in the control of respiration is based on a large number of experiments with uncouplers, and also on data on glycolysis activation by anaerobiosis (Pasteur effect). The important work by Beevers showing that DNP increases, besides $O_2$ uptake, the R.Q. and the alcohol and acetaldehyde production has been confirmed and extended. Beevers' suggestion, that the DNP-induced increase of fermentation is an indirect effect of the uncoupling of phosphorylation at the mitochondrial level, is further supported by studies showing that in pea internode segments DNP induces, together with the increase of alcohol and acetaldehyde production, a strong decrease of high energy phosphate and of hexose phosphate. It is interesting that the observed decrease of Norit A charcoal-adsorbable nucleotide (high energy phosphate) is significantly larger than what could be accounted for by the amount of ATP (enzymatically determined) present in the controls. This seems to indicate not only that all of the ATP initially present was converted to ADP, but also that some ADP was converted to AMP, presumably because of the activity of adenylic kinase. If this were true, one could imagine that the inhibition of glycolysis observed at very high (superoptimal) DNP concentration could at least in part depend on the decrease of ADP occurring when ATP concentration approaches zero. On the other hand, this hypothesis does not explain the inhibitory effect of high DNP concentrations on $O_2$ uptake, as oxidative metabolism of isolated mitochondria is strongly inhibited by DNP concentrations higher than $10^{-4}$ M even in the presence of a phosphate acceptor system. It seems reasonable to ascribe this effect to the depletion of the endo-mitochondrial store of ATP, as ATP is an efficient productive agent for mitochondrial structures. Eliasson has recently suggested that the ATP deficiency induced by uncouplers could explain the progressive inhibition of $O_2$ uptake in intact tissues treated with DNP. It could be observed, however, that in whole cells many factors are simultaneously affected by uncoupling at the mitochondrial level ; among others, the marked decrease of the rate of phosphorylation of hexoses could play a determining role.

An important corollary of accepting the view that the DNP effect on fermentation is an indirect consequence of the uncoupling of mitochondrial phosphorylations, then, is that it would be possible to obtain strong activation of glycolysis—or of other ADP and inorganic phosphate dependent respiratory mechanisms—even under physiological conditions during which oxygen uptake is unaffected. It has recently been found by Sugiura in Oota's laboratory that while pea cotyledons respond to DNP in the usual way—that is by activation of both $CO_2$ production and $O_2$ uptake—in the pea germ axis the uncoupler stimulates only $CO_2$ production while $O_2$ uptake steadily decreases for increasing concentrations of DNP. The interpretation proposed it that in the germ axis the efficiency to ATP-utilizing mechanisms is high enough to allow the maximal rate of mitochondrial oxidations, while the strong DNP-induced increase of ADP—and one would add, in view of the above discussed data by Wu and Racker, perhaps also of inorganic phosphate—may still affect glycolytic $CO_2$ production.

An interesting analysis of *in vivo* changes of adenine nucleotides and phosphate esters in relation to the Pasteur effect in pea seeds has been recently reported by Rown *et. al.* These authors found that anaerobiosis induces in this material, together with the usual increase of $CO_2$ output, a definite increase of inorganic phosphate and ADP, and a decrease of hexosephosphate. This picture is similar to that observed in the above quoted work by Wu and Racker during anaerobiosis in the ascites tumor cells, where the effect on glycolysis is interpreted as a consequence not only of ADP increase, but also of inorganic phosphate increase. The point seems worth emphasis, as the importance of phosphate level changes for metabolic control is usually underestimated, relative to the importance ascribed to phosphate acceptor availability.

An interesting alternative mechanism for the interpretation of the Pasteur effect has been recently suggested by Hatch and Turner, who found that in a dialyzed extract from pea seed acetone powder glycolysis was strongly inhibited by the presence of oxygen. Inhibition by oxygen was accompanied by a rapid decrease of protein sulfhydryl groups, and it was completely reversed by high concentrations of either DPN or glutathione. This leads these authors to suggest that even *in vivo* the oxidation of the essential sulfhydryl of triose phosphate dehydrogenase, in air, and the reductive reactivation of the enzyme under anaerobic conditions might be put in relation with the Pasteur effect. These data are certainly interesting from several aspects, including that of the control of this key step in carbohydrate breakdown ; however, the participation of such a mechanism in the Pasteur effect seems doubtful. The oxidative inactivation of triosephosphate, dehydrogenase observed in these experiments was largely prevented by Versene, and appears a non-enzymatic, metal ion catalyzed process, qualitatively similar to the often occurring and most annoying inconveniences one meets in enzyme work ; thus the occurrence of the phenomenon *in vivo* seems highly hypothetic. Moreover, the strong activation of glycolysis by DNP, usually found *in vivo* under aerobic conditions, seems to indicate that the activity of triosephosphate dehydrogenase per se is not a severely limiting factor for this process ; in fact, there is no reason to suppose that DNP could increase in the cell the amount of DPNH or other reducing agents capable of bringing the triosephosphate dehydrogenase to its fully active state.

*Physiological uncoupling*. The important idea of the possibility of a physiological uncoupling of mitochondrial oxidations from phosphorylations has been postulated by Hock and Lipmann, on the basis of the finding that thyroxin-treated mitochondria are completely insensitive, as far as their oxidative capacity is concerned, to the absence of phosphate acceptors while in the presence of ADP and inorganic phosphate they show phosphorylative and oxidative capacity identical to that of normal, tightly coupled mitochondria. Thyroxin thus induces, under these experimental conditions, a facultative uncoupling of phosphorylation. A similar possibility of physiological, facultative uncoupling has been recently postulated by Gatt and Racker. They suggested that the inhibition of mitochondrial ATPase by ADP and its requirement for a very high ATP level for maximal activity could result, even *in vivo*, in a situation in which, when the level of ATP rises and that of ADP falls to a very low value, the activation of mitochondrial ATPase would partially uncouple phosphorylation thus allowing the production of metabolically important substrates by oxidative reactions.

Several data indicate the presence in higher plants of physiological uncoupling mechanisms of this type. Forti showed that plant mitochondrial ATPase is also strongly inhibited by ADP, and requires a high ATP concentration for significant activity. Other, presumably more important mechanisms of facultative uncoupling might be deduced by the important studies on cyanide and carbon monoxide resistant respiration, reported by Hackett, Chance and Hackett, Ducet and by a number of other authors. It has been shown in many cases—including those of aged slices from storage tissues and of leaf disks—that even when respiration is not inhibited, the inhibitors still block almost any process that, directly or indirectly, depends on a high level of phosphorylative efficiency. This is the case with cell extension, phosphate uptake and incorporation into organic compounds, protein synthesis, and sugar uptake in the aged potato disks, and of salt and sugar uptake by leaf disks. Quite interestingly, a common feature of the materials showing this effect is insensitivity to CO, even when cytochrome oxidase is present in the isolated mitochondria and also insensitivity to uncoupling agents such as DNP. The discovery that in two of the tissues giving this response, the skunk cabbage spadix and the aged potato disks, mitochondria contain an apparently autoxidizable cytochrome, cytochrome-$b_7$, besides the normal cytochrome chain, and the finding that in both cases cyanide does not inhibit $O_2$ uptake, though it blocks phosphorylation led Hackett to suggest that two different electron pathways are operating for the oxidation of mitochondrial substrates. The first of these, cytochrome-$a_3$-mediated and phosphorylating, would operate in the absence of inhibitors ; the second, utilizing cytochrome-$b_7$ as terminal oxidase, would become active only when the cytochrome oxidase pathway is blocked.

The importance of these data and the hypothesis in regard to the problem of facultative uncoupling of phosphorylation is obvious. It seems probable that under conditions in which a high ATP/ADP level would inhibit oxidative metabolism in a tightly coupled system, the existence of an uncoupled, auxiliary pathway would provide a possibility for oxidations to proceed. This pathway would become active as soon as an ADP shortage would close the $a_3$ system. Also, the fundamental difference between this type of uncoupling and that induced by DNP is evident; in the former case ATP is regularly synthesized in the amount required by Ptp dependent reactions, while in the latter case mitochondrial ATP synthesis is suppressed. It is interesting to observe that the reported increases of $CO_2$ output—and perhaps also of $O_2$ uptake—in tissue treated with cytochrome-$a_3$ inhibitors could also be explained through this mechanism. The rise of the level of ADP and inorganic phosphate would presumably stimulate glycolysis and also supply a larger amount of respiratory substrates to the carboxylic cycle.

The more obvious physiological interest of such a mechanism seems to consist in the release of synthesis of anabolites from the restraint imposed on oxidative metabolism by phosphate-acceptor availability. It seems interesting that among the materials showing facultative uncoupling one finds aged slices from storage tissue and green leaves. In the former material, protein synthesis is usually observed. This requires a corresponding rate of synthesis of amino acids and some ATP; but formation of amino acids such as glutamic and aspartic acids from hexoses would imply the synthesis of more high energy phosphate bonds than would be utilizable for the synthesis of the corresponding peptide bonds. Thus, in the absence of a facultative uncoupling mechanism, protein synthesis could be—at least under some conditions—significantly slowed down by the dependence of oxidative breakdown on ADP availability.

The case of mature green leaves is perhaps even more interesting. Oxidative breakdown in these tissues is required for the elaboration of photosynthetic products, and active protein synthesis in the light is often observed. Now it has been reported that—especially under conditions of low $CO_2$ availability—the ATP/ADP+inorganic phosphate value in illuminated green tissues reaches very high values because of photosynthetic phosphorylation. Even if one assumes that some cell compartmentalization limits the rate of intracellular redistribution of adeninenucleotides, it appears quite probable that under these conditions the amount of ADP available for glycolysis and mitochondrial oxidations is drastically decreased. The physiological, facultative uncoupling of phosphorylations in these tissues it seems could thus be a requirement to allow oxidative carbohydrate breakdown to proceed even when photosynthetic phosphorylation tends to reduce the level of phosphate acceptor to a very low value. In this connection also, the otherwise rather puzzling report by Arnon of the specific localization in green leaves of an inorganic phosphate and ADP independent triosephosphate dehydrogenase could find a physiological interpretation.

# 9

# Photoperiodisum and the Flowering Process

What makes a plant suddenly turn from producing leaf buds and begin to form flowers instead? There are at least two good reasons for exploring this mysterious process of nature. Firstly, the question goes to the heart of the origin of form, one of the most interesting and fundamental problems in modern biology. Secondly, if we learned how to control flowering, we could probably find radical and wonderful new ways of growing crops to our desires—in short, we could revolutionize agriculture.

A great deal of thought and experiment has been devoted to the attempt to understand the flowering process, and several years ago the findings up to that time were reviewed in an article in this book [see "The Control of Flowering," by Aubrey W. Naylor; Scientific American Offprint 113]. Since then there have been some rather exciting further developments.

As a preface to this sequel we must recall some of the basic facts that had been established at the time of Naylor's article. Experiments had shown that a plant begins to flower only after it has reached a certain maturity, known as "ripeness-to-flower"; that the main environmental factors which trigger or control flowering are temperature and light, particularly the length of day ; that some plants need a long day to start to flower while others require a short day (*i.e.*, a certain extended period of darkness per day) ; and that the change-over which causes a plant's stem tips to begin forming flowers instead of leaves is controlled by some substance produced in the plant's leaves and transported to the growing tips. This substance has not yet been identified. But there is conclusive evidence that it exists. One of the clinching pieces of evidence is that a plant incapable of flowering (because it has been deprived of the necessary dark period) can be made to flower by grafting onto it a plant already in flower ; clearly the graft must produce this effect by supplying some hormone.

Following up these findings, a number of investigators have tried to analyze the flowering process by breaking it down to a series of steps in the transformation of the plant.

A favorite experimental plant of plant physiologists working on this problem is the cocklebur. This common weed is beautifully convenient for studying flowering. It is easy to cultivate ; it grows well even when defoliated to a single leaf for experimental purposes ; and it is highly sensitive to light conditions. It will flower only when its period of night is more than eight and a half hours long. A single exposure to a continuous dark period of the critical length will trigger it to come into bloom. Its growing points develop into clusters of little green flowers and finally into burs.

Apparently the first step toward flowering is accumulation of a sufficient food reserve by the plant. Karl C. Hamner at the University of Chicago discovered this by an elegantly simple experiment. A plant of course, builds its material by photosynthesis, for which if needs light—the more intense, the better. Hamner starved cocklebur plants by greatly reducing their rations of light, allowing them only flashes at two hour or three-hour intervals. Then he gave them the long night of more than eight and a half hours, which causes normal plants to start forming flowers. But these specimens refused to bloom. However, three hours exposure to sunlight was enough to restore them so that they were able to flower after the necessary dark period. Later James Liverman and James Bonner at the California Institute of Technology showed that the plants' deficiency was indeed one of food substance. They found that they could endow light-starved cockleburs with the ability to flower simply by supplying sugar to the plants.

The first step, then, is the build-up of material in the plant. The second step toward flowering must take place in the required eight-and-one-half-hour dark period. What processes go on during this time? And how does a cocklebur measure eight and one-half hours?

Workers at the U.S. Department of Agriculture research station in Beltsville, Md., succeeded in tracking down one clear clue. It was known that the dark period must be continuous: if it is interrupted by even a single flash of ordinary light, the cocklebur will not flower. The Beltsville workers proceeded to experiment with light of various wave-lengths (colors). They discovered that light at the far-red end of the spectrum did not inhibit flowering: on the contrary it counteracted the effect of inhibiting wavelengths. Red light at the shorter wavelengths (*i.e.*, orange-red) was most effective in preventing flowering. But if, after interrupting the dark period with a flash of this light, the experimenters then followed with a flash of the longer-wavelength red light, the cocklebur would flower. In other words, the farred light reversed the effect of the orange-red light. The plant's behaviour depended on which wavelength it was exposed to last: if orange-red, the plant could not flower; if far-red, it did flower.

The Beltsville group developed the following theory. Light can produce chemical effects only through some absorbing pigment. Therefore a pigment must be involved in the chemical reactions that lead to the plant's flowering. Apparently this pigment, upon absorbing orange-red light, is converted to a different form which is sensitive to far-red light, and *vice versa*. The form produced by exposure to orange-red light must be the one that inhibits flowering. Since sunlight contains more orange-red than far-red light, when a plant is removed from ordinary light to darkness most of its pigment must be in the inhibitory form. This means that one function of darkness in preparing the plant to flower must be to allow the pigment to change spontaneously to the noninhibitory form.

The next question was : How long does the spontaneous conversion of the pigment take ? Does it account for the full critical period ; is this the clock that tells the cocklebur when eight and a half hours have elapsed ? The Beltsville group investigated by exposing cockleburs to intense far-red light before retiring the plants to darkness, the object being to convert all of the pigment to the noninhibitory form. They found that at most this treatment shortened the critical dark period by only two hours—that is, the plant still needed six and a half hours of darkness to flower. Thus the experiments indicated that complete conversion of the pigment takes about two hours, and this conclusion has been supported by other evidence.

This leaves six and a half hours to account for. What are the further preparations for flowering that go on during that time ? Unfortunately we do not yet have any idea : this is still a dark chapter (figuratively as well as literally) in the story of the flowering process. For want of any definite information this unknown step is merely labeled the "preparatory reaction."

The next step is somewhat clearer. The cocklebur needs *more* than eight and a half hours of darkness to flower normally. If it is exposed to light shortly after the end of that critical period, it will flower eventually, but the flower buds develop very slowly. If the dark period is extended further, the buds then develop more rapidly, in fact, the rate at which they will develop after exposure to light depends on the length of the dark period beyond eight and a half hours. This was shown by systematic experiments in which groups of plants were submitted to dark periods of various lengths and then examined under a microscope nine days later. The progress of each group's flower-bud development was directly proportional to the excess of its dark period over eight and a half hours, up to a certain limit.

What does this mean? The simplest and most plausible explanation is that the plant synthesizes the flowering hormone in the late stages of the dark period and the more time it has for synthesizing the hormone, the faster the flower bud will develop. However, diminishing returns set in after about 12 hours of darkness; in fact, it appears that when the dark period is prolonged to 20 hours or so, the hormone begins to be destroyed.

This brings us to the somewhat controversial next step. The cocklebur will often flower even if it is left in the dark for nine days. But James Lockhart and Hamner, working at the University of California at Los Angeles, have found that for full flowering development the plant seems to need a re-exposure to high-intensity light, *e.g.*, sunlight, after the dark period. Consequently it may be that this exposure represents another chemical step in the normal flowering process: the light may be needed to "stabilize" the flowering hormore or to stop some process in the leaves that destroys the hormone.

A very simple experiment has disclosed another distinct stage in the requirements for flowering: namely, transport of the flowering hormone from the leaves to the growing stem tip; where the flowers form. If a cocklebur's leaves are removed at the end of the critical dark period, the plant will not flower. This must mean that the flowering hormone is still in the leaves and none has yet reached the stem tips. (Fig. 9.1).

Finally, there are obviously two more steps in the flowering process: the flowering hormone acts upon the cells in the growing tips in some way to transform, or differentiate, them from the vegetative to the flowering form of growth, and the bud then develops into the specific structures that make up a flower.

To sum up, the flowering of a plant has been resolved so far into about eight steps: (1) the build-up of needed sub-

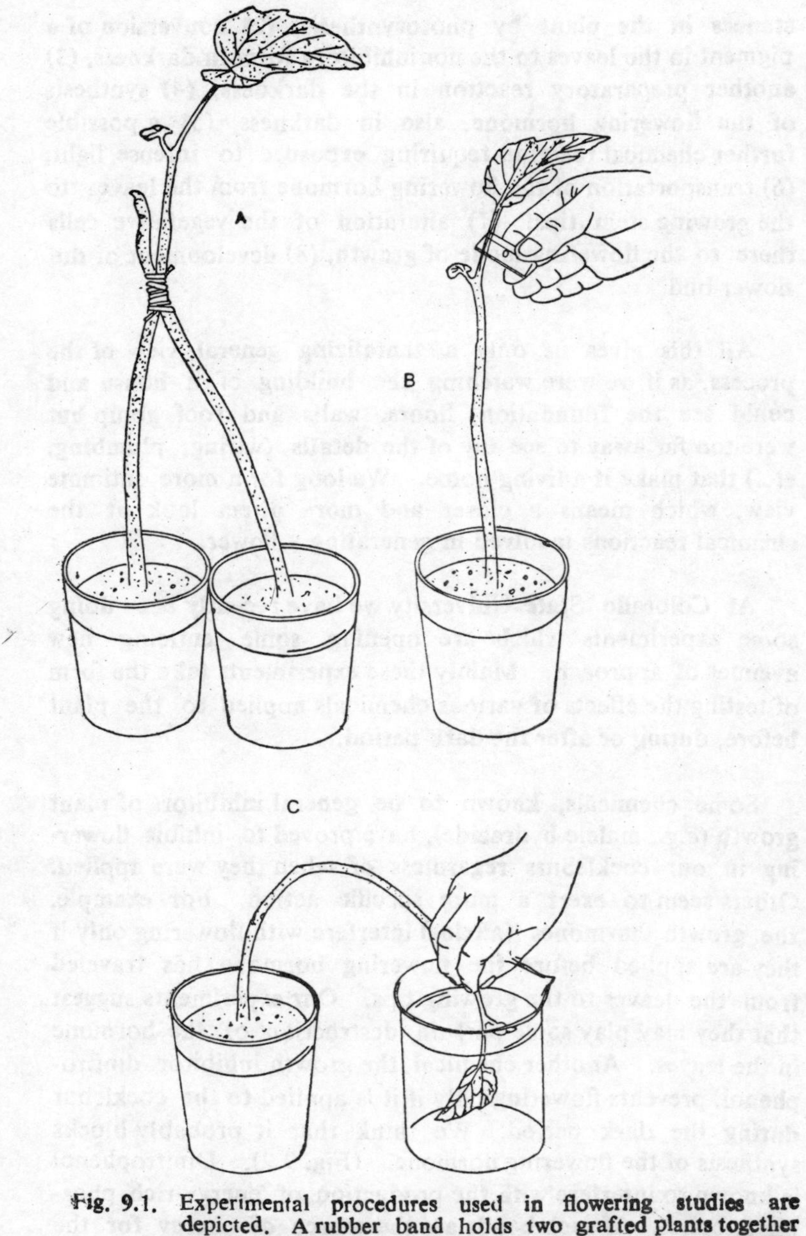


**Fig. 9.1.** **Experimental procedures used in flowering studies are depicted. A rubber band holds two grafted plants together (A) so that hormone can pass from one to the other. The one leaf on an experimental cocklebur plant is cut off with a razor blade in order to stop translocation—the passage of hormone from the leaf to the bud (B). Cocklebur plants are treated with chemicals by dipping a leaf into a solution of the Chemical (C).**

stances in the plant by photosynthesis, (2) conversion of a pigment in the leaves to the noninhibitory form in darkness, (3) another preparatory reaction in the darkness, (4) synthesis of the flowering hormone, also in darkness, (5) a possible further chemical reaction requiring exposure to intense light, (6) transportation of the flowering hormone from the leaves to the growing stem tips. (7) alteration of the vegetative cells there to the flowering mode of growth, (8) development of the flower bud.

All this gives us only a tantalizing general view of the process, as if we were watching the building of a house and could see the foundation, floors, walls and roof go up but were too far away to see any of the details (wiring, plumbing, etc.) that make it a living home. We long for a more intimate view, which means a closer and more direct look at the chemical reactions involved in generating a flower.

At Colorado State University we have recently been doing some experiments which are opening some enticing new avenues of approach. Mainly these experiments take the form of testing the effects of various chemicals applied to the plant before, during or after the dark period.

Some chemicals, known to be general inhibitors of plant growth (*e.g.*, maleic hydrazide), have proved to inhibit flowering in our cockleburs regardless of when they were applied. Others seem to exert a more specific action. For example, the growth hormones (auxins) interfere with flowering only if they are applied before the flowering hormone has traveled from the leaves to the growing tips. Our experiments suggest that they may play some part in destruction of the hormone in the leaves. Another chemical, the growth inhibitor dinitrophenol, prevents flowering only if it is applied to the cocklebur during the dark period. We think that it probably blocks synthesis of the flowering hormone. (Fig. 9.2). Dinitrophenol is known to interfere with the production of energy-rich phosphate bonds. If such bonds are the source of energy for the synthesis of the flowering hormone, dinitrophenol may exert its inhibiting effect by cutting off the power supply, so to speak.

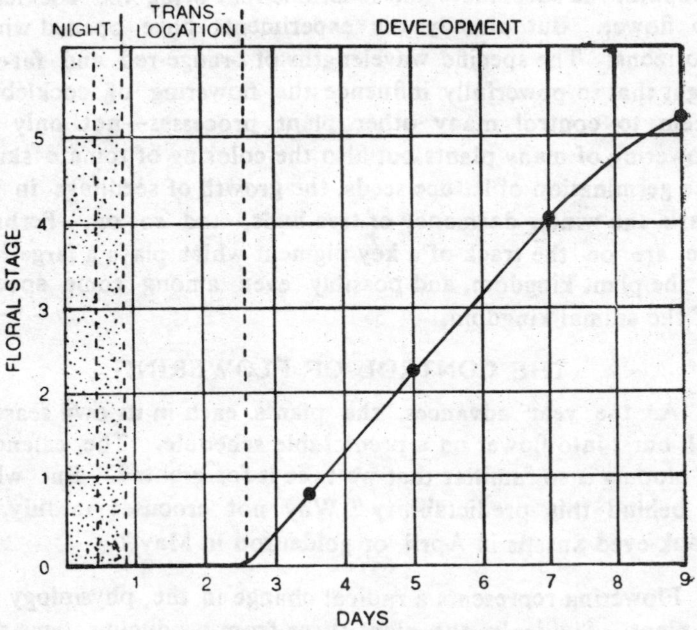


**Fig. 9.2.** **Development of the flower begins after hormone reaches the bud, and continues at a fairly constant rate. In a typical experiment the cocklebur bud reaches the first floral stage in about four days, fourth stage in seven days and fifth stage in nine days.**

One of our most useful tools for dissecting the flowering process has been the ion of cobalt. This chemical inhibits flowering only if applied during the early hours of the dark period. Thus we can narrow down its action to interference with one of two processes: the conversion of the pigment involved in flowering of the "preparatory reaction." Furthermore, among all the chemicals tested so far the cobalt ion is the only one that affects the length of the dark period needed for flowering. In some experiments plants treated with cobaltous chloride did not flower unless their dark period was at least 11 hours long. In other words, cobalt slows down the clock that ticks off the critical night for the cocklebur.

With refined probing tools such as cobalt and other specific growth regulators we have hopes of eventually being able to pinpoint the substances and reactions that bring the cocklebur to flower. But already the experiments have opened wider horizons. The specific wavelengths of orange-red and far-red light that so powerfully influence the flowering of cockleburs seem to control many other plant processes—not only the flowering of many plants but also the coloring of tomato skins, the germination of lettuce seeds, the growth of seedlings in the dark, the winter dormancy of tree buds, and so on. Perhaps we are on the track of a key pigment which plays a large role in the plant kingdom, and possibly even among some species of the animal kingdom.

## THE CONTROL OF FLOWERING

As the year advances, the plants, each in its own season, will burst into flower on a predictable schedule. The calendar of blooms is so familiar that we take it for granted. But what is behind this predictability ? Why not crocuses in July, or black-eyed Susans in April, or goldenrod in May ?

Flowering represents a radical change in the physiology of a plant. Suddenly the plant turns from producing stems and leaves to making the blooms that will yield fruit, and the seeds of reproduction. What master switch sets off this change ? If we understood the mechanism that initiates plant flowering, we could envision some really breath-taking advances in agriculture and our control over nature.

Any study of the subject must begin with the environmental factors that govern a plant's development. Temperature, light, water and nutrition all play their parts, and clever gardeners and greenhousemen have long known how to force plants into bloom to meet deadlines by skillful manipulation of these influences. Their empirical methods have shed little light, however, on the physiological process that transforms a plant from the vegetative (foliage-producing) to the flowering state.

We know that a plant, like an animal, must reach a certain stage of maturity before it is ready to flower and reproduce.

It takes fruit trees, for example, several years to begin bearing blossoms and fruit. A corn plant will not flower until it has produced a certain minimum number of leaves, the number depending on the variety of corn. There is a remarkable bamboo plant native to the mountains of Jamaica that carries the process of growing up to a strange extreme : 32 years after the plant is born, it flowers once and then dies. Its life cycle seems to be independent of the environment ; transplanted to any other part of the world, the plant still blossoms on schedule at the age of 32 years, no earlier and no later.

Investigations to track down the physiological mechanism whereby a plant starts producing flowers have been carried out mainly along two lines, using two different tools to control flowering. The first is temperature. About 35 years ago the German plant physiologist Gustav Gassner discovered that he could influence the flowering of cereal plants by controlling the temperature of germinating seeds. One of the plants on which he experimented was winter rye. Winter rye is planted in the autumn, germinates during the winter and flowers the following summer. If it is planted in the spring, it fails to flower, remaining vegetative throughout the growing season. Gassner found, however, that by keeping the seed at near-freezing temperatures during germination he could make winter rye flower even when he planted it in late spring. This procedure was later adopted by plant breeders to transform winter cereals into spring types, and it became known as "vernalization.'

At first it was thought that the low-temperature treatment changed a plant's general metabolism. But it developed that vernalization was a reversible process ; that the change took place during a critical four-day period, after which it could not be reversed, and that rye seedlings could not be vernalized if they were undernourished. All this indicates that flowering depends on the formation of certain specific substances in the plant, rather than on an alteration in its general metabolism.

The second tool used to investigate flowering is the response of plants to the length of day—what is known as photoperiodism. This root has been far more fruitful than temperature ; it has been utilized by a great many investigators in a great variety of experiments, and it has already yielded some important practical results in agriculture.

The fascinating story of photoperiodism goes back to a historic investigation more than 60 years ago by two workers in the U.S. Department of Agriculture—W.W. Garner and H A. Allard. They wondered why the variety of tobacco called Maryland Mammoth was delayed in flowering when it was grown near Washington, D.C. It flowered so late in the season that its seeds did not mature. The two investigators tried growing the plant under various conditions and explored many blind alleys before they discovered the answer : at the critical time the days were too long and the nights too short for these plants at Washington. Maryland Mammoth tobacco is a short-day plant, meaning that it begins to form flowers when the days are between 10 and 12 hours long. In the Washington growing season the days do not become as short as this until late in the summer. Consequently the tobacco plant, though ripe to flower long before, is delayed in receiving the necessary stimulus and so flowers very late (Fig. 9.3).

Carner and Allard went on to test and confirm this discovery with many other species of plants. They found that plants fell into three general categories : short-day, long-day and indeterminate (not choosy as to length of day). Photoperiodism at once accounted for many things which had long puzzled botanists. It provided a reasonable explanation of why plants of a given variety, even though planted at different times, will nevertheless all flower at the same time, and why certain plants flourish in some latitudes and are practically absent from others.

Ragweed, for instance, starts making flowers when the day is just about 14.5 hours long. At Washington this length of day occurs around July 1, and the plant flowers and sheds its pollen by the middle of August. It has ample time, in other words, to form and scatter its seeds before frost comes. But there is little or no ragweed in northern Maine. There the long summer days do not shorten to 14.5 hours until after

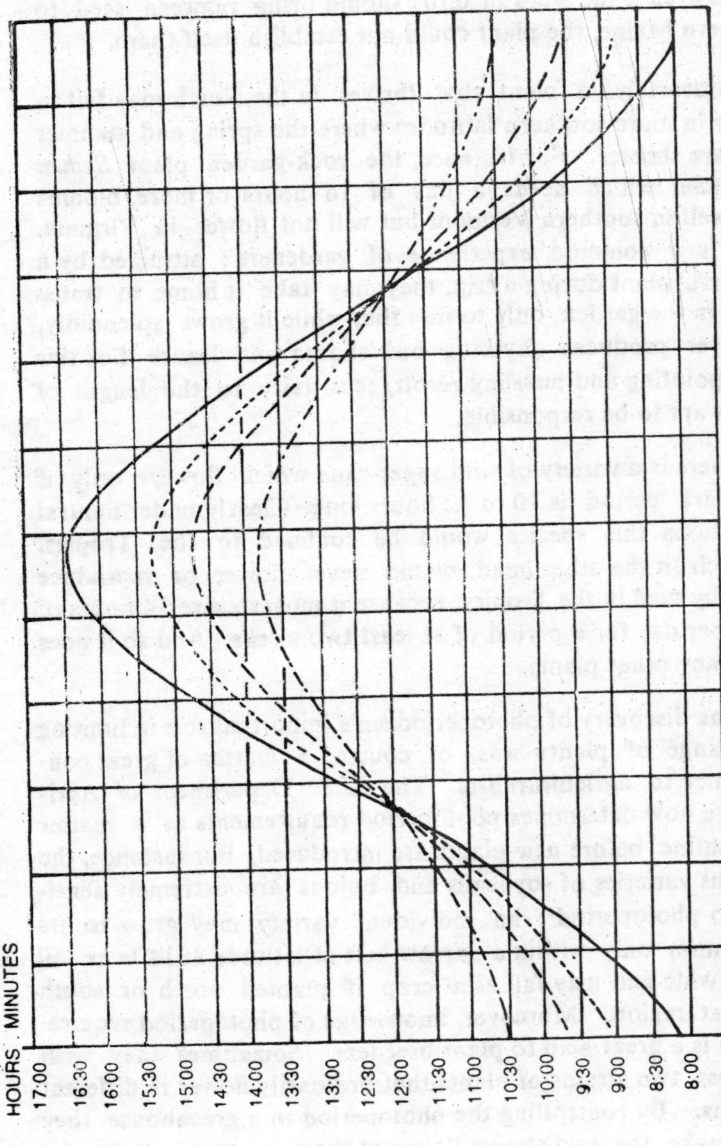


**Fig. 9.3. Length of the day determines when plants flower. The vertical coordinate of this diagram is the length of the day in hours and minutes ; the horizontal coordinate, the months of the year.**

August 1. Ragweed, starting to form flowers after that date, would generally be killed by frost before its seed matured. Hence even if the wind or birds should bring ragweed seed to northern Maine, the plant could not establish itself there.

Conversely, a plant that thrives in the North may fail to flower in more southern latitudes where the spring and summer days are shorter. For instance, the rock-garden plant *Sedum telephium*, which needs a day of 16 hours or more, blooms very well in southern Vermont but will not flower in Virginia. This is a common experience of gardeners ; attracted by a colourful plant during a trip, they may take it home to transplant in the garden, only to find that while it grows splendidly, it never produces anything but shoots and leaves. For this disappointing and puzzling result, sensitivity to the length of day is apt to be responsible.

There is a variety of wild sugar-cane which flowers only if the dark period is 10 to 12 hours long. Clearly under natural conditions this species would be confined to the Tropics. Spinach on the other hand, would never flower or reproduce itself by seed in the Tropics, because it must receive 14 hours of light per day for a period of at least two weeks. And so it goes for many other plants.

The discovery of photoperiodism's important role in limiting the range of plants was, of course, a matter of great consequence to agriculturalists. The U.S. Department of Agriculture now determines photoperiod requirements as a matter of routine before new plants are introduced. For instance, the various varieties of soybeans and onions are extremely sensitive to photoperiod ; an individual variety may grow to its maximum only within a narrow belt of latitude as little as 150 miles wide and may fail as a crop if planted north or south of that region. Moreover, knowledge of photoperiod requirements is a great help to plant breeders. Sometimes they wish to cross two strains of plants that ordinarily flower in different months. By controlling the photoperiod in a greenhouse they can make the two strains flower at the same time and thereby fertilize each other. In this way breeders have obtained some valuable hybrid crop plants which could not be produced otherwise.

But photoperiodism has been most rewarding as a research tool. Very early the experimenters discovered a surprising fact. It seemed reasonable to suppose that if a plant required a certain length of day to flower, darkening the plant for part of the day should interfere with its flowering. This experiment was performed repeatedly, but nothing happened ; the plants flowered just as if they had had a full day of light. Interruption of the nighttime period of darkness, however, told another story. Even a few minutes of illumination affected the plant's flowering. Thus when a short-day plant such as the chrysanthemum is illuminated for a few minutes in the middle of the night during the season when it normally would flower, it fails to bloom. On the other hand, a long-day plant such as pyrethrum can be made to flower in a short-day season when it would not normally do so, simply by exposing it to light for a short time at night. The critical factor in photoperiodism, then, is not the length of day but the length of night ; strictly speaking, plants should be classified as long-night and short-night rather than short-day and long-day.

This information obviously has its practical uses. Flower growers, who used to delay chrysanthemum flowering for the late fall market by extending the daylight period with several hours of artificial light, now can achieve the same end and save a lot of current by illuminating the plants for a few minutes in the middle of the night.

More important, however, is the lead that this discovery has given in the investigation of the flowering process itself. Evidently the chemical reactions responsible for flowering go on in a plant at night. These reactions are sensitive to light. Other experiments show that they require carbon dioxide (or sugar sprayed on the leaves as a substitute) and depend on the quality (*i.e.*, wavelengths) of the light—all of which indicates that the photoperiodic reactions are in some way mixed with photosynthesis. However, they appear to depend on some pigment other than chlorophyll as the light receptor.

We come back, then, to the main question : What is the chemical basis of flowering ? Nearly a hundred years ago the great German plant physiologist Julius von Sachs suggested that the leaves of a plant produce "flower-forming substances" which travel to the growing points where flowers are generated. The first laboratory evidence to support this theory was discovered in 1934 almost simultaneously by several different groups of workers—J.E. Knott at Cornell University, J. Kuijper and L.K. Wiersum in the Netherlands and M.C. Cajlachjan in the U.S.S.R. Knott experimented with spinach, a long-day plant. He exposed the leaves to long photoperiods and covered the stem growing-point for part of the day so it was on a short photoperiod. The plants flowered as readily as if the whole plant had been on long photoperiods. When he reversed the procedure, giving the growing points long photoperiods and the leaves short photoperiods, the effect was the same as if the whole plant had been on short photoperiods—that is, the plant failed to flower. These experiments, confirmed and extended to other plants by the other workers, made it clear that the stimulus for flower formation must originate in the leaves. This strongly supported the idea that the stimulus was some kind of chemical regulator that moved from the leaves to the stem growing-points where flowers are formed.

Karl Hammer and James Bonner at the University of Chicago soon found even more convincing evidence. They did their work with cocklebur, a common noxious weed. This short-day plant is an ideal experimental tool for studying photoperiodism : it is sensitive to differences of less than 30 minutes in the dark period ; a single nine-hour dose of darkness can induce it to flower, and it will survive considerable mutilation. Hamner and Bonner first stripped all the full-grown leaves from the plant and exposed the defoliated stem to nine-hour periods of darkness to see whether the plant would flower. It did not. But they found that when they left as little as one eighth of a fully expanded leaf on the stem, the plant did flower after exposure to the proper dose of darkness. By further experiments—forcing the plant to produce two branches, grafting shoots with leaves on the stem, and so on—they went on to show that the flower-inducing stimulus can travel up and down in the stem and across graft unions. They were also able to make a rough estimate of the rate of movement of the stimulus.

Many other investigators now took up the scent. Experimenting with a wide variety of plants, they demonstrated the activity of the flower-inducing stimulus in some striking ways. A nonflowering plant can be made to flower by grafting on to it a single leaf from a flowering plant, or even an isolated leaf that has been exposed to the proper amount of light. The stimulus can travel a considerable distance through several grafts. In one stunt six double-branched cocklebur plants were grafted together in series and all the leaves were removed except one at the tip of the last plant. After this leaf was exposed to dark periods of the appropriate length, the whole series of grafted plants proceeded to flower in orderly sequence down the chain !

The flower-promoting stimulus is not special and different in each species of plant ; apparently it is some general factor common to all plants. This is shown by the fact that a plant can be induced to bloom by a leaf graft from another variety, species, genus or even from another family of plants. There is a certain parasitic plant, growing underground on the roots of red clover, that apparently obtains the flower-inducing factor from its host !

How can the same factor cause flowering both in long-day and short-day plants, which respond so differently to light ? One working hypothesis is that the factor is a substance which promotes flowering when it reaches a certain minimum concentration and inhibits flowering when it accumulates to a higher concentration. According to this theory, suggested by Harry Borthwick and Marion Parker of the Beltsville Experiment Station, a plant that is ripe to flower produces the active substance continuously, but the substance is unstable to light. In a long-day plant, the substance is somehow protected from destruction by light so that it does not fall below the minimum concentration needed for flowering ; such a plant will flower even if exposed to light continuously. In darkness, however, the substance accumulates in the plant, and if the night is too long, it accumulates to the point where it inhibits flowering. When the plant is illuminated in the middle of the night, the light destroys the excess substance and enables the plant to flower. In a short-day plant, on the other hand, the flower-promoting substance is not protected from the destructive effects of daylight. Hence the plant needs a long period of uninterrupted darkness to accumulate enough of the substance to flower.

All in all the evidence seems overwhelmingly in favour of the idea that a flower-promoting hormone does exist. The Russian worker Cajlachjan succumbed to the temptation to name the suspected substance "florigen." The potential value of this powerful substance, if indeed it exists, is so obvious that many attempts have been made to extract and isolate it from plants, in the U.S. and abroad. So far none has succeeded, though hopes have been raised several times. One of the difficulties is that no one knows just how such a substance should be introduced into a plant to promote flowering ; hence experimenters cannot be certain whether they have failed to find the correct substance or simply failed to apply it properly.

Some investigators believe that plant hormones already known, rather than special "florigens," may be the agents responsible for the control of flowering. They argue that the growth-promoting hormones known as auxins may regulate flowering as well as vegetative growth.

One of the auxins is indole-acetic acid. It has been found that when the cocklebur plant is grown on long photoperiods, unfavorable for flowering, it produces much more indole-acetic acid than when it is on short photoperiods. This would indicate that the auxin inhibits flowering. There are some obvious ways to test that conclusion : anything that neutralizes or destroys auxin in the plant should promote flowering, and an artificial increase in the plant's supply of auxin should suppress flowering. Experiments indicate that this is indeed the case. When cocklebur plants are treated with X-rays or certain chemicals that are known to destroy or counteract auxin. The plants flower more profusely than they would normally. On the other hand, treatment with auxins interferes with flowering. James Bonner and John Thurlow at the California Institute of Technology have prevented flowering in the cocklebur by treating the plant with indole-acetic acid, and the writer has greatly delayed flowering in plants by treating them with other auxins.

There is now good evidence that plants themselves produce natural substances which counteract auxins. These indirect indications point to the conclusion that flowering is regulated by the balance between the auxins and their antagonists in the plant.

We are still far short of solving the mystery of what makes plants flower. It will take a great deal of investigation to determine just what substances are involved, how they are formed in the plant, and how they interact. Nevertheless, the information so far obtained has already made it possible to control flowering by chemical methods in a few plants.

One of these is the pineapple. This crop used to be expensive to produce, because the plants flowered and set fruit erratically, so that pickers had to come back to a field again and again to collect fruit as it ripened. Now all this has changed, thanks to the discovery that various substances, notably acetylene and naphthalene-acetic acid, can control the flowering of pineapple. Six to eight weeks after treatment a whole field flowers with amazing uniformity. As a result pineapple crops can now be harvested with all the techniques and machinery of mass production. The plants are usually treated with naphthalene-acetic acid, but this treatment has a drawback : the fruit-bearing stalks grow too weak and topple over. This can be corrected, however, by spraying the plants at the proper time with B-naphthoxyacetic acid. One of the leading investigators in the field. J. van Overbeek, observes that the pineapple growers still are not satisfied : they want treatments that will make the fruit grow tailor-made to fit the cans !

Chemical control promises to make another dream come true. Hawaiian growers have long wished they could achieve large-scale production of the litchi nut, one of the most delicious of all tropical tree fruit. The litchi tree has been a poor performer in Hawaii ; only about four per cent of the trees flower and bear fruit. Now it seems that naphthalene-acetic acid may be the answer. In recent experiments it has been found that when litchi trees are sprayed with this growth substance at the end of September, about 88 per cent of the trees bear fruit. Development of this discovery may create a new industry in the Hawaiian Islands.

Some work has also been done with vegetable and grain crops. Among the promising candidates for control are lettuce and celery. The problem here is to prevent flowering, because lettuce and celery are grown for their vegetative parts (leaves and stalks) and lose their market value when they begin to go to seed. In the case of lettuce, experimenters have succeeded in delaying the development of seed stalks and flowers by treating the seeds of the plant with cold and a chemical at the time when they are germinating. In the case of celery, they have found one chemical treatment that retards flowering and another that accelerates it. Celery growers who grow celery for the table can use one, and those who grow it for seed can use the other. Tobacco is another plant in which it is desirable to prevent flowering, and some promising results have been obtained with that crop.

But of all plants, perhaps the most important candidate for research on the control of flowering is corn, the No. 1 U.S. crop. Such research has already started, with two ends in view. One is to regulate flowering so that desirable strains which do not ordinarily flower at the same time can the crossed. The other is to sterilize corn tassels by chemical means. Growers of hybrid seed corn must sterilize the tassels to prevent plants from fertilizing themselves and reproducing their own kind ; to produce the hybrid seed it is necessary to fertilize the plants with pollen from selected strains. At present the plants are generally detasseled by hand—an expensive operation. Chemical sterilization obviously would be much easier and less costly.

For both of the control, purposes, promising experimental leads have already developed. Spraying with 2, 4-D, the well-known weed killer, delays flowering in corn without harming the ear. And the tassel can be sterilized by treatment with the plant-growth inhibitor maleic hydrazide.

The results so far in the chemical control of flowering have been highly promising, in some cases even spectacular, and the future is bright. When the flowering process is fully understood, progress will, of course, be much faster. We may even be able to force into bloom plants that now flower rarely.

## FLOWER PIGMENTS

The rich variety of colours displayed by flowers has presented a challenge to both the chemist and the biologist. It has proved no easy matter to answer all the questions that may be asked. Why do the brilliant colours of the plant tend to be concentrated in the flower ? Are many pigments involved or only a few, modified in various ways ? How are the pigments synthesized in the plant ? What function do they serve ? Questions such as these have provided new insights into the intricacies as well as the beauties of nature.

Current studies of flower pigments and their control by genes can be traced to two seemingly unrelated researches of about 100 years ago. One was the experiments of Gregor Mendel, who worked out the basic laws of inheritance in his classic studies of the garden pea. The other was the synthesis by a young English chemist, William Henry Perkin, of the first coal-tar dye, mauveine. Perkin's success inspired an intensive investigation of pigments, both natural and aynthetic. In the late 19th century and the early years of the 20th chemists such as Richard Willstatter and Stanislaus von Kostanecki in Germany and Robert and Gertrude Robinson in Britain identified many flower pigments and worked out their molecular structure.

The rediscovery of Mendel's work in 1900 stimulated research on the inheritance of flower colour. A group of geneticists at the John Innes Horticultural. Institution near London were early leaders in this field. Although many plant geneticists had examined the inheritance of flower colour. Muriel Wheldale Onslow of the University of Cambridge was the first to realize that here might be a way to investigate the chemical events controlled by Mendelian factors She reasoned that the presence of a simple biochemical substance—such as a pigment—might be correlated with a change in a single gene. Working in association with H.L. Bassett, she showed that this correlation did indeed exist in snapdragons, in spite of the fact that the identification of the pigments was incomplete. This work initiated the biochemical-genetic approach to the problem of plant pigments. It was also the first clear-cut demonstration of a link between a single gene and a biochemical entity.

Later in the 1930's Rose Scott-Moncrieff, a student of Muriel Onslow's, noted certain basic patterns of inheritance in the course of working with genetic varieties of many different plant species. Building on these classic studies, with the advantage of new analytic techniques such as paper chromatography, plant biochemists and geneticists have in recent years learned much more about the nature of the pigments and their production by the living cell.

Flower pigments are easily divided into two groups : the water-soluble flavonoids found in the cell vacuole, which give rise to red, blue and yellow colours, and the fat-soluble carotenoids located in the plastid bodies of the cell, which are red, orange or yellow. Although the carotenoids make a substantial contribution to the colouring of flowers, this article will deal only with the flavonoids. These pigments have a basic chemical structure consisting of two six-carbon-atom benzene rings linked together by a chain of three carbon atoms. This basic molecular structure of 15 carbon atoms is sometimes written $C_6-C_3-C_6$.

The various groups of flavonoid pigments are differentiated by the level of oxidation of the three-carbon chain. Two of the more common flavonoids are the anthocyanins (from the Greek for "blue flowers"), which come in shades of blue, purple or red, and the flavonols (from the Latin for "yellow"), which are yellow or ivory. The anthocyanins have fewer chemical variations than the flavonols.

In the cells of the flower the pigment is usually bound to one or more units of sugar, which make the molecule more soluble in water. When the sugar is attached, the pigment is known as an anthocyanin ; the pigment without the sugar is called an anthocyanidin.

Six common and seven rare anthocyanidins have been isolated from flowers (Fig. 9.5). It will be seen that they differ in the number and arrangement of chemical groups attached to the "B" ring of the molecule. This ring may carry one, two or three hydroxyl groups (OH). Occasionally a methyl group ($CH_3$) replaces the hydrogen atom of a hydroxyl group.

Cyanidin, the most common anthocyanidin, was first isolated from *Centaurea cyanus*, the blue cornflower, which gave the pigment its name. Other anthocyanidins were similarly named. The red pigment pelargonidin was originally found in a bright red geranium of the genus *Pelargonium*. The deep purple pigment delphinidin was isolated from a member of the genus *Delphinium*. These three pigments differ from one another by the number of hydroxyl groups found on the B ring.

Another related group of pigments differ in the number of methyl subunits on the B ring. This group includes the purple pigment petunidin (isolated from petunias), the rosy red pigment peonidin (found in peonies) and the mauve pigment malvidin (discovered in a member of the mallow family, *Malvaceae*). An increase in the number of hydroxyl groups on the B ring makes the pigment bluer, whereas an increase in methyl groups makes the pigment redder.

The wide variety of flower colours is due not only to the different pigment molecules but also to the different cellular environments in which the pigments are found. Anthocyanins,

| GENOTYPE | PIGMENTS | | FLOWER COLOR |
|---|---|---|---|
| | ANTHOCYANIDINS | FLAVONOLS | |
| IlhhPp | PELARGONIDIN | KAMPFEROL | PINK |
| IIHhPp | PELARGONIDIN<br>LEUCOPELARGONIDIN | KAMPFEROL | RED |
| LlhhPp | MALVIDIN | KAMPFEROL<br>MYRICETIN | PURPLE |
| LIHhPp | MALVIDIN<br>PELARGONIDIN<br>LEUCOPELARGONIDIN | KAMPFEROL<br>MYRICETIN | MAGENTA |

Fig. 9.4. Flavonoid Pigments found in the petals of four genotypes of the flower *Impatiens balsamina* include anthocyanidins (from the Greek for "blue flowers") and flavonols (from the Latin for "yellow"). Anthocyanidins can colour petals blue, purple or red depending on their chemical environment. Flavonols contribute yellow or ivory shades.

for example, change colour when they accept or release hydrogen ions, depending on the environmental conditions. In a slightly acid environment, which contains an excess of hydrogen ions, the pigment is reddish. As the hydrogen-ion concentration falls, the pigment becomes bluish, then colourless and finally decomposes in a basis solution. The colour may also change when the anthocyanins form complexes with tannins and other pigments. Metallic ions, such as iron or aluminum, can also form complexes with the pigment to produce different colours. Thus a few pigments under a variety of environmental conditions are able to produce a wide spectrum of flower colours.

Often associated with the anthocyanidin pigments are the colourless compounds known as leucoanthocyanins. Only leucopelargonidin, leucocyanidim and leucodelphinidin occur frequently, and these are more often found in the woody portion of the plant than in the more highly coloured portions.

Since flower colour is determined not only by the pigments present but also to some extent by the cellular environment, it is possible for different pigments to produce the same colour. Although some pigments produce characteristic colours, the only sure way to identify the pigments present is by extraction and separation. Pigments can be extracted from flowers with alcohol that has been slightly acidified to keep the pigment from decolorizing. The sugar groups are removed by boiling the pigment with dilute acid. To this solution is then added a small amount of isoamyl alcohol, which rises to the top, carrying the pigment with it.

The mixture of pigments can then be separated by paper chromatography. In this technique pigments of different chemical compositions migrate at different speeds when placed on absorbent paper and exposed to the action of solvents. During the first run of a chromatogram the pigments are spread out along one line and may be overlapping, which makes identification difficult. After the chromatogram has dried it is run in the second direction with the line of pigments along the bottom of the sheet. The resulting pattern of spots is spread over two dimensions and is usually quite faint, but it can be intensified by the use of appropriate sprays. Several pigments fluoresce characteristically under ultraviolet radiation and this also aids in the identification. A certain amount of information can be derived about an unknown anthocyanidin by comparing its performance in a two-dimensional chromatogram with that of pigments of known configuration. Thus the isolation, separation and identification of flower pigments can be achieved with a high degree of accuracy.

To determine the forms of the pigments as they occur naturally in the cell, a simple alcohol extract of the pigments is applied to chromatography paper.

Although it is not easy to isolate and identify the various plant pigments, it is a far more difficult problem to discover how the plant produces them. Green plants can synthesize all organic compounds—both simple and complex—from a few inorganic substances. The metabolic pathway of any one compound is very hard to follow. Some success in elucidating the synthesis of the flavonoid pigments has, however, been gained by two methods, one biochemical, the other genetic.

In the first method the metabolic pathway is followed with the help of molecules labeled with radioactive carbon atoms. Using this technique J.E. Watkin, E.W. Underhill and A.C. Neish of the National Research Council of Canada and others have established the broad outline of the biosynthetic pathway of the flavonoids. The Canadian group studied the formation of the flavonol quercetin in buckwheat. They found that the A and B rings of the molecule are synthesized by different pathways. Ring B and the three-carbon chain are formed by the condensation of a seven-carbon sugar with a three carbon molecule of pyruvic acid. After several changes, in which one carbon atom is lost, the resulting product is phenylpyruvic acid, a $C_6-C_3$ molecule. This molecule then condenses with three molecules of acetic acid, which are the precursors of ring A, to yield the $C_6-C_3-C_6$ backbone of the flavonoid pigments.

It is not clear at what point in the synthesis the hydroxyl and methyl groups are introduced into the structure. Judging from the relative distribution in nature, the more primitive, or basic, structure carries two hydroxyl groups on the B ring. This form or its precursor is modified by the addition or removal of a hydroxyl group. Probably one of the late steps is the addition of methyl groups to form the methoxyl groups. Often delphinidin, petunidin and malvidin are present together in a flower ; this would be expected if malvidin is formed by the stepwise methylation of delphinidin or its precursor.

The second—the genetic—approach to studying pigment biosynthesis is a method that has been successful in elucidating biosynthetic processes in microorganisms. The method involves the use of mutant forms and is therefore more difficult to apply to higher plants, which have longer generation times and a more varied metabolism. Nevertheless, the method has contributed significantly to the knowledge of pigment synthesis.

A genetic study of this type has been carried out with members of the genus *impatiens*. The genetic name of these plants is derived from the characteristic fruit, which opens violently when mature and in the process projects the seeds some distance. Because this can happen if the capsule is merely touched, the plant is known as touch-me-not in many lands and in many languages. Among the members of the genus are the jewelweeds of eastern North America and the popular house plant often called sultana after the Sultan of Zanzibar, in whose country the plant was collected.

A group of plant physiologists at Indiana University headed by Charles W. Hagen, Jr., has been studying the genetic control of the flower pigments of *Impatiens balsamina*, the common garden balsam. Its flowers vary in colour from white and pasted shades to deep purple or bright red. The mode of inheritance of these flower colours was determined several years ago by D.W. Davis and his co-workers at the College of William and Mary. The identification of the anthocyanidin pigments present in the different hereditary strains was made recently by Ralph Alston and Hagen and the flavonols were identified by elevenger.

The formation of flower colour in balsams is controlled by three separate sets of genes. Two genes, L and H, determine the kinds of pigment produced ; the third gene, P, determines the amount. One form of this third gene suppresses pigment production and the flowers are very pale. Another form of the gene acts as an intensifier and the flowers are brightly coloured.

Analysis of the petal pigments present in the various strains has provided some interesting correlations Fig. 9.5. When the gene Lis present, the flower contains malvidin and myricetin, a flavonol resembling delphinidin. Since both pigments carry three groups on the B ring, it seems that L is associated with the addition of the third hydroxyl group. A similar correlation between the gene H and the presence of pelargonidin and leucopelargonidin suggests that H is associated with the formation of a single hydroxyl group on the B ring. Inasmuch as the synthesis of anthocyanidins, flavonols and leucoanthocyanins are all affected by the two genes, it is probable that

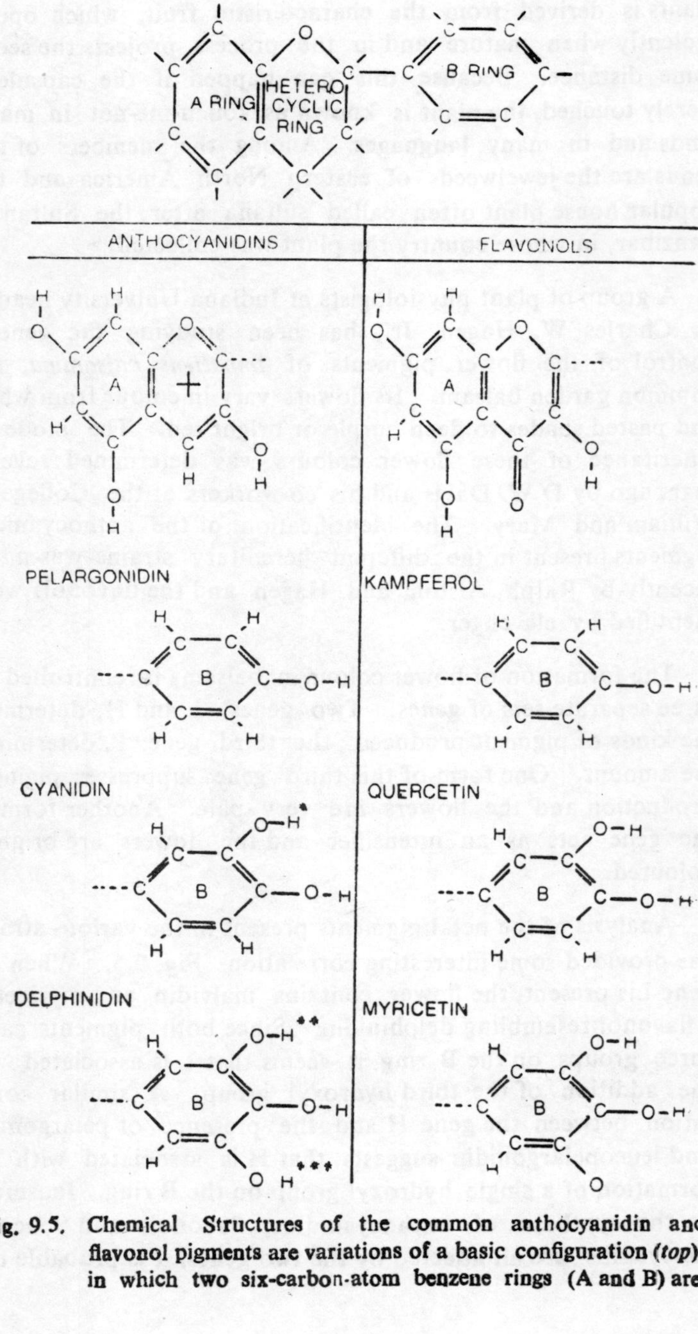


**Fig. 9.5. Chemical Structures of the common anthocyanidin and flavonol pigments are variations of a basic configuration (*top*), in which two six-carbon-atom benzene rings (A and B) are**

these compounds are synthesized along similar pathways. The pigmented sepals contain a different complement of pigments from that found in the petals.

In order to study pigment development under more uniform conditions than those found in the greenhouse, Attila O. Klein of the Indiana group developed a method of growing detached petals in a nutrient culture solution. Light is needed both for the production of the anthocyanins and the expansion of the petals. One unexpected result of this work revealed that the petals in culture are able to synthesize certain pigments (cyanidin, peonidin and quercetin) that are normally found in the coloured sepals of these strains but not usually in the petals. This is further evidence that the action of a gene depends on the chemical, physical and genetic environment in which it acts. The cells of the cultured petals are able to produce substances that intact petals normally do not synthesize.

By pooling such information from various laboratories, one can speculate concerning the possible pathway for the synthesis of the flavonoid pigments-a multibranched pathway leading to many variations of the basic $C_6—C_3—C_6$ skeleton. One branch leads to the formation of the anthocyanidins, perhaps by way of the colourless leucoanthocyanins. Other branches lead to other flavonoids, such as the chalcones, the flavonols and the aurones. The differences between these various groups are slight ; they involve the level of oxidation of the three-

linked by a smaller three-carbon-atom structure known as a heterocyclic ring. What distinguishes the anthocyanidins from the flavonols is the level of oxidation of the heterocyclic ring ; the flavonols have a doubly bonded oxygen atom there. Differences in hues within each family depend mainly on groups of atoms attached to the B ring. The pigments produced by the addition of one, two and three hydroxyl groups (OH) to certain positions on the B ring are shown at bottom. If a methyl group ($CH_3$) were substituted for the hydrogen atom in the hydroxyl group indicated by an asterisk, the result would be a rosy red pigment called peonidin. A similar methyl substitution at the double asterisk would produce petunidin and at the double and triple asterisks malvidin.

carbon chain and the number, kind and position of additional groups. Yet these slight variations contribute to the full spectrum of colours displayed by the plant world.

The comparison of the pigment complement present in different species of the same genus gives valuable information concerning the evolution of species within the genus as well as about the genetic control of pigment synthesis. Within any genus the differences between species have evolved gradually over the years through mutation and selection of genes. Undoubtedly some of these genetic changes have affected the synthesis of pigments. Consequently a comparison of the pigments in closely related species should shed light on the fine-scale evolution of pigment synthesis.

The genus *Impatiens* is well suited to such a study because it contains a great many species with a wide variety of flower colours. The greatest number of species are found in the tropical areas of the Eastern Hemisphere. Two some what similar plants scientists have investigated provide an excellent illustration of the possibilities offered by the comparative study of pigments. Both plants are native to Indonesia. *Impatiens platypetala*, found in Java and Sumatra, has large, showy magenta flowers. On the nearby island of Celebes grows an *Impatiens* with bright yellow-orange flowers. Because the two plants are so similar some taxonomists consider the second to be a variety of the first and call it *I. platypetala var. aurantiaca*. Others maintain that the second is a separate species and should be recognized as *I. aurantiaca*.

An analysis of the flower pigments strengthens the second viewpoint. The magenta petals of *I. platypetala* contain mainly malvidin, but the orange flowers of *I. aurantiaca* contain an apparently unrelated anthocyanidin not previously found in nature. The structure of the new pigment, called aurantinidin, is now being worked out by J.B. Harborne of the John Innes Institute. The pigment seems to resemble another rare anthocyanidin called luteolinidin. Since the two groups of plants synthesize such widely differing pigments, it seems unlikely that they are closely related. and so one can conclude that their resemblance is superficial rather than fundamental. A mixture of pigments from the flowers of these two species was used to make the chromatogram shown on the cover.

So far only a small percentage of the known species of the genus *Impatiens* has been analyzed. Cyanidin and malvidin are the most common pigments in the petals ; cyanidin is the most common in the sepals and stems. From a morphological point of view one would expect the petal to be more highly evolved than the sepal, and sepals, in turn, to be more complex than the stem. It appears that this ordering may also be true for the pigments if one assumes that the nonmethylated pigments are the more primitive and the methylated pigments the more elaborate.

The suggests that the biosynthetic sequence leading to the production of anthocyanidins, at least as far as the B ring is concerned, proceeds as follows : first, two hydroxyl groups are added as in cyanidin, then a third as in delphinidin ; finally stepwise methylation produces first petunidin, then malvidin. At times methylated pigments appear with small amounts of their unmethylated counterparts, indicating that such a stepwise procedure is plausible. These separate steps appear to be under the control of different genes ; the pigment produced in any given cell depends on whether a particular gene functions or is prevented from functioning.

In spite of the abundance of the flavonoids in nature, they represent an unresolved area of plant metabolism. One of the more important unanswered questions about anthocyanins in particular and flavonoids in general concerns their function in the living plant. Many roles in metabolism have been postulated but none has actually been shown to obtain in the plant. These substances may be end products of metabolism that gradually build up in the cell ; they may act as agents of oxidation and reduction ; they may be waste products that serve little or no use, or perhaps they act as antifungal agents.

Undoubtedly one of the functions of the pigments in the flowers is to attract the pollinating agents. Although it is

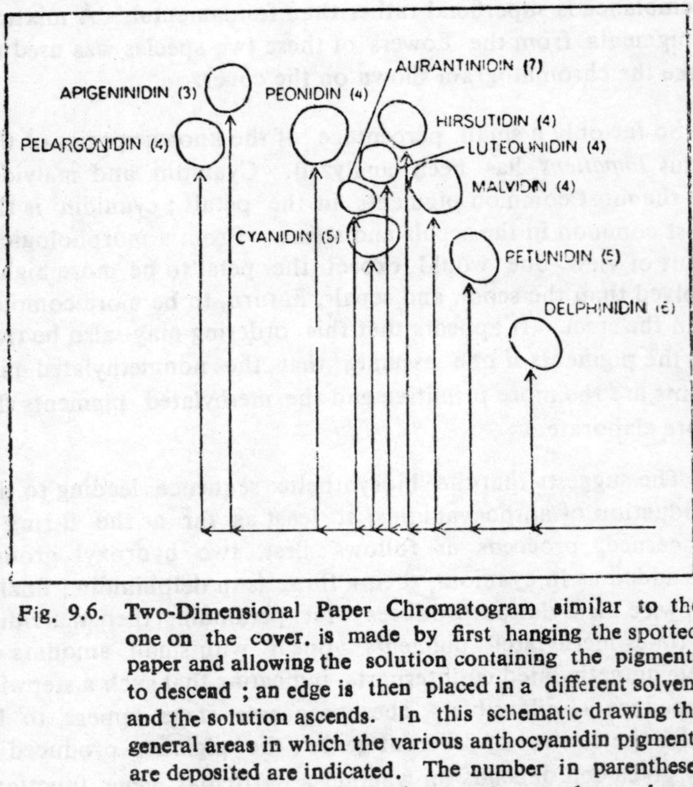


Fig. 9.6. Two-Dimensional Paper Chromatogram similar to the one on the cover. is made by first hanging the spotted paper and allowing the solution containing the pigments to descend ; an edge is then placed in a different solvent and the solution ascends. In this schematic drawing the general areas in which the various anthocyanidin pigments are deposited are indicated. The number in parentheses after the name of each pigment denotes the number of free hydroxyl groups on a molecule of that pigment.

quite clear that flower pigments do perform this function, it is probably a function that developed only secondarily : anthocyanins are found in plants that are not dependent on attracting pollinating agents. Bright red and redorange flowers, particularly prevalent in the Tropics, attract birds. Bees favour yellow and blue flowers. Day-flying butterflies pollinate brightly coloured flowers, whereas their night-flying counterparts, the moths, choose the white or pale flowers, which are more noticeable during the poorly lighted hours. In addition to attracting pollinators by the conspicuousness of their colouring, most flowers have streaks, dots or lines known as pollen guides to help orient the alighting visitor in search of nectar.

Since flower pigments are so widely distributed, one feels that they must have other important functional roles. With the wide appeal that this subject has for plant physiologists, geneticists, biochemists and taxonomists, it is certain that a great deal of new information on the subject of flower pigments will be produced in future. However one looks at these pigments, they are a most attractive subject for investigation.

# 10

# Photoperiodic Effect

Several excellent reviews on the physiology of flowering have appeared within recent years. The review of Lang initiated the series and was one of the most complete, covering both photoperiodism and vernalization.

Because of this profusion of recent excellent reviews, complete coverage of all of the literature should not be essential, and Chouard's thorough discussion of vernalization obviates the necessary of including this topic. The approach is similar to Liverman's in that the component processes of the flowering response in short-day plants (especially *Xanthium*) are emphasized. Since his review, however, increasing emphasis has been placed upon the general significance of these processes to biologv. Virtually all the steps of the flowering process have their counterparts in other plant and animal functions ; photoperiodism implies a response to light and a measurement of time, both of which phenomena are now known to be widespread (measurement of time may prove to be a characteristic of living things). Synthesis and translocation of the flowering hormone are less general in nature, but response to the hormone is an example of what might be the most fundamental phenomenon of biology : differentiation or morphogenesis. And if morphogenesis is commonly a response to specific chemical substances, then knowledge concerning the synthesis, movement, and mode of action of the flowering hormone is certainly of basic biological significance. The following discussion considers these fundamental problems as they apply to the flowering process. Topics such, as the pigment system and measurement of time are discussed as they apply to plants in general, but space limitations seldom allow any mention of how these basic processes operate in the various specific long- and short-day plants that have been studied.

## GENERALIZATIONS ABOUT THE FLOWERING PROCESS

The details relating to flowering are extremely involved and complex, and contradictions are not uncommon in the mass of literature that has accumulated on the subject. As J. Bonner has said, "Perhaps it is impossible for one individual even to remember simultaneously all the relevant facts." Thus the basic facts should be summarized before recent developments are considered. To this end eight generalizations have been formulated. Obviously, future research may prove them to be less general than is implied here, but at present they appear to be the "typical" responses of the plants that have been most intensively studied. Choice of what to include must be a very arbitrary matter, and the author assumes complete responsibility.

I. Short-day plants flower (on a 24-hour cycle of light and darkness) if the day-length is less than some critical amount. Experiments utilizing cycles greater or less than 24 hr. have shown that short-day plants flower more in response to dark periods that exceed a minimum critical length than to the length of the light period. The light and dark periods may interact, however.

II. Long-day plants flower in response to day-lengths exceeding a certain minimum critical value (cycle lengths of 24 hr.). Again, the response to dark periods of less than a maximum value appears to be more critical than the respon :

to the light period, and an interaction may complicate this situation.

III. Short-day plants are inhibited and long-day plants are promoted in their flowering by an interruption of the dark period with light.

IV. The action spectrum for light interruption of the dark period in both short- and long-day plants shows a broad peak in the orange-red between 620 and 665 m$\mu$, and the effects of this orange-red light are reversed by far-red light (peak between 710 to 740 m$\mu$). These action spectra place the light interruption responses in the realm of other light phenomena in plants which are controlled by the so-called photomorphogenetic pigment system. Effects of blue are also encountered in these systems as well as in flowering.

V. There is strong evidence (see below) for a flowering hormone (stimulus) that moves from the leaves to the apical meristems. This hormone has been called florigen.

VI. Flowering in the majority of photoperiodically sensitive plants is an inductive process. That is, plants given the photoperiodic treatment that promotes flowering will continue to produce flowers even though they are returned to a photoperiodic regime that would not lead to flowering. One or more, and sometimes many, photoperiodic cycles are required to bring about this condition.

VII. A combination of responses is commonly observed within a single plant. Many plants are known in which the response may be dayneutral at one temperature and sensitive to day-length at another temperature ; species are being discovered that require a combination of day lengths in a particular order ; and a few plants are known to respond to light intensity changes. Indeed, plants are relatively rare in which the day-length response is only slightly influenced by temperature. Even the test plants of photoperiodism studies are so influenced to a greater or lesser degree : *Xanthium pennsylvanicum* Wall. (cocklebur). *Glycine max* (L.) Merr. (Biloxi soybean), *Hyoscyamus niger* L. (henbane), *Perilla ocymoides* L., *Chenopodium amaranticolor* Coste *et* Reyn.,

*Kalanchoe blossfeldiana* v. Poelln., etc. Recently, a few others have been added to this group. Of these, *Pharbitis nil* Chois. (Japanese morning glory) is the most important. It is a very sensitive short-day plant that responds to a single inductive dark period. Seedlings are usually used, and the characteristics of their response have beed worked out in careful detail by the Japanese. *Pharbilis* has recently been shown to be day-neutral at low temperatures if plants are given sucrose. *Lemna perpusilla* Torr. 6746 is also of special interest because experiments with this plant may be carried out in only a week and its response to day-length is influenced by its nutrient medium (it responds to a single dark period but only when grown on media containing iron chelating agents).

Plants that require short-days followed by long-days were described as early as 1930 by McKinney and Sando [winter wheat and again by Loomis and co-workers in 1949 [certain grasses. Resende and others have also described such plants, as well as those that require long-days followed by short-days. The present interest in short-long-day and long-short-day plants, however, stems from the detailed work of Sachs on *Cestrum nocturnum* L., a woody plant. Among other things, Sachs showed that light interruption of the dark period inhibited the short-day phase of induction and promoted the long-day phase, as in plants with only a single day-length requirement. He suggested that all plants have a series of requirements for flowering that, in the dual day-length plants, may be met only by the proper-sequence of day-lengths. For example, a "normal" short-day plant may require a "long-day" of only two or three hours preceding the short-day phase of induction. Considering the long-short- and short-long day plants together, he concluded that one should not expect the results of long-day treatment to be metabolically the same in all long-day plants, nor of short-day treatment in short-day plants. Considerable evidence is accumulating in favour of this suggestion.

VIII. In short-day plants a number of component processes have been recognized. High-intensity light must precede the dark period if the latter is to be effective in inducing flowering. Timing mechanisms in the leaf that involve an internal conversion of the photomorphogenetic pigment from one form to another, and probably other unknown mechanisms, constitute the first part of the dark period. The second part of the dark period is concerned with a synthesis of flowering hormone. This hormone must then be translocated from the leaf to the apical or lateral meristems or both ; these respond by producing floral organs instead of leaves. In long-day plants the component processes may be similar, but evidence for their existence has been difficult to obtain because of certain inherent properties of the plants that are normally used. Most of the recognition of the component processes in short-day plants depends upon work with *Xanthium*, frequently used because it responds to a single inductive cycle. Only recently have long-day plants of such sensitivity been described and as yet comparable work has not been done with these plants, although the component process concept is being extended.

## THE PIGMENT SYSTEM

Three principal questions relating to the photoreceptor pigment will be considered. The first question concerns the nature of the pigment ; the second, the kinetics of its action in the flowering process ; and the third, the biochemistry of its action in flowering.

### WHAT IS THE PIGMENT

Since light must be absorbed to bring about a photochemical reaction, a pigment system must form an integral and essential part of the flowering mechanism. Action spectra (generalization number IV) indicate a photoreversible pigment system that may be represented as follows :

$$\text{R-Phytochrome} \underset{\text{Far-Red (730 to 735 m}\mu\text{)}}{\overset{\text{Red Light (660 to 665 m}\mu\text{)}}{\rightleftharpoons}} \text{F-Phytochrome}$$

(or more slowly in darkness)

R-Phytochrome refers to pigment in the red-absorbing form, produced by illumination with far-red light or equilibration in darkness. F-Phytochrome is the far-red-absorbing form of the pigment. A prime object of research has been the isolation and characterization of this system. Much of the physiological work has been carried out by workers at the Plant Industry Station at Beltsville, Maryland [including Hendricks, Borthwick, and others. It is appropriate then that Butler *et al.*, working with this group, were the first to succeed in obtaining *in vitro* preparations of the photomorphogenetic pigment system.

The reversible nature of this system suggested an assay. Absorption peaks of the pigment should be dependent upon the quality of light last used to illuminate the pigment. The obvious difficulty lies in the fact that light used to measure optical density at either absorption peak would tend to destroy that peak. The Beltsville group solved this problem with a spectrophotometer system possessing high sensitivity and capable of measuring transmission through dense samples. A reaction-forcing illumination of bright light (e.g., red) is given, and optical density at the desired peak is measured. Optical density is again measured following a second reaction-forcing illumination with the "opposite" wavelength (e.g., far-red). The difference between the two optical densities is a measure of the presence and quantity of the pigment system. The original equipment was assembled from component parts at Beltsville, and commercially available equipment has also been used. Thus work with the pigment system can be carried out in any laboratory having sufficient interest.

The Beltsville workers have obtained the pigment *in vitro* from a number of dark-grown plant tissues. Corn seedlings have proved to be a rich source, and B. Bonner in another laboratory, has used etiolated pea tissue. The pigment has been obtained at Beltsville from green spinach leaves, but the presence of chlorophyll complicates the procedure.

The Beltsville workers named the pigment system "phytochrome." It is of considerable interest that phytochrome proved to be protein in nature since chlorophyll is thought to be closely associated with protein in the chloroplasts, and the phycobilin pigments must also depend upon protein for their photoactivity. Initial attempts to separate a prosthetic group from phytochrome have been unsuccessful. Indeed, attempts have been sufficiently extensive to permit the tentative conclusion that the chromatophore is not a dialyzable group. Complete purification of the pigment system had not been accomplished at the time of this writing although its concentration *in vitro* had been increased at least tenfold. At any rate, study of the photoreceptor pigment system *in vitro* has opened a great many doors to our potential understanding of photomorphogenetic processes in plants.

## WHAT ARE THE KINETICS OF ACTION OF THE PIGMENT SYSTEM IN FLOWERING ?

The deductions based upon experiments with whole plants now await confirmation in the isolated phytochrome system. An elegant series of experiments by the Beltsville workers led to the conclusion that the photoreaction was first order in both directions. This has been confirmed in the *in vitro* system. Another early conclusion of these workers, that the conversion of the pigment from one form to another involved a reaction (probably oxidation-reduction) with another compound was not substantiated in work with the *in vitro* pigment. However the majority of the early conclusions, including the idea of reversibility itself, were substantiated. Yet those of us who indulge in the essential practice of deducing ideas about the chemical nature of the flowering mechanism from the results of physiological experiments with whole plants, should again be reminded that these conclusions must remain tentative until they are supported by more direct evidence.

A key problem, discussed rather extensively by the Beltsville group concerns the action of the pigment when illuminated by relatively high intensity light which contains various ratios of red and far-red. They demonstrated that a number of photomorphogenetic responses could be understood if this problem was properly taken into account ; and it is also of importance in flowering. Recently, they have shown that under incandescent illumination (slightly more far-red than red) about two-thirds of the pigment consists of F-phytochrome Thus it is significantly easier to drive the reaction from R-phytochrome to F-phytochrome than it is to drive it in the reverse direction. Stated another way, red light is more effective than far-red light at equal intensities (R-phytochrome has a higher absorption coefficient than F-phytochrome; quantum efficiencies are near 1.0).

The workers at Beltsville as well as Bonner have studied the destruction of the pigment *in vitro*. The conversion of F-phytochrome to R-phytochrome in the dark does not occur *in vitro*. The Beltsville group has also studied its synthesis and conversion *in vivo*. Present results are too preliminary to permit any far-reaching conclusions, but developments in this field are anticipated.

Although we are just beginning to learn about the kinetics of the light response *in vitro*, considerable research has been done with whole plants. The studies on germination, dormancy, stem extension, leaf expansion, and other photomorphogenetic responses are probably pertinent, but space limitations will not allow a discussion of this work here. A few general principles will be discussed as they relate to the flowering process.

*Reciprocity.* Withrow has considered the question of reciprocity. If a photochemical reaction is stoichiometric, only the quantity of light is important. That is, light of high intensity applied over a short duration should be equal in effect to the same quantity of light applied at lower intensity over a longer duration. Withrow has pointed out that this reciprocity relationship seldom holds in photobiological processes, since short duration times are relatively more effective than longer times. (Reciprocity failure is also common in photochemistry—including photography). We might conclude on the basis of relatively early work in photoperiodism that reciprocity does not fail in this process. Since that work, however, the methods used in flowering research have been improved so that smaller differences in response may be observed. In the early work, results were expressed as a percentage of the plants treated alike which ultimately flowered.

Thus, expression of the data assumed nonuniformity of the material. If the uniformity of response were extremely high, either all plants would flower or none would flower, and percentage data would have little meaning. During the past decade most workers have overcome this difficulty by expressing their data according to some measure of the degree of flowering. Early workers did not use this method because it was felt that conversion to the reproductive state was the most important aspect of flowering, and confusion with development was to be avoided. Yet the rate of development is strongly dependent upon conditions during induction (the dark period in short-day plants) and relatively refractory to the usual conditions during development. All plants in an experiment develop under the same conditions anyway. Sometimes these latter conditions will cause borderline stages to revert to the vegetative state. In a percentage system these early stages contribute as much to the per cent flowering as the more advanced stages, but in a stage system the numbers assigned to the low stages are usually in accordance with their borderline nature.

The problem of reciprocity in flowering has yet to be studied in sufficient detail and by use of proper methods. The experiment should be done with an accurate control of light (intensity, quality, and duration) and using highly uniform plants trimmed to a single leaf (to avoid problems of shading and the sensitivity changes dependent upon age); the light interruption should be applied at the proper time during the dark period (discussion below); and data should be expressed using a system of floral stages. In a classroom demonstration the author obtained results, using an electronic flash (designed for photography), that tentatively show reciprocity failure in flowering. It has also been reported that reciprocity fails in *Pharbitis*, but in the opposite sense to that mentioned above.

*The time of effectiveness of an applied light interruption.* This has been a subject of experimentation by many workers during the past two decades. In general it may be concluded that maximum inhibition of flowering in short-day plants (or promotion in long-day plants) occurs when the light interruption is applied near the end of a duration of darkness nearly equal to the critical dark period. With "normal" dark periods (within a 24-hour cycle) maximum effectiveness often comes near the middle. The per cent flowering should be influenced only by light interruptions that divide the dark period into periods close to the critical, and this is generally the case. Thus, experiments on time of effectiveness of a light interruption must use a quantitative measurement of flowering. If the numbered stages can be shown to have any sort of linearity, kinetic consideration of the data will be more reasonable.

*Effective quantities of light used in interruptions.* This problem has also received considerable attention since it could indicate the quantity of photoreceptor pigments present in the leaf as a function of time during induction. Unfortunately, many of the results are conflicting and difficult to interpret. There are at least three immediate reasons for these deficiencies. First, before any work of this nature can be meaningful, the question of reciprocity must be resolved. Second, many of the data have been expressed on a percentage flowering basis which subject to the difficulties mentioned above. Third, it is quite conceivable, and results seem to indicate, that it may not be necessary to convert the pigment system completely to F-phytochrome to cause complete inhibition of flowering during the most light-sensitive portion of the inductive dark period. A fourth, and probably equally important, point is that it is experimentally difficult to control the quality, quantity and duration of light that simultaneously reaches the leaf surfaces of all the test plants in a given treatment.

Nevertheless, experiments with *Xanthium* in which intensity was held constant and duration was varied seemed to indicate that the amount of light required for maximum response at a given time (saturation quantity) remains constant during most of the inductive dark period. From these results it was concluded that the photoreceptor pigment completed its conversion to R-phytochrome within the first two or three hours of the inductive dark period, and that the amount of R-phytochrome remained relatively constat during the rest of the dark period. Konitz, in Bunning's laboratory, came to opposite conclusions (using *Chenopodium*). He reports that red light interruptions during the dark period could always produce vegetative plants, but that the amount of light required varied drastically during the dark period—the least amount near the middle, much more towards the beginning or end. It was concluded that the amount of R-phytochrome changes as drastically during the inductive dark period. This conclusion is, however, subject to serious question. Since Konitz's plants were not defoliated to a single leaf, one might expect shading and differing sensitivities among the various leaves to make it virtually impossible to obtain a saturation value. Furthermore, saturating quantity cannot be determined if the interruption causes plants to remain completely vegetative. This always occurred in Konitz's experiments, probably because the longest portion of any interrupted dark period was only about 1.5 hr. longer than the critical night. It is not valid to speculate about the quantity of photoreceptor pigment if saturating quantities of light are not determined. His experiment seems to be a further demonstration of the idea that plant response to F-phytochrome is a function of time during the inductive dark period.

*Light quality during the light period.* Flowering was not only inhibited by interruption of the inductive dark period with red light, but also by interruption of the intervening light periods with far-red light. This new phenomenon immediately suggests that flowering is inhibited both by F-phytochrome during the dark period, and by R-phytochrome during the light period. Bunning and Joerrens have since described an analogous situation in the diapause induction of an insect. Violet and blue wavelengths promote during the first half and inhibit during the second half of the daily cycle, while yellow or red light acts in an opposite fashion.

The inhibition brought about by far-red light applied in the middle of the light period could be reversed by red light. In a sense, this result makes interpretation of the initial observation much more difficult. As mentioned above, phytochrome responds to white light primarily as though it were a red source. This should be especially true of fluorescent light, rich in red and poor in far-red. If red will reverse the far-red effect during the light period, why doesn't the normal white light of this light period reverse the effect as well and make the phenomenon impossible to observe ? In case of *Salvia* a light period of the proper mixture of wavelengths will allow a red light interruption of the dark period to promote instead of inhibit flowering.

*The pigment system in* Pharbitis Nil. Takimoto and Ikeda and others in Japan have been carrying out an intensive study of light interactions with the inductive dark period. Although Imamura initially seemed to feel that *Pharbitis* was almost an exact Japanese counter-part of *Xanthium*, it has become abundantly clear that some of the responses of *Pharbitis* to light interruptions of the dark period are unprecedented by work with any other plant.

Two figures from one of the recent papers summarize many of the early results. These two figures have been generalized and combined into Fig. 10.1. The curves show flowering response to red or far-red or either one followed by the other, applied at various times during a single inductive dark period. A curve showing response to different night-length has also been added. In the reported experiments, the dark period was 48 hr., but other papers seem to indicate that the shape of the curves would be essentially the same with any night-length exceeding 16 hr. It can be seen that both red and far-red inhibit flowering maximally when they are applied about 8 hr. after the beginning of the dark period, but far-red is also effective during the hours preceding this. Indeed, a dark period preceded by far-red (before time zero in Fig. 10.1 is less effective than one preceded by white light. This early far-red inhibition can be reversed by red (which in turn can be reversed by far-red, etc.), but red interruption effects at ca. 8 to 12 hr. are not reversed by far-red.

Borthwick, Nakayama and Hendricks have obtained essentially the same results at Beitsville, except that they observed, in the first part of a 16-hour dark period, a promotion with red that was reversed by far-red. Thus in their experiments red light (F-phytochrome) was promotive and reversible in the first part of the dark period, and inhibitory

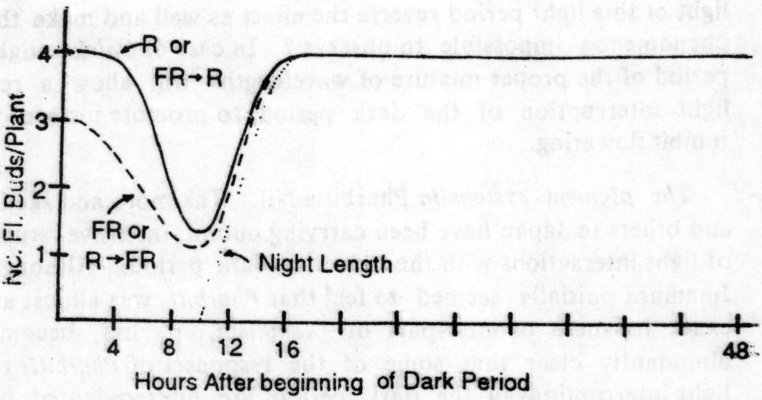


**Fig. 10.1. The effect on flowering of *Pharbitis Nil.* by a light interruption at various times during a dark period of 16 hr. or more (interruptions were not given at 9, 10, or 11 hr. ). R=red ; FR=far-red. For comparison, the response to various night lengths is also shown.**

and not reversible in the middle of the dark period. They suggest that failure of far-red to reverse red is due to some short of quenching reaction of F-phytochrome with a substrate or an inhibitor. They point out that *Pharbitis* acts in a sense like a long-day plant during the first part of the dark period and a short-day plant near the middle of the dark period.

Another interpretation of why far-red fails to reverse red seems possible. Optimal flowering in *Pharbitis* may require a mixture or balance of the pigment forms near the middle of the dark period. This might be indicated by the fact that both red and far-red inhibited flowering, but neither inhibited it completely in the experiments of Takimoto and Ikeda (Fig. 10.1). Furthermore, they were able to obtain slight reversals of red with far-red or *vice versa* if the reversing light was applied at high intensities for short durations. Thus it seems possible that shifting the pigment system in either direction in the middle of the dark period might result in something incompatible with flowering. Only a very rapid, partial shift back to the dark condition would halt this process slightly. The Beltsville workers suggest however, that the far-red inhibition at the middle of the dark period is a result of absorption by R-phytochrome rather than F-phytochrome. The relatively high energies required would be due to low absorption by R-phytochrome at these wavelengths. Yet energies were about the same in the Japanese work.

Takimoto and Ikeda could find no evidence for a second high intensity light process in *Pharbitis*. Indeed, if plants were given a fairly short dark period (12 hr.), a light break, and another short dark period [4 hr., as in comparable experiments with *Xanthium*], they flowered more than if they were simply returned to continuous light. This might be expected from the curves of Fig. 10.1 since a light interruption at 12 hr. does not inhibit flowering as much as does returning the plants to full light at this time. If a balance in the pigment system is required at this time, one might imagine that a return to darkness would be more effective in allowing re-establishment of this balance than would leaving plants in the light. Short dark periods following a light break at 16 hr. or later were without effect, also to be expected on the basis of Fig. 10.1.

Many of the Japanese papers report that far-red (or low intensity light) preceding a long dark period (16 hr. or more) inhibits flowering [opposite to the finding of Borthwick *et al.* with *Xanthium*]. They also state that far-red preceding short dark periods (12 hr. or less) promotes flowering. The critical dark period is shortened, but flowering is less than that of controls receiving a dark period equal to the time of far-red illumination plus the following dark period. The Japanese have stated that the first part of the dark period is relatively insensitive to light and that far-red or low intensity light preceding the dark period extends its length. But since this could only be true for 12-hour dark periods and not for 16-hour dark periods (far-red inhibits in this case), it seems to imply that the plant reads the mind of the experimenter and responds according to whether it is his intention to use a long or a short dark period ! If far-red acts like darkness before a 12-hour dark period, why should it act differently before a 16-hour dark period ? Actually, the promotion by far-red preceding short dark periods is only relative, as indicated above. It may be considered as an inhibition if the total time is considered rather than just the length of the following dark period. At any rate, the intricate arguments of the Japanese are illustrated, and we must again appeal to future research and clear thinking for clarification of the picture.

*Flowering systems responding to a balance in phytochrome forms.* Several authors have shown that light containing a mixture of red and far-red (incandescent light) is most effective in lengthening the day and causing long-day plants to flower. Takimoto even reports that *Silene armeria* L. flowers best under continuous far-red. The Beltsville group explains this by pointing out that far-red causes stem elongation, a process common to all long-day plants used in the above studies. Indeed, in some plants stem elongation may lead to flowering. More commonly, flowering is a direct response to duration of light or darkness (a response to red light—F-phytochrome), but it may be aided by elongation in response to far-red (R-phytochrome). A balance in pigment forms may lead to optimal flowering. Only one non-Japanese paper known to the reviewer claims that incandescent light is more effective in a dark period interruption (long-day plant) than fluorescent light, which may indicate that maximum inhibition of a dark-period depends upon a balance of pigment forms. If proven to be generally true, this suggestion might explain why a short duration of far-red will reverse a red interruption of a short-day plant, while a long duration of far-red is itself inhibitory. Pure R-phytochrome maintained for a long period may be inhibitory, but returning plants to darkness may allow the proper balance to be established in time to reverse the red light effects.

## WHAT IS THE BIOCHEMISTRY OF THE PIGMENT'S ACTION IN FLOWERING?

In certain systems it has been possible to postulate an action of phytochrome in a specific step of a biosynthetic pathway (*e.g.*, anthocyanin synthesis in apple skin), but its action in flowering is essentially unknown. Only a few general suggestions are available. For example, the inductive nature of the flowering process as contrasted to the non-inductive nature of many other photobiological responses (including anthocyanin synthesis) must have some bearing upon the problem. Downs compares the inductive response to an on-off switch, while non-inductive processes compare to a modulating switch. In the "modulating" case it is possible to imagine that pigment in the proper excited form leads to an observed response that is directly related to amount of pigment. In the "on-off" case of flowering, however, the pigment system acts as a trigger, perhaps as a catalyst. It is as if some reaction leading to flowering is catalyzed by photoreceptor pigment in the proper form, or it is blocked or reversed by pigment in the improper form. In such a system, extent of the reaction may be highly dependent upon the concentration of reactants and products, the temperature, the time (*e.g.*, night length), etc., as well as upon the catalytic pigment. The Beltsville workers have concluded that the biologically active form of the photoreceptor pigment is F-phytochrome. Downs showed that far-red, to be effective, had to be applied within a short time after a red light interruption. Thus F-phytochrome seems to bring about some condition in *Xanthium* that is not reversed by reversing the pigment. In a later paragraph we will conclude that F-phytochrome is related to the flowering clock and that it will halt synthesis of flowering hormone. Direct biochemical research on the nature of phytochrome may be our main hope for a satisfactory solution.

## TIMING IN THE FLOWERING PROCESS

Until the last decade the problem of biological time measurement has been either ignored by physiologists or relegated to a position of rather minor importance. Only since 1950-54 has the word "clock" been used in the physiological literature, yet the very term "photoperiodism" implies an ability of the plant to measure the duration of the photoperiod. It would probably be justifiable to assign the credit for the discovery of biological time measurement to Garner and Allard who used the term "photoperiodism" in 1920. Zoologists commonly assign this discovery to Beling for his work on honey bees nearly a decade later.

Time measurement allows a plant to meet the requirements of natural selection and occupy a niche in space and time. A plant at a given latitude may time its growth and reproduction to the season if it is able to measure the length of day or night with accuracy. Withrow has pointed out that an organism must be able to measure time with a precision on the order of 1 to 3 per cent if it is to measure seasonal time to one week, and 4 to 12 per cent for an accuracy to one month. Since the rate of change in day length is greatest in the spring and fall and least in the summer and winter, the absolute precision required depends upon time of year. The unpredictable temperatures of natural environments require a high degree of temperature compensation if organisms are to measure time with the requisite accuracy. Response to the light intensities of twilight is also important. Because of the sensitivity to low light intensities, the photoresponses of the plant go from total darkness to saturating light intensities within a few minutes as the sun rises in the morning, and the extreme daytime fluctuations in light intensity brought about by clouds are "ignored" by the plant. Nevertheless, clouds during twilight may influence the measurement of day-length.

In spite of the early lead by plant physiologists, most of the work done after the discovery of photoperiodism seemed to consider the problem of timing as of secondary interest, and research was concentrated upon other aspects of the flowering process. The problem was finally brought most directly to the attention of plant physiologists by Bunning with his hypothesis relating flowering to his previous studies on endogenous rhythms. Persistent rhythms of leaf movement, flower opening and closing, and a great many other phenomena in both plants and animals, had been studied since at least the middle of the last century. These were shown by Bunning and other workers to be in many cases at least, endogenous in nature—*i.e.*, the rhythms continued even though the plants were held under constant external conditions (within experimental limitations). Recently the term "circadian" has been applied to rhythms of this type that approximate 24 hr. in their period.

On the basis of leaf movement observations, Bunning proposed that plants alternate every 24 hr. through a "photophil" and a "scotophil" phase. He applied the theory to flowering by proposing that light applied during the photophil phase promoted the flowering process and light during the scotophil phase was inhibitory to flowering. Light following a dark period set the phases (as could be shown with leaf movements). In long-day plants the rhythm was delayed so that an extended day-length or light interruption of the dark period fell during the photophil phase, while in short-day plants the rhythm was not delayed and a short-day allowed the onset of darkness to coincide with the scotophil phase.

The theory agreed fairly well with most of the facts of photoperiodism known by the middle of the 1940's, and some new evidences were marshaled as well. For example, experiments by Carr in 1952 seemed to strongly support Bunning's hypothesis. *Kalanchoe* plants were given 72-hour dark periods. Light interruptions were applied at various times during these periods and produced peaks of inhibition spaced 24 hr. apart as well as 24 hour peaks of promotion. Carr concluded : "This experiment constitutes a critical test of the theory of Bunning which must therefore be regarded as finally and decisively proved. It is desirable however that the experiment should be repeated with other short-day plants and with long-day plants, and it is hoped that this will be done in the near future."

Most plant physiologists, in the United States at least, were not so anxious to accept the theory. In retrospect, the principal reason for this reluctance appears to have been a feeling on the part of these workers that the theory of Bunning was something of an exercise in mysticism. Biological phenomena were to be explained in physiochemical terms. Thus unknown precursors, intermediates, and final flowering substances were postulated. Bunning, on the other hand, spoke only of phases that loved either the light of the darkness. The feeling was that even if Bunning's hypothesis were true, it had little meaning in terms of understanding biochemical mechanisms in flowering. Timing in the flowering process resulted from the rates of certain critical reaction, and the proper duty of researchers was a discover these reactions and their rate constants.

Experimental evidence and interpretations were accumulating that failed to agree with Bunning theory. Carr had already done experiments using 48-hour cycles (38 hr. darkness) and had found that light interruptions near the end or beginning of the dark period were most effective in the inhibition of flowering. Claes and Lang had suggested that light interruptions may interact with the main light period if they are close to it. In extended cycle experiments with two long-day species. Hussey concluded in 1954 "that so far as *Anagallis* and *Arabidopsis* are concerned, the persistent alternation of photophil and scotophil phases as postulated by Bunnning cannot be the basis of the photoperiodic reaction." Experiments by Kribben also failed to support Bunning, and Wareing obtained data clearly opposed to the basic theory of Bunning. Light breaks near the beginning or the end of extended dark periods inhibited flowering whether they fell in the photophil or the scotophil phase, and the relationship of the initiation of the dark period to the beginning of the previous light period seemed unimportant. Bunning's theory failed in other ways as well. It should be noted that the explanation of many of these workers seem nearly as mystical as those of Bunning, but a concerted effort to postulate "substances" is usually evident, and reaction rates were frequently discussed. Many workers felt that Bunning's theory had collapsed and, unfortunately, there was also a feeling that since Bunning's theory had collapsed, no relationship existed between endogenous rhythms and timing in the flowering process.

Bunning then modified his theory even to the extent of admitting that long- and short-day plants both exhibit the same scotophil or photophil characteristics at the same relative times of day. He recalled that leaf movement rhythms would persist in some plants for only one or two 24-hour cycles under constant conditions, providing a possible explanation for the failure of extended cycle experiments, and that the cycle of some plants is set by going from light to darkness rather than from darkness to light. But Bunning still maintains that timing in flowering is related to endogenous rhythms, and indeed, he has recently reported on a similar situation in the diapause induction of an insect.

In the past few years the problem has been reexamined. Hammer and his students have carried out an extensive program of experimentation in which plants were given a series of day and night cycles of various total lengths. His results had been foreshadowed by an experiment of Wareing in which it was shown that, using such cycles, if the nights are very long the days must be relatively shorter. In Hamner's much more detailed experiments it was evident that optimal flowering occurs when the light and dark periods total some multiple of approximately 24 hr. This does not prove Bunning's hypothesis, but it emphasizes the possible relationship between persistent circadian rhythms and the flowering process. The experiments of Konitz, in which it was shown that far-red inhibits when applied during the light period, also emphasize this possible relationship.

As may be evident from the above paragraphs, the fundamental question of timing in the flowering process is often buried in the details of a theory. Participants in an argument about endogenous rhythms often finally discover that no one knows what the issues are. Thus it might be well to summarize the problems that await solution.

*The endogenous or exogenous nature of timing.* Most biologists believe that timing is endogenously controlled, in a manner somewhat analogous to a wind-up clock. Pittendrigh is a spokesman for this viewpoint. Brown almost singlehandedly defends the viewpoint that time measurement is exogenously controlled, as an electric clock, by some subtle fluctuation in the external environment (cosmic rays, for example). This is a basic aspect of the problem of biological timing, but it is probably of secondary importance to timing in the flowering process.

*The pendulum or hourglass nature of the flowering clock.* This is the basic question raised by Bunning. Timing controlled by the rate of a simple chemical reaction would be analogous to an hourglass in that only one cycle could be timed, after which the reaction would have to be reinitiated (the hourglass turned over). An oscillating system such as a pendulum, however, must account for persistent circadian rhythms under constant environmental conditions. The Beltsville workers are proponents of the reaction clock in flowering. They argue that the dark conversion time of F-phytochrome to R-phytochrome (half time of one or two hours) provides a simple solution to the problem of timing in the flowering process. Reduction of F-phytochrome to 3 or 4 per cent of its original level would take 8 to 12 hr. A pendulum theory would assume that there is a timing mechanism with a circadian period but that the pigment system is not this mechanism. Phytochrome would be only the linkage between the flowering clock and the ligh environment. For example, the reaction of F-phytochrome with its substrates might depend upon the status (condition) of the clock. On the basis of the present marginal evidence, a decision between the two theories cannot be made. Two lines of reasoning seem to favour the pendulum theory, however.

First, the assumption that a light interruption of the dark period resets the flowering clock is implicit in the hourglass (pigment system) theory. (Such an assumption is not excluded by the pendulum theories.) Thus it was argued that the effectiveness of a light interruption in the first part of an inductive dark period depended upon the length of the remaining dark period. Yet in plants treated with cobaltous ion, timing is slowed so that an interruption in the middle of a 16-hour dark period allows flowering, while a somewhat later interruption will not. Since the 8-hour interruption divides the dark period into periods less than the critical night, it is difficult to reconcile this with the idea that the clock is reset by a red light interruption of the dark period. Nevertheless, the clock that controls leaf movements is under the control of the phytochrome system and is reset by red light.

Second, it was shown that nearly equal quantities of light were required to saturate the pigment when applied any time after the first three hours of an inductive dark period. This seems to indicate that pigment conversion is complete before the completion of time measurement, which agrees with older work and with experiments of Borthwick *et al.*, (in which they were able to shorten the critical dark period only about two hours by converting the pigment with far-red light). When timing was slowed by the cobaltous ion the quantity of light required for saturation did not seem to be influenced. Of course, experiments that relate flowering to circadian rhythms also favour the oscillator model.

Third, a decay curve based on half times is nearly flat by the time only 3 to 4 per cent of the original quantity is left. Such a flat curve seems incompatible with the sharp critical dark periods that are observed. Could critical night be only the difference between 3 and 4 per cent of the original F-phytochrome ?

*One or more physiological clocks.* The parsimonious conclusion is in favour of a single master clock for the control of flowering and the various circadian rhythms, but the problem might be a very difficult one to resolve. Using models, it is quite possible to imagine a single master clock provided with a number of cams, each of which might be connected with the master clock through a different gear ratio so that all did not measure time according to the same period. Of course, it is well known that a number of superimposed rhythms may be observed in a given organism. The important question here is : if the flowering clock is an oscillator, is it the same one that controls other plant rhythms ?

*The basic clock mechanism.* The nature of the flowering clock is of interest and importance regardless of its relationship to other rhythms. Bunning hypothesizes about a relaxation oscillator, and Pittendrigh has developed a fairly complex hypothetical system. Such mechanical models, with their springs, pendulum wheels, and gears, are easy to devise, but chemical oscillators are less obvious. The restriction of temperature compensation makes this especially difficult even though there may be some temperature dependence in plants. Thus the special time-measuring phenomena of living organisms have no satisfactory biochemical models, let alone direct biochemical understanding. It seems quite likely that new approaches will be required to bridge the gap between physical models and biochemistry.

Two lines of research offer some promise. Bunning's group has established a possible relationship between biological timing and the nucleus. Bunning reported in 1957 that urethane (ethyl carbamate) and colchicine sometimes influence the exhibition of biological rhythms (leaf movements). Recently one of his students extended the observations to a number of other compounds. Some were ineffective (or caused "transients"), but a few such as KCN, $Co(NO_3)_2$, ethyl and methyl alcohol, and theobromine changed the period of the rhythm. Some of these, at least, are known to influence the nucleus (*e.g.*, cell division). Furthermore, the volume of the nucleus varies with a circadian rhythm. The second line of research is based upon the observation that cobaltous ion slows timing in the flowering process, and sulfhydryl protectors such as cysteine and glutathione will reverse this action. Since sulfhydryl groups are known to react with cobaltous ion in solution, they are not positively implicated in the flowering clock. Cobaltous ion does seem to be implicated, however, and sulfhydryl groups may be. Although both of these lines of evidence are marginal at best, any biochemical information is of interest.

## SYNTHESIS OF FLOWERING HORMONE

*Evidence for flowering hormones (promoters and inhibitors).* The evidences for existence of a flowering hormone have been reviewed in detail by other authors. Flowering may be brought about in a graft partner held under strictly non-inductive conditions of day-length (or temperature) by an induced graft partner of the same, or in some cases not the same, species. Grafts between long-and short-day plants have been successful in both directions, and day-neutral plants grafted to non-induced short-day plants will cause these to flower. A few failures have been encountered. An essentially sufficient evidence for a flowering stimulus is the response of the leaf to photoperiod conditions while it is the bud that differentiates to produce the floral organs. Assuming the absence of nervous or electrical influences, a chemical interchange is implied, and if this is not a nutrient or energy supplying substance, then by definition a hormone is involved. Since flowering of photoperiodically sensitive plants is not induced by nutrients or energy supplying substances, we should make no apologies about use of the term "hormone." There are a number of possible complications, however. The hormone (chemical interaction) might be a flowering inhibitor produced in the buds and removed to the leaves that are being induced, but the idea seems unreasonable, and there is no evidence to support it. Another possibility is that the leaf produces an inhibitor of flowering under non-inductive conditions, and production of this inhibitor is stopped by the act of induction. Guttridge has obtained very convincing evidence that this is the case in strawberry, a short-day plant. Perhaps his most impressive evidence is the induction of flowering by removal of leaves. This was considered as proof of the inhibitor concept by Lang and Melchers working with the long-day plant *Hyoscyamus*, but defoliation does not lead to flowering in most short-day plants. Guttridge was unable to find evidence for a flower promoting substance in strawberry but other workers compromise the ideas of inhibitors and promoters in flowering. Schwabe using the short-day plants *Kalanchoe*, *Perilla*, and *Chenopodium*, obtained good evidence for inhibitors produced on long days that are interspersed between short inductive days (destruction of flowering hormone did not seem likely) ; yet, a positive-acting flowering hormone is also required in his scheme. He postulates the formation of an adaptive enzyme during successive inductive cycles and has recently added evidence to this concept. It is interesting to note that most research with *Xanthium* is best interpreted on the basis of a positive flowering hormone, but Bogorad and McIlrath have shown that an inhibitor of lettuce seed germination is produced by *Xanthium* leaves on long days. The relationship of this observation to flowering is not clear.

Further impressive evidence for the existence of the flowering hormone is provided by experiments in which leaves of sensitive short-day plants are removed at various intervals following a single inductive dark period. If plants are defoliated immediately following induction, they remain vegetative ; a later defoliation allows flowering. Such experiments were designed primarily to study the translocation of this material, but their success argues in favour of a flower promoting substance and against the removal of a flower inhibitor by induction. Certainly species differences are to be expected, and it is likely that inhibitors play their most important role in certain long-day plants.

*Time of hormone synthesis.* In short-day plants the flowering hormone is apparently synthesized during the part of the inductive dark period that follows the critical night—the time-measuring reactions of the dark period. This is evidenced by the observation that the rate of development of the floral bud following a single inductive dark period is a function of how much this dark period exceeds the critical night, although a "saturation" night length is usually reached. A red light interruption during the period of hormone synthesis acts much as would returning plants to the main light period. We may imply that F-phytochrome halts hormone synthesis.

Most workers in photoperiodism have considered the idea that some hormone is synthesized before the end of the critical night but that this amount is insufficient to cause flowering and may be destroyed in less than 24 hr. There is some evidence for this idea and it helps to explain certain observations. A destruction of flowering hormone seems indicated by the inhibitory effect of extra-long dark periods and it has been suggested that applied auxin results in a destruction of flowering hormone. If synthetic auxins will result in a destruction, it seems reasonable to assume that the natural auxins of the leaf under normal conditions might also result in the destruction of a certain amount of flowering stimulus. Thus, on a regime of subcritical dark periods, some flowering hormone may be synthesized in the hours preceding the end of the critical night and then be destroyed by natural auxins before the end of the next critical night. If this is the case, any treatment to prevent this natural destruction, would, in essence, shorten the critical night. At least three such experiments with *Xanthium* have been reported. Plants grown under threshold conditions (dark periods slightly less than the critical or dim light of intensities slightly too high for flowering) were caused to form buds by application of antiauxins. *Xanthium* plants given low temperatures or high pressures of carbon dioxide immediately following a subcritical dark period were also induced to form buds. All of these experiments might be explained if plants normally produce a small amount of flowering hormone on subcritical dark periods, if this hormone is normally destroyed in the plant, and if the above mentioned treatments in some way prevent its destruction. Alternatively, one or more of these treatments might prevent F-phytochrome from halting hormone synthesis.

It is possible that time measurement in flowering may be a matter of phytochrome conversion followed by synthesis of small amounts of flowering hormone, amounts too small to last until the end of the next sub-inductive dark period. Results of experiments in which chemicals were applied to *Xanthium* plants at various times in relation to a single inductive dark period are in opposition to this idea. Cobaltous ion inhibits flowering only when it is applied before the end of the critical dark period. In addition, it extends the critical night and influences the kinetics of light interruption of the dark period. A number of compounds discussed below inhibit flowering when they are applied during the last part of the dark period but not when applied thereafter. Most of these have no effect on the critical night. The (ethionine and picolinic acid) that lengthened critical night do not influence the kinetics of light interruption and seem to influence critical night through a stoichiometric inhibition of flowering hormone synthesis. Thus two phases of the inductive dark period—time measurement and hormone synthesis—seem to be adequately separated.

*Leaf sensitivity and temperature response.* Hormone synthesis appears to be most active in the most rapidly expanding leaves of *Xanthium*. Since these leaves are probably most active in synthesizing metabolites in general, this is not surprising. Many plants are strongly influenced by temperature, but *Xanthium* does not appear to be highly temperature sensitive. Temperature has been studied as a variable in many experiments but only recently have short periods of extreme temperatures been used to study specific steps in the flowering process. Schwemmle found that flowering was inhibited most in long-day plants when the dark period was "interruped" near the middle with low temperatures (5°C) : and least when the interruption occurred near the ends of the dark period. Short-day plants reacted least when interrupted near the middle and most when near the ends. High temperature (45°C) interruptions act in an opposite way.

*Biochemistry*. Synthesis of the flowering hormone implies chemistry. In contrast to the question of timing which seemed to be neglected for some time, the question of synthesis of flowering hormone has been of primary concern since it was formulated in the 1930's. Yet actual information is not much more extensive than that concerning timing. The problem has been the lack of a suitable assay for the flowering stimulus. The only measure of its presence so far is the flowering of a mature test plant, and this is complicated by a requirement that there be a phloem connection between the source of the flowering hormone and the bud that differentiates into a flower. So far, the relatively few instances in which applied chemicals have caused flowering, rather than solving the problem, have tended to increase its complexity. Synthetic auxins and other chemicals applied to the pineapple plant caused it to flower but no one felt that any one of these chemicals was itself the flowering hormone. Gibberellins will cause many species of long-day, cold-requiring plants to flower under non-inductive conditions and promotions of flowering in short-day plants under certain inductive conditions have been reported. Chailakhyan suggests that flowering may require two substances. Gibberellin is one of these and limits flowering in long-day plants ; the other limits flowering in short-day plants. Yet most workers feel that the gibberellins do not constitute the elusive flowering substance although they do meet some of the originally established criteria. Extracts are obtained from higher plants and such extracts may induce flowering. Indeed, plants induced to flower by the proper environmental conditions have sometimes yielded more of these gibberellin-like substances than comparable control plants. The only persistent claim for extraction of the flowering substance (a lipid) comes from Roberts but his methods have not been described in sufficient detail for duplication in other laboratories. Furthermore, it is very possible that his flowering substances are of the same category as the gibberbellins, or the auxins that promote flowering of pineapple—they may influence the flowering process without themselves being the flowering substance.

Many early attempts were made to learn about the biochemistry of hormone synthesis by studying the changes in metabolites that accompany flowering. Changes in growth promoters have also been measured recently. In all cases there are so many changes that their meaning is not clear. Cause and effect cannot be separated. A sort of last resort has been the application of chemicals to whole plants.

Earlier reviews have discussed the results obtained by the application of auxins. To summarize : the flowering of many plants is inhibited by applying auxins at relatively high concentrations, while flowering is promoted in a few plants, including pineapple and the litchi tree. Other plants have been promoted in flowering by application of auxins under special conditions [low temperatures, low light intensity, low concentrations to older leaves and threshold day length. Since auxin inhibition may be overcome by simultaneous application of antiauxins it is probable that the action of auxin in flowering is very similar to its action in other growth responses such as cell elongation. Evidences that the natural auxins of plant tissues influence flowering include not only promotion of flowering under threshhold conditions by antiauxins but promotion of flowering by ethylene and maintaining plants in an inverted position. All of these treatments were interpreted as lowering the natural auxin concentrations within the plant leaf. Clear cut inhibitions can be obtained only when quantities of auxin are applied that will cause quite serious side effects to the plant. The primary effect must not be upon flowering hormone synthesis because applied auxin inhibits flowering in short-day plants in the hours following the dark period as well as during the dark period. Preliminary experiments indicate that auxin also does not influence the kinetics of hormone translocation.

When it became evident that auxin could not be implicated in the synthesis of flowering hormone, other chemicals were investigated. Liverman and Bonner showed with short-day plants and Fredericq with longday plants that the requirement for high intensity light could be satisfied by applying sucrose and other energy substrates to the plant. Lang has shown in a preliminary way that reducing substances including glutathione, cysteine, and ascorbic acid will also fill this requirement. A requirement for sucrose or energy substrates implies that part of the dark process involves conversion of these substrates to energy—respiration. For a time respiratory energy exists in the form of reducing power, which may help explain Lang's observations. Respiration was also implicated by Nakayama, Fukamizu and Sei who showed that cyanide, azide, fluoride, and arsenate will inhibit flowering of *Pharbitis* when applied during the second half of the dark period. 2,4-Dinitrophenol will also inhibit flowering of *Xanthium* when applied during the last half of the dark period. In the Japanese experiments malonic acid did not inhibit flowering, and they concluded that only the glycolitic pathways are essential for flowering and that the Krebs cycle phase of respiration was not. Results with dinitrophenol would indicate that ATP is essential.

Experiments with applied chemicals can be interpreted with some degree of confidence only if time of application data are available. A substance that inhibits flowering only in the early part of the dark period may affect only the timing mechanism, as with cobaltous ion. Compounds that influence flowering after the end of the inductive dark period may have little or nothing to do with synthesis of flowering hormone. Maleic hydrazide, 2,2-dichloropropionic acid (dalapon), and 2,4-dichlorophenoxyacetic acid (2,4-D) probably inhibit development of the floral bud, because they all inhibit flowering regardless of when they are applied.

Current research by the author with W. Collins and J. Bonner concerns the effects of a number of antimetabolites upon flowering of *Xanthium*. Of twelve anti-amino acids that were tried only one inhibited flowering. Thus protein synthesis may not be a part of flowering hormone synthesis. The one effective compound was ethionine, an inhibitor of methionine. Ethionine may inhibit donation of methyl groups from methionine. Quercetin, an inhibitor of rutin metabolism in animal biochemistry, and picolinic acid, which is known to inhibit thiamine metabolism in certain animal systems, inhibit flowering when they are applied during the period of hormone synthesis. The action of these two compounds in the flowering process is unknown.

A number of compounds that inhibit various phases of nucleic acid metabolism in certain systems have been tried with *Xanthium*. Ineffective compounds include adenosine, 2-amino-4-methyl pyrimidine, barbituric acid, 5-hydroxyuridine, and 5-mercaptouracil. Compounds that inhibit flowering include 5-fluorouracil, benzimidazole, thiouracil, 2,6-diamino purine sulfate, and 8-chloroxanthine. As yet no reliable conclusions can be drawn from these lists since inhibitors of both RNA and DNA metabolism are found among the successful and among the unsuccessful groups. All of the above compounds have been tested in experiments designed to study the effect of a range of concentrations, the effective time of application, and effects on the critical night, but only 5-fluorouracil has been studied in any further detail. Of the above compounds that were successful in the inhibition of flowering, none influenced the length of the critical dark period in a clear cut manner (thiouracil and 2,6-diamino purine sulfate had slight effects), and all except one inhibited flowering only if applied before the end of the dark period (thiouracil also inhibited flowering when applied during the translocation period).

A test of whether an antimetabolite is acting in flowering in a manner similar to that in other biochemical systems is to see if simultaneous application of the corresponding metabolite will reverse the effect upon flowering. Of the above compounds only ethionine and 5-fluorouracil have been adequately tested in this manner. Ethionine inhibition is completely reversed by simultaneous application of methionine and 5-fluorouracil inhibition is reversed by orotic acid. Reversals of inhibitions by some of the other compounds have been accomplished in preliminary unpublished experiments. Attempts to reverse the effects of 5-fluorouracil with thymidine or uracil have been consistently and conclusively negative. Sucrose is also unable to affect the 5-fluorouracil inhibition (uracil is known to take part in sucrose metabolism). These experiments are being continued and the results so far seem to merit a more direct biochemical approach, which is also being undertaken.

A most interesting aspect of the work with 5-fluorouracil was the finding that under certain conditions the inhibition caused by application of this compound was most complete when the material was applied only to the bud and not to the leaf. Subsequent experiments by Bonner and Zeevaart indicate that 5-fluorouracil may be translocated from the leaf to the bud but that it is never translocated from the bud to the leaf. Thus this compound must be effective in the bud during the period of flowering hormone synthesis.

Three possible conclusions come to mind. (*a*) Some process essential to flowering but sensitive to 5-fluorouracil is taking place in the bud during the last part of the inductive dark period. No evidence for such a process has ever been reported, but perhaps these experiments constitute sufficient evidence to warrant a detailed search. (*b*) Some material other than the flowering stimulus (which is typically translocated following the dark period) is translocated from the leaf during the dark period. Action of the material in the bud is inhibited by 5-fluorouracil. This explanation postulates two flowering hormones. (*c*) The effective time of application of 5-fluorouracil is a coincidence caused by a delay (slow reaction rates) in its inhibition of processes that normally occur in the bud. Thus it must be applied a day or two before the flowering stimulus reaches the bud. Such a delayed reaction seems unlikely.

To summarize : experiments with anti-purines and pyrimidines have implicated nucleic acid metabolism in the flowering process, and results of the time of application experiments have tied this effect to the period of flowering hormone synthesis. The results with 5-fluorouracil have located a response in the bud rather than the leaf, the known site of hormone synthesis. There is every reason to believe that perusal of these anomalies, especially by direct biochemical research, will lead to considerable further clarification of the flowering mechanism ; but at present the perplexities are much more obvious than the explanations. At least localization in the bud of the nucleic acid effects saves us from the necessity of immediately postulating translocatable nucleic acids.

An exciting aspect of the antimetabolite work is that when the inhibition due to an antimetabolite can be reversed by its corresponding metabolite, this is evidence for participation of the metabolite in the flowering process. Attempts have been made in the past to label the flowering stimulus by allowing the plant to photosynthesize in the presence of radioactive carbon dioxide. These have been uniformly unsuccessful because labeled carbon dioxide ends up in nearly all of the compounds of the plant. If specific precursors are identified, however, labeling them and introducing them into the plant at the proper time might produce more acceptable results.

## THE SECOND HIGH INTENSITY LIGHT PROCESS

In relatively recent years the work of Lockhart and Hamner has renewed the discussion of the effects of light conditions following the inductive dark period. Early work indicated that low intensity light either preceding or following the inductive dark period reduces subsequent flowering. Lockhart and Hamner studied this phenomenon in detail by giving *Xanthium* plants an inductive dark period, a relatively short light break, and then a short period of darkness (usually four hours). This post inductive dark period inhibited flowering. Unfortunately, they confused the issue somewhat by assuming that applied auxin (IAA) had the same effect as the post-inductive dark period, and thust in most of their experiments auxin was applied before this short dark period. They concluded that the period of hormone synthesis during the last part of the inductive dark period produced only a precursor of a flowering hormone which then had to be stabilized by high intensity light applied to the leaf. Their alternative explanation seems more logical to the reviewer some leaves (depending upon age or as yet unknown environmental conditions) may cause the destruction of flowering hormone in darkness, and high intensity light changes the condition of such leaves so that they are no longer capable of destroying flowering hormone. The data fit either explanation almost equally well, but the second one does not require a second synthesis step. It is not known why some leaves have this ability while others do not, but a post-inductive dark period is not always effective.

Some problems were evident at the time of Lockhart and Hamner's experiments, and others have become apparent since that time. To begin with, it was known that even *Xanthium* will flower if it is left in total darkness for about a week (plants die after an extended period). Considerable data have been accumulated to indicate that many plants will flower under conditions of total darkness. These include short-day plants, long-day plants and day neutral plants. Both etiolated and green plants seem to respond in this way although some do not. If only *Xanthium* were being considered, one might say that these results certainly eliminate the rigorous requirement for a second high intensity light process. Since so many plants flower in extended darkness, however, it is tempting to imagine that this is a special situation, one that may bypass in certain respects the normal steps in the flowering process. If this were the case, high intensity light might still be required under "normal" conditions. As discussed above, however, a second high intensity light process cannot be demonstrated in *Pharbitis*.

Carr, using *Xanthium*, was able to overcome the requirement for a second high intensity light period by treating the plants with sucrose. Previous attempts at this had failed. Carr concluded that the second high intensity light process was not a stabilization of flowering hormone precursor in the leaf but simply an effect upon translocation. Photosynthate was required to move the flowering hormone out of the leaf either by mass flow or by providing energy for an active movement. There must also be a destruction of flowering hormone in the leaf, otherwise it should simply be moved out at the close of a post-inductive dark period when plants are finally returned to high intensity light conditions ; and post-inductive inhibition could not be observed. Metabolism of sucrose may change the leaves so that they are incapable of hormone destruction, or the hormone may be moved out of the leaf before any destruction can occur [the second possibility seems less likely since hormone is usually not moved out so soon]. If the destruction in the post-inductive dark period is because of relatively high auxin levels in the leaf, as suggested by Lockhart and Hamner, then sucrose metabolism might be responsible for the lowering of these auxin levels. It might also result in stabilization of a hormone precursor. At any rate, Carr's experiments seem to indicate that the second high intensity light process is related to photosynthesis.

## THE INDUCED STATE

So far the discussion has considered only the act of induction : the events occurring within the plant that result in the formation and translocation out of the leaf of a flowering stimulus. Many plants are capable of attaining the induced state – that is, flowers will continue to develop even though the plant is returned from the inductive day-length conditions to a non-inductive photoperiod regime. Plants that require only a single inductive cycle are the most extreme examples of this category. Other authors have used different terms in the discussion of these concepts.

*Translocation of the flowering stimulus from the leaf.* Initial translocation (see discussion of defoliation experiments above) must be a part of the act of induction, but certain aspects of the subsequent movement of hormone within the plant seem to be characteristic of the induced state. Therefore, translocation will be discussed under that title. It was shown almost two decades ago by Withrow and Withrow that the flowering stimulus would move only through living tissue in *Xanthium* and soy bean. It will move in either direction. Recently Imamura and Takimoto have demonstrated the same things in *Pharbitis.* Selim claims xylem movement in *Perilla*, but this is disputed by Zeevaart. Dissipation of the hormone in the stem of *Pharbitis* has been studied and the rate of movement was found to be 2.6 to 3.8 mm. per hour. These rates are in agreement with rough estimates made by less ingenious methods and indicate that movement of the flowering hormone is one or two orders of magnitude slower than movement of other organic materials through the phloem elements. This observation might be a clue to a means of attack of the flowering stimulus in isolation as suggested by Bonner, and it also may indicate a high molecular weight for the flowering hormone.

Many authors have found that transmission of the flowering stimulus may be inhibited by leaves maintained on a non-inductive day length. Various explanations of these results have been forthcoming : the leaves may "absorb" the flowering hormone as it goes by ; leaves on the non-inductive day lengths may produce inhibitors of flowering ; or such leaves may otherwise prevent the hormone from reaching the bud.

Some recent studies by Carr are most stimulating. On the basis of leaf graft experiments and *in vitro* culture of induced and non-induced apical buds, he concluded among other things that the flowering stimulus is initially moved completely out of the leaf, leaving the leaf devoid of this substance. Then after a few days the hormone reappears, after which the leaf is truly in the induced state. It is well known that young *Xanthium* leaves that were not subjected to the inductive cycle but grew out after the plants had been induced, also are converted to the induced state. Lona and Zeevaart have shown that young *Perilla* leaves are not so converted, although the leaf initially subjected to the inductive dark period is in the induced state from the act of induction until it dies.

*Necessity for active buds.* Although this topic was reviewed in detail by Doorenbos and Wellensiek, the subject should be briefly outlined again for the sake of completeness. In early experiments Lona claimed that leaves that had been detached from the plant were still capable of the act of induction. Carr was unable to induce detached leaves, but Lona's group again succeeded and Zeevaart induced detached *Perilla* and *Xanthium* leaves and disbudded plants by giving at least ten inductive cycles. Nevertheless, in experiments using disbudded plants and only one or a few inductive cycles, it was shown that plants disbudded at the right time were incapable of transmitting the flowering stimulus when grafted to plants having buds. "Inactive" buds were used to obtain essentially the same results without the necessity of grafting to a receptor plant. It was shown by experiments of this type that if active buds were removed just previous to or any time following the act of induction, inactive buds would be vegetative when they became active. If they were to receive the flowering stimulus and become reproductive, they would have to become active before the stimulus was "dissipated" from the plant (about four to six days). It was shown that the requirement for active buds could be replaced by auxin, or by young leaves recently shown to produce auxin [or something that maintains axillary buds in a dormant condition]. It was concluded that flowering stimulus was "preserved" by the presence of auxin until inactive buds can become active. Gibberellin may help young tissues preserve the stimulus. Carr and Zeevaart suggested that the effect of auxin was merely to make the buds become active, sooner, before the hormone had a chance to become "dissipated." It is well known, however, that auxin usually inhibits the growth of axillary buds rather than promoting it, and, indeed, unpublished observations by the author during the original experiments bore out this idea.

It was observed that when active floral buds were removed up to at least seven days after induction, inactive buds were vegetative when they became active. Thus it is possible to cause a plant to revert from the flowering to the vegetative condition by the removal of the flowering buds. Young leaves tend to prevent this reversion. Recently, this phenomenon has been reinvestigated by Lam and Leopold.

## THE ACTION OF FLOWERING HORMONE

Probably the most fundamental question in biology concerns the nature of morphogenesis. It is a phenomenon common to all living things (even the virus particle is organized after the synthesis of its constituents) and is responsible for the myriad forms of life that surround us. Beyond the idea that it may be controlled by certain chemical entities, we know essentially nothing about the process. Indeed, the very problem of how a chemical can control the formation of a three dimensional structure seems to be beyond the approach of our present research methods. The flowering hormone is apparently such a chemical, and flowering should be an ideal situation for study of morphogenesis. In sensitive plants such as *Xanthium* the time of differentiation can be determined and controlled within a matter of hours. Morphogenesis encompasses an even more perplexing problem than the one mentioned above : how do the genes control specific species characteristics of a morphological nature ? In flowering, the releasing trigger that allows the species-chracteristic floral genes to act seems to be the arrival of the flowering hormone at the meristematic site. Thus virtually all phases of morphogenesis could be studied with the flowering process as a basic frame of reference.

Studies in this field, however, have been extremely limited. Although there has been a great deal of general descriptive work relating to the development of floral organs, etc. work with photoperiodically sensitive species directed towards understanding the cellular and tissue changes that take place at the apical meristems during floral differentiation have been very few in number. F.L. Naylor published a description of the surface changes that take place in developing *Xanthium* inflorescences, but no attention was paid to changes at the cellular level. Millington and Fisk published a paper on the vegetative characteristics of the *Xanthium* apical meristem, but a promised paper on the floral apex has never appeared. Popham and Chan described the developing inflorescence of

*Chrysanthemum morifolium*, a short-day plant. Thein studied the flowering shoot of *Xanthium*, including some chemical changes. Sachs and co-workers have studied cell divisions in the stems of bolting long-day plants, especially as gibberellins influence the process, but this is not a study of floral differentiation itself although it does approach a truly experimental study of the problem. Wetmore and co-workers have published a recent description of the cellular changes taking place in the differentiating apical meristem. Little information exists on the biochemical changes that take place in the differentiating meristem. Some pioneer work has been done on the biohistochemistry of vegetative apices, but intensive histochemical work on differentiating apices of photoperiodically sensitive plants remains to be done.

The work of Wetmore *et al.* was admittedly preliminary, and the experiments could easily be improved. They were overcomplicated by use of various numbers of inductive cycles samples were taken only in duplicate and only every other day, while the initial changes that occur must take place within a matter of hours. Nevertheless, their description of what happens is the best available at this time. Their observations were summarized in five points. (*a*) The cells just below the "central zone" in the apical meristem were the first to become active in the differentiating bud. This activity consisted of cell divisions and some enlargement and was found in all species studied by this group, as well as in the literature cited by them. This might be the most important finding of the entire study. Perhaps research efforts should be directed towards biochemical understanding of this phenomenon. (*b*) Activity then spread to the other cells, including those in the central zone. Cells in this zone become smaller by dividing and eventually there is a layer of meristematic cells two or three cells deep covering the entire bud. The bud at this stage is an easily recognizable floral primordium and has passed far beyond the initial changeover from vegetative to reproductive state. (*c*) Enlargement of the pith rib meristem cells causes swelling of the bud. (*d*) The layer of meristematic cells covering the surface forms the bracts, flower parts, etc (*e*) Extension of the axis stops, and apical dominance is lost.

To the present reviewer one of the more unfortunate aspects of work in this field concerns the time wasted in quite philosophical discussions on the controversy of the meristem *d'attente*. The French school of morphologists under Gregoire and Buvat suggested that the central zone of the apical meristem consisted of inactive cells simply lifted into space as the stem grew, and that such vegetative stem growth originated from division of cells below this group. At the time of differentiation of the floral bud these cells, the meristem *d'attente* (the meristem in waiting), became active and developed into the flower. If this theory were true, it would, of course, have implications for the physiologist – only the cells of the meristem *d'attente* must respond to flowering hormone. Experiments with photoperiodically sensitive plants should settle the controversy once and for all. Indeed, the five points of Wetmore *et al.* based on such actual observation fail to support the theory, and the French school has retreated considerably.

At any rate crucial problems in the physiology and morphogenesis of flowering still await the efforts of future research. The basic experimental procedures are quite obvious. They need only to be done by someone trained in the proper techniques. Refinements of the obvious techniques have been suggested by Green and Erickson at the University of Pennsylvania. These include the use of identical plants, so that identification markers such as axillary bud primordia might be chosen. Such markers would be very valuable for interpreting the changes in anatomy that accompany floral differentiation. Identical plants could probably be obtained, even with *Xanthium*. Plants might be chosen very carefully according to leaf size, height, age, and direction of phyllotaxy, and grown under carefully controlled conditions. Any work in this field should provide a groundwork for solving some of the problems of morphogenesis. And in view of the suggestions made above in the section on synthesis of the flowering hormone, a most exciting subject for histochemical study might be the DNA and RNA status within the various cells of the differentiating apical meristem.

# 11

# Seed and Bud Dormancy

## The Structure of Seeds

All seeds contain an embryo plant which is enclosed by one, or more commonly by two, seed coats. The seed coats originate from the integuments of the ovule and often exhibit external structural evidences of this origin even in the mature seed. Among these are the *hilum*, which represents the place where the seed was attached to the ovule stalk (*funiculus*), and *micropyle*, which frequently persists in the mature seed, and the *raphe*, a remnant of ovule stalk which in certain kinds of seeds is adherent to the seed coats. When a single seed coat is present it is usually hard and woody, but when there are two seed coats the inner is almost invariably thin and membranous.

The embryos in seeds of different species of plants differ markedly in size and appearance, but all mature embryos are composed of one or more *cotyledons*, a *plumule* and a *hypocotyl*. The cotyledons are the seed leaves and vary in number from one in the monocotyledons to as many as fifteen in the embryos of some conifers. The embryos of dicotyledons have two cotyledons, as the name implies. Structurally the cotyledons are modified leaves, but usually they differ greatly in appearance from the foliage leaves of the same species. The cotyledons (or cotyledon) are attached near the upper end of the short thick stem-like axis of the embryo, the hypocotyl. The plumule or bud of the embryo is usually located just above the point at which the cotyledon or cotyledons are attached to the hypocotyl. The plumule consists of a meristem with several rudimentary foliage leaves. The primary root of the plant develops from the lower end of the hypocotyl. The rudimentary root at the lower end of the hypocotyl is often called the *radicle*.

An *endosperm* is also present in the seeds of many species. This tissue develops from the endosperm nucleus and usually contains considerable quantities of accumulated foods. In the seeds of those species which contain no endosperm, such as the legumes, the cotyledons are usually enlarged and contain considerable quantities of reserve foods. Some seeds contain a *perisperm* which represents remnants of the nucellus. The so-called "endosperm" in the seeds of gymnosperms is not a true endosperm but represents the female gametophyte.

**Germination of Seeds**

The resumption of active growth on the part of the embryo resulting in the rupture of the seed coats and the emergence of the young plant is known as *germination*. The seeds of many plants will germinate as soon as ripe if environmental conditions are suitable. Pea seeds, for example, sometimes germinate within the pod, corn grains may sprout while still attached to the parent plant, and the seeds of some citrus species frequently germinate while still within the fleshy fruits. Seeds of many other species, however, will not sprout until after an interval of weeks, months, or years, even if environmental conditions are favourable for germination. The causes of this condition of *dormancy* in seeds will be discussed later.

In nature, germination of seeds usually occurs either at or just below the surface of the soil. The latter is more apt to happen in forests where seeds, especially smaller varieties, often fall into interstices in the decaying detritus which composes the top layer of many forest soils and are later covered by falling leaves. In the laboratory, nondormant seeds will usually sprout if they are brought into contact with any moist substratum or even simply exposed to a saturated atmosphere, provided other environmental conditions are also suitable.

The initial step in germination is the imbibition of water by the various tissues within the seed. This generally results in an increase in its volume. The increase in the hydration of the seed coats usually causes a pronounced increase in their permeability to oxygen and carbon dioxide, which is very low in the dry seed coats. The swelling of the seed often ruptures the seed coat, but in some species this does not occur until the emergence of the primary root.

With an increase in the hydration of the cells, enzymes become activated. In seeds possessing an endosperm, enzymes apparently move into that tissue from the embryo as described. Stored foods, whether they occur in the endosperm or cotyledons, are digested and the soluble products of the digestion process are translocated toward the growing points of the embryo. If chemical analyses are made of samples of seeds at successive stages during their germination, it is found that the quantity of starches, oils, and proteins in the seed decreases markedly. A large proportion of the fats present are usually converted, after digestion, into soluble carbohydrates. The soluble carbohydrates are not present during the later stages of germination in amounts quantitatively equivalent to the starch or other storage carbohydrates digested during the process, indicating that a large proportion of these compounds is consumed in respiration or assimilated in the construction of the carbohydrate constituents of cell walls. In oily seeds the soluble carbohydrates utilized in respiration result largely from chemical transformations of the products of the digestion of fats. Digested proteins are usually represented in the seeds by quantitatively equivalent amounts of amino acids, asparagine, etc. This indicates that proteins are not consumed in respiration but are utilized in the synthesis of the organic nitrogenous compounds of the growing embryo.

Insofar as the actual mechanics of seed germination are concerned two principal groups of seeds may be recognized : (1) those in which the cotyledons emerge from the seed and (2) those in which the cotyledons remain permanently within the seed. Most seeds of dicotyledons and seeds of some monocotyledons such as onion belong to the first group while the seeds of grasses and of some dicotyledons such as peas and oaks belong in the second.

1. *Seeds in Which the Cotyledons Emerge*. The sequence of events that takes place during the germination of the seed of the lima bean (*Phaseolus lunatus*) will be described as a type example in this group (Fig. 11.1). Germination is initiated by a marked swelling of the seed which usually ruptures the seed coat. This is followed by the emergence of the primary root which develops from the lower end of the hypocotyl and is the first structure of the embryo to make contact with the external environment. The primary root grows downward in the soil producing lateral roots and root hairs. The hypocotyl then elongates rapidly, pulling the cotyledons upward out of the soil into the air where they separate into an approximately horizontal position on either side of the plumule. The plumule then begins active growth giving rise to the stem and foliage leaves of the seedling. Since the bean is a seed without an endosperm, the food used during germination is largely derived from the accumulaonts in the thick cotyledons.

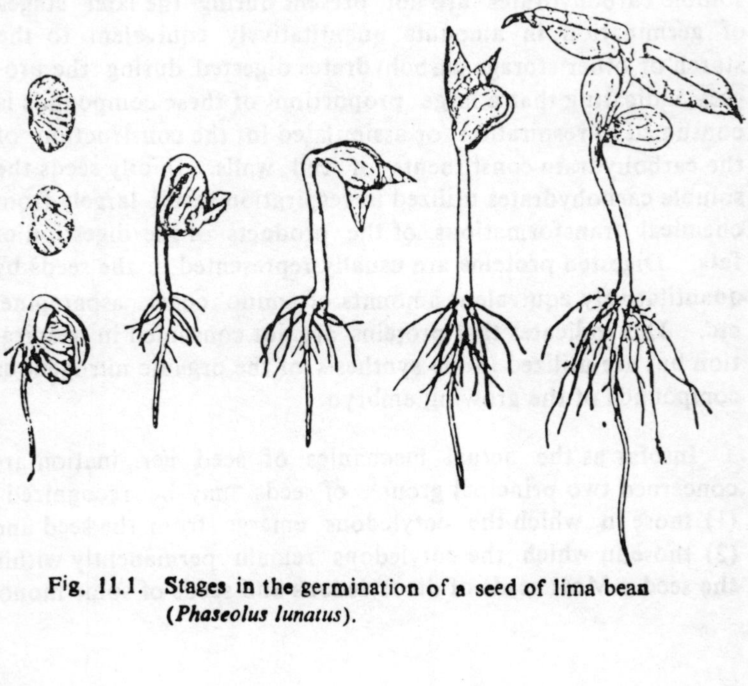

Fig. 11.1. Stages in the germination of a seed of lima bean (*Phaseolus lunatus*).

2. *Seeds in Which the Cotyledons do not Emerge.* The seed of the pea is structurally very similar to that of the bean, but its germination behaviour is very different. Elongation of the hypocotyl does not occur, and the cotyledons remain in the seed. The primary root elongates early in the process of germination much as in the bean. The plumule is elevated through the soil by rapid elongation of the *epicotyl*, which is the stem region between the cotyledons and the first true leaves—in other words, the first internode.

This type of germination is also exhibited by oak acorns (Fig. 11.2).

Many monocotyledons also show this type of germination behaviour. In the germination of the corn grain, for example, the primary root develops from the lower end of the axis of the embryo, growing through the *coleorhiza* and the wall of the grain (pericarp). As the primary root elongates, lateral roots soon appear and root hairs begin to develop just back of the elongating regions on all of the roots. The single cotyledon (*sculellum*) remains within the seed and acts as an absorbing organ through which soluble foods in the endosperm move into the tissues of the rapidly enlarging embryo. Soon after the appearance of the primary root the plumule and the *coleoptile*, which completely encloses it, grow out through the walls of the grain and upward as a result of elongation of the region of the axis just below the plumule. About the time the coleoptile breaks through the surface of the soil, or soon thereafter, the first foliage leaf grows through the tip of the coleoptile and emerges into the light and air.

## Environmental Conditions Necessary for Germination

The seeds of all species of plants require at least three external conditions before germination can occur : (1) water, (2) a suitable temperature, and (3) oxygen. A fourth factor, light, is of importance in the germination of the seeds of some species.

1. *Water.* A low water content is one of the prominent characteristics of the dormant seeds of most plant species, although some seeds become injured and lose the capacity to

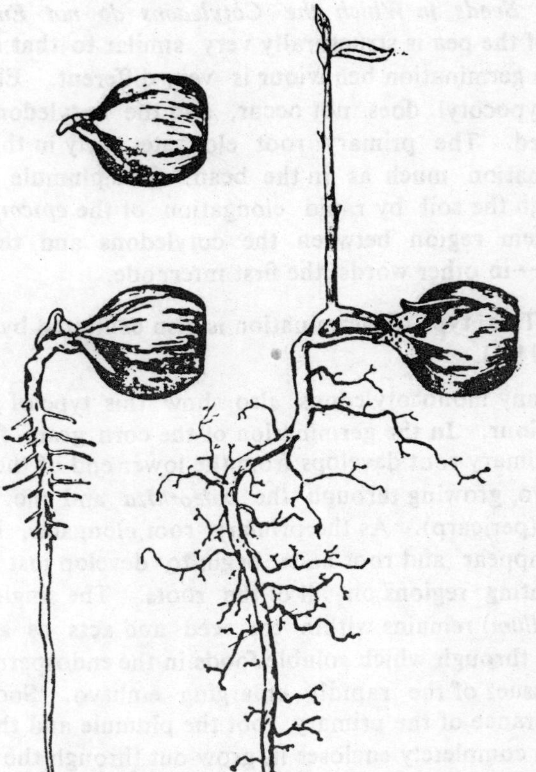

Fig. 11.2. Stages in the germination of a red oak (*Quercus borealis*) acorn. Redrawn from Korstian (1927).

germinate if their water content falls to low levels. The germination of grapefruit seeds, for example, was reduced 30 per cent by drying at room temperature to 52 per cent moisture, although much lower moisture contents did not reduce germination if the temperatures were reduced to 5°C.

The physiological processes of living cells occur largely in an aqueous medium, and germination cannot occur unless the seed can absorb water from its environment. The absorption of water initiates a series of physical and chemical processes which, in the absence of any limiting factor, results in the emergence of the embryo from the seed. Soil-moisture contents need not be high, however, for germination to occur. When soil moisture levels are at or slightly below field capacity, germination of seeds is usually most rapid, but good germination of many species of seeds and subsequent emergence of the young seedlings may occur when the soil moisture level is at the permanent wilting percentage or even slightly lower.

The quantity of water absorbed during germination by seeds of different species or even by varieties of the same species varies within wide limits. Seeds of a drought resistant variety of cotton, for example, have been found to absorb only about one-half as much water during germination as seeds of a variety less resistant to dry weather.

Water-vapor as well as liquid water can be imbibed by seeds. Most seeds, therefore, will pass through the earlier stages of germination in an atmosphere which is saturated, or nearly so, but if the vapour pressure of the atmosphere is appreciably below the saturation value germination will be checked or inhibited.

2. *Oxygen.* The respiration of germinating seeds proceeds at a rapid rate especially during the early stages of germination. The partial pressure of oxygen in the atmosphere can be reduced considerably, however, without greatly interfering with the rate of respiration. In fact, the seeds of some water plants such as cattail germinate better under low oxygen pressures than in air. Seeds of many terrestrial plants can germinate under water where the concentration of oxygen often corresponds to a partial pressure of oxygen very much less than that of the atmosphere. Seeds of many common weeds may lie buried in the soil for years without germinating ; yet when the soil is plowed or otherwise disturbed so that these seeds are brought to the surface germination usually occurs

promptly. The cause for this delay in germination of viable seeds appears to be the low oxygen content or high carbon dioxide content, or both, of the soil in which they are buried. Exposure to the air, which occurs as a result of plowing, so increases the oxygen supply in the environment of the seeds that germination usually takes place within a few days if other conditions are not limiting.

During the early stages of germination of seeds of pea and some other species, respiration is largely or almost entirely of the anaerobic type because of the relative impermeability of even hydrated seed coats of such species to oxygen. As soon as the seed coats are ruptured, however, aerobic respiration replaces anaerobic oxidative processes even in seeds of this type.

3. *Suitable Temperature*. In the absence of other limiting factors the seeds of any species will germinate within a certain range of temperatures, but a temperatures above or below this range no germination will occur. As a rule, the seeds of species indigenous to temperate regions germinate in a lower range of temperatures than seeds of species whose native habitat is in tropical or subtropical regions. Wheat seeds, for example, germinate at temperatures only slightly above 0°C. and at temperatures as high as 35°C., whereas the range of temperatures for germination of seeds of maize lies between a lower value of 5°-10°C. and an upper limit of about 45°C. The optimum temperature is usually about midway between the two extremes of temperature at which germination will occur. It is not possible to designate any exact temperature as the optimum for germination, because this varies with the other prevailing environmental conditions and also with the exact criterion selected as an index of germination. The most favourable temperature for the elongation of the primary root, for example, does not always correspond to the most suitable temperature for the development of the plumule.

4. *Light*. A few species of plants including the strangling fig mistletoe, and some other epiphytes have seeds which fail to germinate unless exposed to the light. Many kinds of seeds germinate better when exposed to the light than when kept in

total darkness. Examples of these light-favoured seeds are those of many grasses, especially species of *Poa*, the evening primrose, and mullein. On the other hand, the germination of the seeds of some species appears to be retarded or even prevented by exposure to light. Seeds of the onion and many other members of the lily family belong in this group. The effect of light upon the germination of seeds is profoundly influenced by other environmental factors. For example, the germination of the seeds of some species of grass is ordinarily influenced by light, but after a period of dry storage this effect disappears. Similarly the ordinarily light sensitive seeds of the pampas grass will germinate readily in complete darkness in an atmosphere of pure oxygen. Freshly harvested lettuce seeds germinate much better in diffuse light than in total darkness. The beneficial effects of light disappear, however, if the seeds are placed in an atmosphere containing 5-20 per cent of $CO_2$ or if the seeds are treated with thiourea.

**Advantages of Dormancy**

In temperate zones, there are large seasonal changes in temperature, ranging from near 100°F in midsummer to well below freezing in midwinter. Obviously, most plants could not survive the cold temperatures of winter in the vegetative or flowering state. Thus, in many plants seed and bud dormancy begins at the onset of winter cold, allowing the plant to pass through the winter with little or no damage. In the grain areas of the United States and Canada, for example, will oat infestation is a serious problem because of the ability of the grains to survive the winter in a dormant state and then germinate the following spring. In contrast, the seeds of many other noxious weeds have only a brief dormant period, germinate in the fall, and are killed off during the severe winters that are common to the northern midwest areas.

The significance of dormancy among plants growing in arid regions is immediately apparent. Certainly, it is advantageous to the plant if germination and growth can be managed during the relatively brief periods of rainfall in these areas. Thus, seeds that can remain viable but dormant until sufficient water is available have a very good chance for survival. An even more bizarre example of the importance of dormancy in the adaptation of a plant to a dry region may be found in a study of the desert shrub quayule. In this plant, the chaff covering the seed contains a germination inhibitor that causes the seed to remain in a dormant state. However, with a rather strong rainfall, sufficient dilution of the inhibitor occurs to allow germination.

While talking about the advantages of dormancy, we should also mention how seed coats impermeable to water help in the persistence of a species. This type of seed coat may be found in some species of *Convolvulus*, growing in arid regions. In order for these seeds to imbibe water and germinate, their seed coats must be mechanically broken. However, permeability to water gradually occurs over a long period of time. The advantage here is that never will all of the seeds germinate at one time, but only a certain number each year. Thus, it is virtually impossible for the entire species to be wiped out during the vulnerable seedling stage due to some adverse environmental condition.

Dormancy in plants is both a convenience and an inconvenience to man. The temporary dormant period experienced by many cereal grains allows for their harvest, dry storage, and ultimate use as food. Otherwise, these grains would germinate in the field and be useless to man. The ability, however, of certain weed seeds to lie dormant for many years in the soil has proved a great inconvenience. During plowing of the soil, dormancy of many of these seeds will be broken, allowing them to compete with any economic crop sown in that area. The eradication or even control of many of these weeds is almost impossible because they can never all be caught in the vulnerable seedling or vegetative state. Although some are triggered to germinate by soil disturbances caused by plowing, there are always some that remain dormant in the soil. Therefore, each year the farmer is presented with the same problem, the germination of some but not all of these weed seeds. He can only destroy those that germinate and has almost no control of these lying dormant in the soil.

*Dormancy of Seeds.* Many kinds of seeds, apparently ripe, fail to germinate even if placed under such conditions that all environmental factors are favourable. In such seeds resumption of growth by the embryo is arrested by conditions within the seeds themselves. The state of inhibited growth of seeds or other plant organs as a result of internal causes is usually called *dormancy* but is sometimes referred to as the "rest period."

Failure of seeds to sprout does not necessarily mean that they are dormant. Environmental conditions may be unfavourable for germination. The water supply may be inadequate, or the temperature may be unfavourable. Deeply buried seeds are often prevented from germinating by an inadequate oxygen supply, or certain kinds of light-sensitive seeds may fail to germinate because of unfavourable light conditions. The term "dormancy" as applied to seeds is generally restricted to those which fail to germinate as a result of internal causes. For convenience we shall use the term *quiescence* as a designation for the situation in which failure of a plant organ to grow is a result of environmental conditions. In practice, it is often difficult to determine whether seeds or other plant organs under natural conditions are actually dormant or merely quiescent without resorting to an experimental test.

Dormancy of seeds results from one or a combination of several different factors :

1. *Seed Coats Impermeable to Water.* The seed coats of many species are completely impermeable to water (and probably also to oxygen) at the time the seeds are ripe. This condition is very common in the seeds of many legumes (clovers, alfalfa, black locust, honey locust, etc.), of the water lotus, and of the morning glory. Germination fails to occur until water penetrates through the seed coats. In many such seeds the permeability of the coats to water increases slowly in dry storage, but it occurs more rapidly when they are exposed to the fluctuations of temperature and moisture that are present in soils under natural conditions. The action of bacteria and fungi also increases the permeability of the seed

coats to water and so shortens the dormant period of seeds of this kind that are buried in the surface layers of soil.

It is probable that the cause of the impermeability of seeds to water will be traced to some feature of the structure or the chemical composition of the walls of the cells in the seed coats. Comparative studies, however, of the cellular anatomy of seed coats of legume species with impermeable seed coats and of legume species with seeds that are permeable to water have failed to reveal the presence of any structural characteristics associated only with the impermeable condition.

2. *Mechanically Resistant Seed Coats.* The seeds of some of the commonest weeds, such as mustard (Brassica), pigweed (*Amaranthus*), water plantain (*Alisma*), shepherd's purse (*Capsella*), and peppergrass (*Lepidium*), remain in the dormant condition because the seed coats are strong enough to prevent any appreciable expansion of the embryo. In the seeds of the pigweed (*Amaranthus retroflexus*), for example, water and oxygen penetrate through the seed coats readily, but the enlargement of the embryo is limited by the mechanical strength of the seed coats. As long as the seed coats remain saturated with water, dormancy persists, and may last for a period of 30 years or even longer. If the seed coats become dry, however, certain changes occur in the colloidal compounds that make up the walls of the cells of the seed coats so that, upon being again saturated with water, they are no longer able to resist the pressures developed by the imbibitional forces in the embryo. The coats are ruptured and germination occurs. High temperatures (above 40°C.) may also induce some germination of pigweed seeds at the time of ripening because the seed coats are less resistant at these temperatures. As the seeds age, the minimum temperature for germination becomes lower. The embryos of these seeds have no dormant periods and will grow readily if the seed coats are removed. Likewise any treatment which weakens the seed coats increases the percentage of seeds that germinate. In general, other seeds with mechanically resistant seed coats exhibit similar behaviour.

3. *Seed Coats Impermeable to Oxygen.* The two seeds in a fruit of cocklebur (*Xanthium*) are not equally dormant. Under natural conditions the lower seed usually germinates in the spring following maturity while the upper seed remains dormant until the next year. The dormancy of these seeds has been demonstrated to result from the impermeability of the seed coats to oxygen. If the seed coats are ruptured, or if the oxygen pressure is increased around infact seeds, germination occurs. The oxygen requirements for germination are greater in the upper seed than in the lower, and this explains the more pronounced dormancy of the former. During dry storage or under natural conditions the seed coats gradually become more permeable to oxygen, and the oxygen requirements of the embryo decrease. Hence the intensity of the dormant condition gradually diminishes. The seeds of a number of grasses and of many Compositae also have dormant periods that seem to result from the impermeability of the seed coats to oxygen.

4. *Rudimentary Embryos.* Many species of plants have seeds in which the embryo does not develop as rapidly as the surrounding tissues, so that when the seeds are shed the embryos are still imperfectly developed. In some species the ripened seeds contain embryos that have grown little beyond the fertilized egg stage, whereas in other species development of the embryos may be nearly complete when the seed are shed. The germination of such seeds is necessarily delayed until formation of the embryo is complete. Examples of species in which dormancy of seeds is a consequence of incompletely developed embryos include ginkgo (*Ginkgo biloba*), European ash (*Fraxinus excelsior*), holly (*Ilex opaca*), and many orchids.

5. *Dormant Embryos.* In many species, although the embryos are completely developed when the seed is ripe, the seeds fail to germinate even when environmental conditions are favorable. Dormancy of such seeds is a result of the physiological condition of the embryo. The embryos of such seeds will not grow when the seeds first ripen even if the seed coats are removed. Among the many species whose seeds

exhibit dormancy of this type are apple, peach, hawthorne, iris, lily-of-the-valley, basswood, ashes, tulip poplar, dogwood, hemlock, and pines. Germination of such seeds occurs only after a period of "after-ripening." In many wild species after-ripening occurs during the winter while the seeds lie on the ground or just under the soil surface. Such seeds will not germinate in the fall just after they are shed, but will germinate the following spring if environmental conditions are favourable. In some species after-ripening occurs over a period of years, some germination occurring each year. After-ripening involves principally a series of changes in the physiological condition of the embryo which gradually converts a dormant embryo into one that can resume growth. The nature of these physiological changes is not clearly understood at the present time. In some species after-ripening also involves changes in the properties of the seed coats. The length of time required for completion of the after-ripening process can be greatly modified by environmental conditions as subsequent discussion will show.

6. *Germination Inhibitors.* Germination of seeds is sometimes checked or prevented by the presence of compounds known as inhibitors. These are often substances produced in one or more plant organs. The juice of tomato fruits, for example, inhibits the germination of tomato seeds and many other seeds as well. Even when tomato juice is diluted in a proportion of 1 : 25 it completely inhibits the germination of seeds of *Lepidium sativum.* Germination inhibitors are known to be present, within the seeds of some species such as those of iris and of at least some varieties of cabbage. It is probable that such germination inhibitors are present in the seeds of many other species and that they represent a widespread mechanism of dormancy. Certain compounds having the structure of unsaturated lactones appear to be widely distributed in plants and to possess the property of inhibiting seed germination and other growth phenomena. Among the compounds in this category are coumarin and parasorbic acid. There are indications that at least some of the naturally occurring inhibitors are compounds of this type.

*Secondary Dormancy.* Some seeds which are capable of germinating as soon as they are harvested lose this capacity after being kept in an unfavourable environment for a while. This induced dormant period is known as *secondary dormancy.* Usually secondary dormancy can develop only when at least one of the conditions essential for germination is unfavourable. For example, if seeds of white mustard (*Brassica alba*) are exposed to high concentrations of carbon dioxide they fail to germinate, even under favourable conditions, for a long period after the removal of the carbon dioxide. Light sensitive seeds may pass into secondary dormancy if kept in the dark, and seeds that germinate only in the dark may become dormant if exposed to light. Likewise, secondary dormancy may be induced in some kinds of seeds by exposures to low temperatures and in others by high temperatures.

Secondary dormancy is often caused by changes in the seed coat since, in some species, the embryos are able to grow immediately if the seed coats are removed. In other kinds of seeds, however, the dormancy is produced by physiological changes that occur within the embryo itself. Secondary dormancy, like primary dormancy, may be interrupted by various treatments.

*Special Types of Dormancy.* The failure of seedlings to develop from seeds is not always traceable to dormancy of the seed itself. In many of the spring wild flowers, for example, germination of the seed may occur, but growth is restricted to the establishment of the young root system, the epicotyl failing to grow. In some species the epicotyls may push through the seed coat but remain dormant after this has been accomplished. Epicotyl dormancy is usually broken by exposure to low temperatures (1°-10°C.).

Young root systems also may exhibit dormancy after emerging from the seed and becoming established in the soil. In some species to *Trillium*, for example, there is present a double dormancy—that of the seed and that of the young root after emerging from the seed, Root dormancy, like dormancy of the epicotyl, is usually broken by exposure to low temperatures.

*Methods of Breaking the Dormancy of Seeds.* The dormancy of seeds presents a practical problem of considerable economic importance. Plant growers are often interested in securing seed that will germinate soon after it is harvested. Ordinarily this would be possible only with seeds that have a short dormant period or none at all. Methods have been devised however, whereby the dormancy of many kinds of seeds can be broken, and whereby the length of the dormant period in many other kinds can be shortened. The methods employed for the breaking of dormancy vary depending upon its cause. Methods which can be used for breaking the dormancy of one species may be totally ineffective when used with seeds of another species, and sometimes may even prolong dormancy.

1. *Scarification.* Whenever dormancy is a consequence of any of the causes inherent in the seed coats it can be interrupted by scarification. We shall use this term to apply to any treatment—mechanical or otherwise—which results in rupturing or weakening the seed coats sufficiently to permit germination. For example, machine-threshed legume seeds usually show a higher percentage of germination than those that have been harvested by hand. The mechanical treatment is sufficiently severe to scratch or crack many of the seed coats, and this permits ready ingress of water. Various types of mechanical treatments have been devised for breaking the dormancy of seeds of this kind. Strong mineral acids have likewise been used successfully to interrupt seed dormancy caused by resistant or impermeable seed coats. It is essential, however, that any method used to interrupt seed-coat dormancy should not be injurious to the embryo. Under natural conditions dormancy of such seeds is broken by the slow decay of the seed coats or by the action of alternate freezing and thawing.

2. *Low Temperatures.* After-ripening of many seeds occurs more rapidly when they are stratified in moist peat at low temperatures than when stored at higher temperatures. Temperatures between 5° and 10°C for 2 or 3 months are

effective with conifer seeds and greatly increase the percentage of germination. Similarly, low temperatures combined with moisture have been found to reduce the period of after-ripening in seeds of mountain ash, basswood, elder, bayberry, and many other species. The effectiveness of low temperatures in breaking dormancy appears to be associated in some species at least with a favourable relation between respiration rates and the rate of oxygen absorption or carbon dioxide liberation. Permeability changes in the seed coats may also be an important factor.

3. *Alternating Temperatures.* In some seed testing laboratories it is common practice to subject seeds alternately to relatively low and high temperatures. The temperature extreme of such treatments may not differ by more than 10° or 20°C., and both are commonly well above the freezing point. The germination of seeds of Kentucky blue grass (*Poapratensis*), for example, is greatly improved by subjecting the seeds alternately to temperatures of 20°C for 16-18 hr. and 30°C. for 6-8 hr. ; and the percentage germination of Johnson grass (*Holcus halepensis*) seeds is increased by alternate treatments at 30°C. for 18-22 hr. and 45°C. for 2-8 hr. The dormancy of some seeds may be interrupted by alternate freezing and thawing, although this is decidedly harmful to other species. The action of the alternating temperatures upon the seeds is not understood. It is entirely ineffective with seeds of some species. Seeds of carrot and timothy, for example, germinate just as well at constant temperatures as when temperatures are varied. In general, this method of treatment is used principally with seeds in which dormancy is inherent in the embryo.

4. *Light.* In the previous discussion light was mentioned as one of the conditions essential for the germination of certain species of seeds. Light may be considered, therefore, as a means of breaking dormancy of such species. In some of the species other environmental factors can be substituted for light. In seeds of *Veronica longifolia* (one of the commonly cultivated speed wells), for example, light improves germination at low temperatures, but the seeds germinate equally

well in total darkness at high temperatures. In seeds of Kentucky blue grass exposure to light is effective in improving germination at both variable and constant temperatures.

5. *Pressures.* Seeds of sweed clover (*Melilotus alba*) and alfalfa (*Medicago sativa*) showed greatly improved germination after being subjected to hydraulic pressures of 2000 atm. at 18°C. When the pressure was applied for periods of from 5-20 min, the germination of the seeds was increased by 50-200 per cent. The effect of the pressures persists after the seeds have been dried and stored and undoubtedly results from changes in the permeability of the seed coats to water.

6. *Growth Regulators.* Growth regulators are widely used to haster the development of roots on cuttings and to increase the number of roots produced. The success obtained by these treatments has suggested the possibility of interrupting the dormancy of seeds or improving their germination by similar means. Considerable study has been devoted to this problem but with very few exceptions neither dormant nor quiescent seeds appear to react the treatments with growth regulators.

*Longevity of Seeds.* The life-span of seeds varies from a few weeks to many years, depending upon the species and the environmental conditions to which the seeds are subjected. The silver maple (*Acer saccharinum*) may be cited as an example of a species which has short-lived seeds. When the seeds of this species are shed in June their water content is about 58 per cent. Once their moisture content drops below 30 to 34 per cent, the seeds die. Since this ofter happens within a few weeks in nature, seeds of this species soon perish. The seeds of the majority of crop plants are relatively short-lived under the usual storage conditions, generally remaining viable for only one to three years. The life-span of such seeds can often be increased several fold by keeping them under suitable storage conditions.

At the other extreme there are a few authentic recods of seeds which have lived for more than a hundred years. Bequerel (1934) succeeded in germinating in 1934 seeds of *Cassia bicapsularis* which had been collected in 1819, and seeds of *Cassia multijuga* which had been collected in 1776. These are both South American species of legumes. Viable seeds of the Indian lotus (*Nelumbo nucifera*) have been found buried under layers of peat and soil in Manchuria of such depth that they must have been at least 120 years old and may have been 200 to 400 years old. With this one exception all the authentic records of seeds living for 75 years or longer are of legumes. Further data on the longevity of seeds are given by Crocker (1938).

At least some of the seeds of a number of species of wild plants will remain viable for 50 years or more. This is especially true of hard-coated species. As a general rule only seeds with a pronounced dormancy remain viable very many years in nature. The seeds of many weed species are notoriously long-lived as compared with the seeds of most crop plants. This is illustrated by an experiment initiated by Beal at East Lansing, Michigan, in 1879. Seeds of twenty herbaceous species were mixed with sand and buried in pint bottles. Twenty such bottles were prepared. Once every five or ten years one of the bottles was excavated and the enclosed seeds tested for their percentage of germination. Fifty years later seeds of five species remained alive and showed the following percentages of germination : yellow dock (*Runmex crispus*), 52 per cent ; evening primrose (*Oenethera biennis*), 38 per cent ; moth mullein (*Verbascum blattaria*), 62 per cent ; black mustard (*Brassica nigra*), 8 per cent ; and water smartweed (*Polygonum hydropiper*), 4 per cent. The 70-years period for this experiment is now past and the results in 1950 showed that seeds of three species (*Oenethera biennis*, *Rumex crispus*, and *Verbascum blattaria*) were still viable. The germination percentage of the *Verbascum* seeds was still high, 72 per cent, while the percentages for the seeds of *Rumex* and *Oenethera* were 8 and 14, respectively.

Similar results were obtained from the buried seed experiment of Duvel which was terminated in 1945. In this experiment seeds of sixteen species were still capable of germination and growth after about 40 years.

*Methods of Detecting Seed Viability.* It is often important to know whether seeds are merely dormant and capable of germination under suitable conditions or whether they are dead. Since dormancy may be prolonged and since the causes of dormancy are sometimes difficult to ascertain, the existence of some reliable method of measuring the viability of seeds is highly desirable. Three of the many methods proposed have received considerable experimental study : (1) the behaviour of excised embryos, (2) the use of certain dyes, and (3) the activity of specific enzyme systems. Of these the first two seem more reliable indices of viability than the last named. Enzyme activity is not closely correlated with seed viability in all species and cannot be used successfully as a general method to distinguished between living and nonliving seeds.

The excised embryo test is based upon the behaviour of embryos when removed from the surrounding tissues of the seed. If carefully separated from the seed and placed upon moist filter paper at room temperature, embryos of viable seeds will show various types of growth behaviour, whereas nonviable embryos deteriorate rapidly. This method of determining the germinating capacity of seeds seems widely applicable and has a high reliability.

Methods involving dyes are of two kinds. One depends upon the impermeability of living cells to certain stains which readily enter and stain dead tissues. The second method depends upon the reduction by living cells of certain colourless compounds which are strongly coloured in their reduced states. With this method the living cells become strongly coloured and the nonliving tissues remain unstained. This method has in general proved to be more satisfactory than the use of dyes which enter and stain only the dead tissue. The two compounds most widely used are 2,3-diphenyl-5-methyl-tetrazolium chloride and 2, 3, 5-triphenyl-tetrazolium chloride. The latter is commonly preferred to the former. Dilute solutions of these salts (0.05-2.0 per cent) are used as solutions in which sections of the seeds are soaked for several hours. The embryos of living seeds become stained a deep red or orange while dead embryos remain unstained. The tetrazolium methods are highly successful with many species of seeds, especially so with seeds of the grasses (maize, wheat, etc.), but must be used with caution with some seed species.

**Methods of Prolonging Seed Viability**

Seeds of many species rapidly loose their capacity to germinate if stored in an unfavourable environment. The critical factors in determining the viable period of most seeds are moisture and temperature, although the oxygen and carbon dioxide content of the surrounding atmosphere are also of importance. In general, seeds remain viable for longer periods of time at low temperatures (5°C.) than at room temperature, and low relative humidities are usually more favourable for prolonging the life of seeds than high ones. Neither factor can be considered independently of the other, however, since the effect of one commonly depends upon the magnitude of the other. Marked fluctuations of temperature or relative humidity in storage rooms are not conducive to prolonging the viability of seeds. Similarly the packaging of vegetable or flower seeds in paper envelopes when displayed for sale frequently exposes the seeds to variations in temperature and humidity which shorten significantly the period in which good germination can be obtained.

Longevity of many seeds seems correlated with respiration rates of the embryo, and it is for this reason that low temperatures and low seed moisture content are usually essential if viability is to be maintained for long periods of time. There are exceptions to this general rule, citrus seeds and coffee seeds, for example, remaining alive longest in a humid atmosphere. It is nevertheless generally true that environmental conditions which reduce the rate of respiration in seeds are also conditions which are most favourable for seed storage.

In addition to the environmental factors just discussed a number of internal factors are known to influence the longevity of seeds in storage. The stage of maturity at the time the seed is harvested, the moisture content of the seed when placed in storage, and genetic factors all play a role, sometimes a decisive role, in determining the viability of the seeds. These factors vary so widely in different species of seeds that any general statement of their relative importance would be difficult or impossible.

**Dormancy of Buds**

Lateral and terminal buds ordinarily develop on the newly elongated shoots of temperate zone woody plants during the spring and early summer months. Such buds are not commonly in the dormant state during this period. Defoliation of the tree or shrub from any cause, for example, often results in the development of new shoots from buds formed during the current season. By the time of, or somewhat prior to, leaf-fall, however, such buds have passed into a state of dormancy. In nature dormancy of buds borne on woody stems is broken by exposure to relatively low winter temperatures. The length of time for which such buds retain their dormancy, however, differs considerably according to species.

Howard (1910) made a comprehensive study of bud dormancy in a large number of species of woody plants, some of American, some of European, and some of Asiatic, origin. This investigation was conducted in Missouri. Of 234 species collected, buds of 125 species developed when branches were brought into a greenhouse between October 28 and November 4. Most of these were European or Asiatic species. In other words, the buds of more than half of the species experimented with were no longer in a dormant state on this date. A second collection of branches from 283 species was brought into the greenhouse on January 8-10. Of these, buds developed on 244 species. On February 23 a third collection of 63 species was made composed largely of kinds on which the buds failed to develop in the preceding two tests. On this date buds developed on all but five of the species collected. The results of this study indicate that the buds of some woody species retain their dormancy much longer into the fall or winter than others and that the buds of many such species are in a quiescent rather than a dormant state during much of the winter.

Dormancy of buds is sometimes a result of correlative effects, many of which probably have a hormonal mechanism. A familiar example is the phenomenon of apical dominance in which lateral buds remain dormant as long as the terminal growing region remains intact, but usually resume growth upon its injury or removal.

Likewise the buds of many kinds of tubers, rhizomes, corms, and bulbs often remain dormant for periods of greater or less duration while environmental conditions are favourable for their development.

**Methods of Breaking of Dormancy of Buds**

As the previous discussion has shown the buds on many species of woody plants remain dormant through the autumn and at least the forepart of the winter. This dormancy is not caused by low temperatures, because it is also exhibited by such species when kept in a greenhouse where the temperature is continuously maintained at levels typical of midsummer weather. The dormant condition often persists as long as the temperatures are high but may be broken by subjecting the plants to temperatures near the freezing point. The buds of some species such as peach will begin to grow, if environmental conditions are favourable, after only a few days of chilling, but the buds of other kinds of plants such as the blueberry fail to grow well unless exposed to temperatures near the freezing point for several weeks. Coville demonstrated that the effect of the low temperatures is restricted to the tissues exposed. When single branches are chilled while the rest of the plant is kept warm, only the buds on the chilled branch are able to grow, all of the others remaining dormant.

Early in the twentieth century Johannsen discovered that dormant buds of many kinds of plants will begin to grow after being exposed to vapours of ether or chloroform for a day or two. The interval of time between the ether treatment and the beginning of growth differs widely with the time of the year at which the vapour is applied. In late summer or early fall, several weeks may elapse between the exposure to ether vapour and the active growth of the buds. Late in the winter, however, the interval is shortened to a day or two.

More recently a number of other chemical treatments have been discovered that are successful in breaking the dormancy of buds. Denny (1926) found that dormant potato tubers sprouted freely when exposed to ethylene chlorhydrin vapour. Sodium and potassium thiocyanate, thiourea, dichloroethylene, carbon bisulfide, xylol, ethyl bromide, and a number of other compounds were also effective. Vapours of ethylene chlorhydrin so hastened the growth of tubers of the Irish cobbler variety that vines 2 feet high bearing young tubers 1 cm. in diameter grew from the treated tubers before the sprouts of the untreated tubers appeared above the surface of the ground. Solutions of sodium and potassium thiocyanate gave almost equally striking results. Thiourea solutions differed somewhat from the other compounds used in that they overcame the inhibiting effect of the terminal bud upon the growth of lateral buds and caused the development of several shoots from each eye on the tuber.

Combinations of two or more compounds, each effective in overcoming the dormancy of buds in potato tubers, will give better results than any single treatment. A mixture of ethylene chlorhydrin, ethylene, and carbon tetrachloride was more effective in breaking the dormancy of potato tubers than the sum of the effects of each compound used separately. These so-called "synergistic effects" are attributed to the increased penetration of the compounds as a result of changes in cellular permeability induced by one component of the mixture.

The vapours of ethylene chlorhydrin and ethylene dichloride also induce the growth of dormant buds of lilac, flowering almond, and some other species of woody plants. The effect of the vapors was so restricted that when one of two paired buds of a lilac was exposed to the vapour, growth occurred only in the treated bud. The untreated bud remained as fully dormant as other buds more distant from the treated area. Apparently the factors causing the dormancy of the buds of these plants reside within the buds themselves and not in the adjacent tissues.

The dormancy of the buds of fruit trees is usually broken in nature by the low temperatures of the winter months. It is not uncommon, however, for entire winters to pass in southern. United States without sufficient cold weather to interrupt the dormancy of buds of fruit trees. When this happens flowering is irregular and uncertain and yields of fruit suffer correspondingly. The discovery of some method of breaking the dormancy of fruit trees is a matter of great practical importance, and the problem has been given much study. The dormancy of the buds of some fruit and nut trees can be broken by the use of certain chemical compounds sprayed upon the trees during the dormant period, but results are not equally successful with different species. Methods of treatment have not been sufficiently well developed as yet to warrant their extensive use in commercial orchards.

### Methods of Prolonging Dormancy of Buds

Often it becomes important to prolong, rather than to shorten, the dormancy of buds. The warm weather of an early spring may cause the buds to open before danger from frost is past. Under such conditions a severe frost can cause enormous damage. The prolongation of dormancy of buds of nursery stock held in storage or prepared for shipment is also very desirable. Dormancy of buds may be prolonged by spraying with certain chemical compounds—some of the growth regulators being particularly effective in this respect. The results of experimental studies of orchard grown fruit trees have not always been consistent, but greater success has been obtained in the treatment of nursery stock.

Probably the most complete and satisfactory control of dormancy through the use of chemical agents is that obtained over the buds of potato tubers. The sprouting of potato tubers in storage may be prevented almost completely by treatment with certain growth regulators. The methyl ester of naphthalene acetic acid is now widely used for this purpose. When shredded paper impregnated with small amounts of this ester is seattered among potato tubers in a storage bin, sprouting of the buds is inhibited. The tubers remain completely dormant and are firm and sound after overwinter storage. The dormancy induced in this way may be interrupted at any time by the treatments described earlier in this chapter.

**Growth-Inhibiting Substances**

In a series of studies on potato bud dormancy, Hemberg (1947 ; 1949 ; 1950) demonstrated that substances extracted from potato peels of dormant tubers are capable of counteracting the effect of IAA in the *Avena* curvature test. The group of inhibitors extracted by Hemberg were made up of acid and neutral substances. The acid inhibitors could not be detected at termination of dormancy. Paper chromatographic analysis showed that the acid inhibitor present in potato peels of dormant potatoes is *inhibitor*-β, a complex of organic compounds first identified by Bennet-Clark and Kefford (1953) from paper chromatograms of plant extracts. It is interesting to note that inhibitor-β has recently been extracted from dormant buds of the silver maple.

A correlation has been shown between increase and decrease of inhibitor-β with respect to onset and release of potato bud dormancy, and inhibitor-β has been extracted from the dormant buds of at least one woody species. In at least one study it was demonstrated that abscisin II, (or dormin), at very low dosages, is capable of almost completely inhibiting the sprouting potato buds. Perhaps the inhibiting component of Hemberg's inhibitor-β is abscisin II.

**Compounds breaking bud dormancy**

The regulation of dormancy release has academic as well as practical importance. Controlling the time of release of dormancy by artificial manipulation of the environment, or by the use of active compounds, many times will reveal some of the mechanisms involved in dormancy, thus contributing to our understanding of the overall process. The release of dormancy by some artificial means often aids the farmer in an economic way. For example, the release of potato bud dormancy of newly harvested tubers will allow some growers to produce a second crop. This, of course, would depend on the length of the growing season. Some of the chemicals found useful in the release of dormancy are ethylenechlorohydrin, thiourea, and gibberellin.

*Ethylenechlorohydrin* ($ClCH_2CH_2OH$). A thorough study of the dormancy-breaking abilities of many different compounds by Denny (1926) brought into prominence one especially active compound, ethylene-chlorohydrin. This compound proved highly effective in inducing sprouting in dormant potato tubers with the added attraction of having a wide safety margin between active and toxic dosages. In addition, ethylenechlorohydrin proved to be very successful in breaking bud dormancy of fruit trees when applied in vapour form.

Several metabolic changes may take place as a result of ethylenechlorohydrin application. A study of its effect on dormant potato tubers showed that there is a rise in respiration, catalase, and peroxidase activity, sucrose, and glutathione. There is a drop in $H^+$ ion and citric acid concentrations. Most likely, citric acid is being utilized as a substrate in respiration, and its decrease in content decreases the $H^+$ ion concentration.

*Thiourea* ($NH_2CSNH_2$). Although not as effective as ethylenechlorohydrin, thiourea has proved effective in forcing sprouting in dormant potato tubers. Thiourea has an unusual effect in that it may cause the growth of several bud primordia in one "eye"., As many as eight sprouts have been observed growing from one eye. In contrast, ethylenechlorohydrin causes the growth of only one sprout in each eye. It is interesting to note that reduced oxygen pressures have an effect somewhat similar to thiourea, causing multiple sprouting.

*Gibberellin.* A special case can be made for gibberellin acting as a bud dormancy breaker. In contrast to ethylenechlorohydrin and thiourea, it is a natural compound and may be involved as a controlling factor in the overall process of bud dormancy. In earlier discussions, we learned that gibberellin has a profound influence on dormant shoots and seeds, promoting the growth of both when applied. One would logically assume that gibberellin would also be capable of breaking bud dormancy. This was eventually demonstrated with excellent success on dormant potato tubers and on dormant peach buds. Generally in those plants requiring a low temperature period to break dormancy, gibberellin will substitute for the cold treatment and force the release of dormancy.

Much work has been done on the effect of gibberellin on potato bud dormancy. Gibberellin can induce sprouting of potato tubers still on the plant and in harvested tubers at any time in their dormant period. Therefore, the induction of sprouting by gibberellin may be accomplished anywhere from the beginning of tuber enlargement to the end of the dormant period. The remarkable ability of gibberellin to induce sprouting when applied as a spray to potato plants 4, 2, and 1 week before harvest can be seen in Table 11-1.

As mentioned before, endogenous gibberellins may play a major role in the control of dormancy. This assumption has been given major support in several studies. For example, in 1959 it was reported that gibberellin-like substances are present in potato tubers and that higher levels can be detected in sprouting tubers than in newly harvested tubers. Subsequently Smith and Rappaport (1961) reported that the concentration of endogenous gibberellin remains low during the dormant period, but rises thirty fold after sprouting begins.

**Table 11-1.** Percentage of sprouted tubers at harvest from plants that received preharvest foliar sprays of gibberellin 4, 2, and 1 week before harvest.*

| *Gibberellin, mg/liter* | *Tubers sprouted at harvest, %* | | |
|---|---|---|---|
| | *4 weeks* | *2 weeks* | *1 week* |
| 0 | 0·0 | 1.4 | 0.2 |
| 10 | 3.0 | 1.5 | 1.5 |
| 50 | 58.3 | 18.0 | 0.4 |
| 100 | 75.6 | 34.3 | 2.1 |
| 500 | 83.6 | 50.0 | 5.8 |

*After L.F. Lippert *et. al*. 1958, Plant Physiol, 33 : 132.

**Gene derepression**

There is evidence that the primary act in the breaking of potato bud dormancy may involve the derepression of genes. The genome of the dormant potato bud is almost completely repressed ; that is, the dormant potato bud lacks the ability to synthesize RNA in vivo, and chromatin isolated from it cannot support DNA-dependent RNA synthesis even though incubated with a reaction mixture containing all the necessary components for this synthesis. On the other hand, nondormant potato buds support an active DNA-dependent RNA synthesis.

Tuan and Bonner (1964) found that ethylenechlorohydrin treatment of dormant potato buds causes a rapid synthesis of RNA in the buds (Table 11.2). They suggested that the mechanism of action of ethylenechlorohydrin in the breaking of bud dormancy may be the *derepression of repressed genes.* The new RNA and protein synthesis resulting from this derepression would then stimulate growth of the bud.

We learned that the addition of GA to certain dormant plant tissues can stimulate the synthesis of new RNA and protein synthesis. Thus, GA may also be able to derepress repressed genes. Since ethylenechlorohydrin mimics the action of GA on potato bud dormancy, it is entirely possible that GA—like ethylenechlorohydrin—breaks the dormancy of these buds by releasing repressed genes.

**Table 11-2.** Effect of ethylenechlorohydrin on growth and RNA synthesis of dormant potato buds.*

| *Parameter* | *Days after ethylenechlorohydrin treatment* | | | |
|---|---|---|---|---|
| | *0* | *2* | *6* | *10* |
| fresh weight of buds (mg) | 0.40 | 0.48 | 12.0 | 65.6 |
| RNA content ($\mu$g/bud) | 4.0 | 5.6 | 15.2 | 64.2 |
| Rate of RNA synthesis (m$\mu$g/2.5 hr/bud) | 0.03 | 0.08 | 0.86 | 2.5 |
| Rate of RNA synthesis per unit bud DNA (m$\mu$g RNA synthesized/$\mu$g DNA/2.5 hr) | 0.020 | 0.048 | 0.37 | 0.23 |

*Data of D. Tuan and J. Bonner. 1964. Plant Physiol 39 : 768. Taken from J. Bonner, 1965. Plant Biochemistry p. 859, Academic Press, New York.

# 12

# Role of Potassium in Plants

In plants the potassium is one of the macronutrients present in quantities larger than those of any other mineral nutrient derived from soil with the exception of hydrogen and nitrogen. Some plant tissues accumulate much potassium. Tobacco leaves, for example, may contain as much as 8% potassium on the dry-weight basis and may show symptoms of potassium deficiency if the content falls much below 3%. Potassium is a relatively abundant and widely distributed constituent of the surface rocks of the earth, making up an estimated 2.6% of the earth's crust by weight. Russell (1954) pointed out that, although potassium is removed from soil continually, most soils continue to supply potassium for a long time. Consequently, relatively few soils are so low in potassium that cropping is entirely dependent on fertilization. In contrast, there are large areas of soils on which no worthwhile crop can be grown without phosphorus fertilization and others on which only low yields are obtained without nitrogen fertilization.

## Potassium Content in Soils

The potassium content in mineral soils is usually much greater than that of nitrogen of phosphorus. According to estimates by *Lipman* and *Conybeare* (1936), the average total potassium content of the plowed layer of the crop land is

0.83%, which is 5.8 times as great as the nitrogen content and 13.4 times as great as the phosphorus content.

The potassium content of soils of the various regions is related to parent material and the degree of weathering. Soils of the Coastal Plain area, for example, have developed on transported materials that were strongly weathered and low in potassium at the time of deposition. The low potassium content of soils of the Pacific Northwest, on the other hand, may be attributed primarily to weatherning in place and not to an initial dearth of potassium in the parent material.

With increasing depth in the soil, the potassium percentage usually remains about the same or increases in soils of initially uniform parent material. The deficit in the upper part of the profile may be attributed primarily to a combination of chemical weathering and leaching. Where strong weathering has reduced the potassium content of the entire profile to a low level, however, the surface soil sometimes has a higher potassium percentage than the subsoil. The action of plants in transporting potassium to the surface probably is responsible.

The potassium percentage varies with particle size in a less consistent manner than do nitrogen and phosphorus percentages. Failyer *et al.* (1908) analyzed separates of 27 soils from different parts of the United States and found that the potassium percentage in the clay fraction exceeded that of the silt fraction in 22 soils and that the potassium percentage in the silt fraction exceeded that in the sand fraction in 24 soils. The tendency, therefore, was for an increase in potassium percentage with a decrease in particle size. Salmon (1964) found from analysis of particle-size fractions of 30 samples of soil from Rhodesia that the highest potassium percentage occurred in the sand fraction (2 to 0.02 mm diameter) of 17 samples and decreased with decreasing particle size. The highest potassium percentage was in the silt fraction (0.02 to 0.002 mm diameter) in 11 samples and in the coarse clay fraction (0.002 to 0.0005 mm diameter) in 2. Thus Salmon's (1964) results differ from and are less consistent than those obtained by Failyer *et al.* (1908). An explanation for the various types of behaviour is found in the potassium-bearing minerals present. The discussion of these minerals and some of their properties in the section on mineralogical forms of potassium in soils will provide some basis for understanding why the potassium percentage varies in such an inconsistent manner with particle size.

## PHOSPHORUS CYCLE IN SOIL

### Crop Removal

Barber and Humbert (1963) published a table of values of potassium removal from soils by the principal world crops. Their figures are for relatively high yields and range from 45 kg/hectare in spinach, coffee, and tea up to 520 kg/hectare in sugarcane. Lipman and Conybeare (1936) estimated that in 1930 the average removal of potassium from the soils of the harvested crop land of the United States amounted to 19.5 kg/hectare. This figure refers only to the potassium in the harvested portion of the crops. The total removal of potassium was probably of the order of 35 to 45 kg/hectare. Forty kilograms per hectare is equivalent to about 0.21% of the average content of potassium in the plowed layer of soil. The total removal of potassium is much higher than removal of phosphorus and about equal to the removal of nitrogen. As a percentage of the total amount present in the soil, however, potassium is removed less rapidly than either phosphorus or nitrogen.

### Leaching

The losses of potassium, nitrogen, and phosphorus by leaching from a fallow, sandy soil has been studied. In this the losses of potassium in both absolute amount and in percentage of the additions in the fertilizer were below the corresponding figures for nitrogen and above those for phosphorus, which is the usual behaviour. Nevertheless, one must recognize that considerable variation in absolute and relative losses of the different nutrients occurs under different conditions. In the experiment the soil was sandy and low in fertility, no crop was present, and the loss of water by drainage was large. All these conditions are of significance in interpreting the results, and their relevance will be discussed in the following paragraphs.

With regard to cropping, loss of water, and soil fertility effects, pertinent information is provided by another treatment. In this treatment, fertilization was the same as shown for the fertilized soil in Table 12.1 but crotolaria, a legume, was grown each year. Drainage decreased to 44 cm annually ; and losses of nitrogen, potassium, and phosphorus decreased to 24, 3, and 0.1 kg/hectare annually. The crop removed both water and nutrients and hence decreased both the amount of drainage water and the concentration of nutrients in it. The much greater decrease in loss of potassium than of nitrogen may be explained on the basis that the crop took up both nitrogen and potassium from the soil but, being a legume, added to the soil some nitrogen, a part of which subsequently appeared in the drainage water. The infertility of the soil is evident in

**Table 12.1.** Annual Loss of Nitrogen, Phosphorus, and Potassium in Drainage Water from 109-Centimeter Columns of Fertilized and Unfertilized Fallow Sandy Soil in South Carolina.

| | *Annual Loss* | | | |
|---|---|---|---|---|
| | *Nitrogen, kg/hectare* | *Potassium, kg/hectare* | *Phosphorus, kg/hectare* | *Water, cm* |
| Fertilized | 40 | 30 | 0.2 | 68 |
| Unfertilized | 30 | 13 | 0 1 | 67 |
| Proportion of added nutrient lost in drainage water, % | 56 | 47 | 1 | — |

the ratio of nitrogen to potassium in the drainage water from the fallow soil. In soils of high fertility, the ratio of loss of nitrogen to potassium usually is much greater than the values shown in the table. For example, in analyses of drainage water from a 102-cm column of an unfertilized, fallow, highly fertile silt loam soil in Illinois, Stauffer (1942) recorded an annual drainage loss of 16 cm of water containing the equivalent of 86 kg of nitrogen and 1.5 kg of potassium per hectare.

The sandy nature of the soil has another effect on potassium loss that requires explanation. To be lost in drainage water, potassium must be present in solution. But potassium moving downward through soil in solution equilibrates continuously with the exchangeable cations in the soil. A given potassium ion will spend most of the time in exchangeable form and only a little of the time in solution as a freely diffusible cation associated with a freely diffusible anion. Accordingly, the downward passage of potassium is much delayed. The extent of the delay may be perceived to increase with the cation-exchange capacity of the soil and with the ease of replacement of the exchangeable cations. The sandy soil used in this experiment had a low cation-exchange capacity. Much of the exchange capacity was probably occupied by aluminum, which would not be displaced readily by potassium Hence loss of added potassium by leaching would take place readily.

Volk and Bell (1945) published the results of an experiment that illustrate the operation of principles of ion exchange in leaching of potassium and other ions from a 122-cm column of a sandy soil in Florida. At the beginning of the experiment, the soil was saturated with tap water and allowed to stand until drainage ceased. Then it received, at the surface, an application of salts containing calcium, ammonium, and potassium cations and nitrate, sulfate, and phosphate anions. The drainage water obtained from the soil as a result of natural rainfall then was analyzed at intervals, and the concentrations of the ions were plotted against the accumulated amount of drainage water.

The results in Fig. 12.1 show that concentrations of all ions were relatively low in the first drainage water, but then nitrate and cations (mostly calcium) emerged in definite peak concentrations. These peaks are due to the fertilizer applied at the beginning of the experiment. Although no doubt the nitrate ions emerging in the nitrate peak came mostly from the fertilizer and appeared early because they moved through

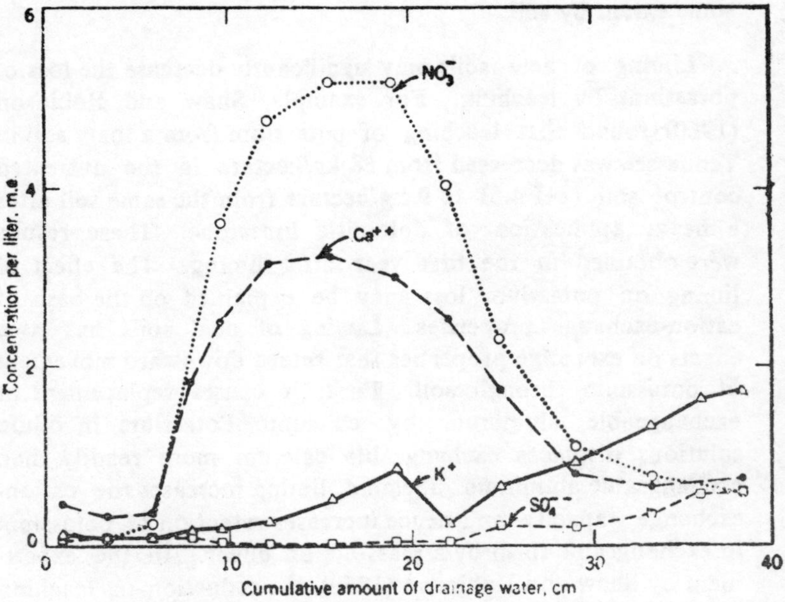


Fig. 12.1. Concentration of ions in increments of drainage water.

the soil without attachment, the peak concentration of cations that appeared with the nitrate peak resulted from the electrical neutrality requirement for cations to accompany the nitrate. The cations emerging in the cation peak with the nitrate no doubt came mostly from the exchangeable form in the lower part of the column of soil, the fertilizer cations added to the surface of the soil being retained for the most part in exchangeable form in the upper part of the soil. Thus, in this first cation peak, the total concentration of cations was determined mostly by the concentration of nitrate, and the proportions of the cations were determined mostly by the proportions and ease of exchange of the exchangeable cations in the lower part of the soil column. At the end of the experiment, the concentration of potassium in the drainage water was increasing toward what would have been a peak concentration of potassium ions derived from the fertilizer. The concentration of sulfate in the effluent likewise was increasing at the end of the experiment toward a peak concentration of sulphate ions derived from the fertilizer. Emergence of sulphate was delayed because, unlike nitrate, sulphate is adsorbed to some extent by soil.

Liming of acid soils may significantly decrease the loss of potassium by leaching. For example, Shaw and Robinson (1960) found that leaching of potassium from a loam soil in Tennessee was decreased from 58 kg/hectare in the untreated control soil (*p*H 4.5) to 9 kg/hectare from the same soil after a heavy application of dolomitic limestone. These results were obtained in the first year after liming. The effect of liming on potassium loss may be explained on the basis of cation-exchange properties. Liming of acid soils has two effects on exchange properties that retard downward movement of potassium through soil. First, it causes replacement of exchangeable aluminum by calcium. Potassium in dilute solutions displaces exchangeable calcium more readily than exchangeable aluminum. Second, liming increases the cation-exchange capacity and hence increases retention of potassium in exchangeable form by a mass-action effect. In the experiment by Shaw and Robinson (1960), the reduction in leaching loss of potassium from liming was far greater in the first year than in succeeding years, as would be expected if the effect of liming were attributable to an alteration in the cation-exchange properties just mentioned. If the effect of liming had been primarily on release of potassium from nonexchangeable to exchangeable form, the effect would be expected to persist over a period of years.

**Erosion**

Considerable potassium may be lost from soils by erosion. Lipman and Conybeare (1936) estimated the annual loss of potassium by erosion from the crop land of the United States at 158 kg/hectare. The selective removal of finer particles during erosion depletes selectively the portion of the soil that is most important in supplying potassium to plants. The same is true of nitrogen, as pointed out previously. Losses of potassium in this way are seldom taken as seriously as those of nitrogen, however, because of the difference in distribution of these two elements in soil profiles. Nitrogen is concentrated near the surface, which results in a decrease in nitrogen availability with increasing erosion. Potassium, on the other hand, is present throughout the profile. Loss of surface potassium by erosion, which may cause plants to develop in a surface soil having lower potassium availability, is compensated in part by addition of a new source of potassium in the deeper soil horizons that are brought nearer the surface.

## FORMS OF POTASSIUM IN SOILS

In soil different forms of potassium have been studied as given under the following heads.

### Mineralogical Forms

Most of the potassium in soils is present in minerals classified as feldspars and micas. The most important of these are orthoclase and microcline feldspar, biotite and muscovite mica, and micaceous clays, known as illite.

The feldspars occur almost exclusively in the sand and silt fractions of soils but are found occasionally in the coarse clay. The feldspars may be of either primary or secondary origin but, if secondary, are thought to be inherited from the parent material and not formed during soil development.

Biotite and muscovite mica also occur mainly in the silt and sand fractions and may be of either primary or secondary origin. The forms of these minerals found in soils have been altered from the original, having lost some potassium and gained some water Denison *et al.*, (1929). Secondary biotite and muscovite are thought not to form under conditions of soils. Illite, the principal potassium-bearing mineral in the clay fraction of soils, is of secondary origin. There is some doubt among mineralogists whether the illite in soil clays is developed there. Evidence indicates that illite in soils is, in general, undergoing degradation. Addition of potassium fertilizer, however, may favour formation of illite, as will be explained further subsequently. Likewise, minerals similar to muscovite and biotite may from form degraded micas where potassium fertilization is practiced.

Mineralogical analyses of bottom sediments in the oceans indicate that illite is formed there by alteration of other minerals, particularly degraded illite (low in potassium) and kaolinite Grim *et al.*, (1949). Formation of illite in sediments on the ocean bottom provides an explanation for the low content of potassium in relation to sodium in ocean water. According to data summarized by Rankama and Sahama (1950), the ratio of potassium to sodium by weight is 0.92 in igneous rocks, 0.37 in river and lake water, and 0.04 in ocean water. Mackenzie (1955) reviewed the subject of potassinm in clay minerals.

The occurrence of various potassium-bearing minerals in soils may be illustrated by Schachtschabel's (1937, 1937a) detailed mineralogical analyses of seven soils of Germany. The samples he analyzed contained an average of 2.3% potassium. On the average, the samples consisted of 27% by weight of potassium-bearing minerals (15.5% potassium feldspar and 11.3% micaceous minerals), 62% of nonpotassium-bearing minerals and large stones, and 5% of unidentified minerals. Of the feldspar, 14.5% was in the sand and silt, and 1% was in the clay. Of the micaceous material, 2.7% was in the sand and silt, and 8.6% was in the clay. The clay fraction of these soils averaged 40% micaceous material, which probably would be classified now as illite. Similar analyses of the silt and clay fractions of several soils of the United States were published by Reitemeier *et al.* (1951).

**Chemical Forms**

From the chemical standpoint, soil potassium often is divided into three categories : non-exchangeable, exchangeable, and water-soluble. Sometimes the nonexchangeable fraction is divided into two fractions on the basis of the potassium extracted by an acid, a cation-exchange resin, or a potassium-precipitating reagent.

In most soils, the great bulk of the potassium is nonexchangeable. As shown in Table 12.4, an average of 99.6% of the total potassium in samples of twenty soils of New Jersey was nonexchangeable. The remaining 0.4% included both the exchangeable and water-soluble forms. The water-soluble potassium usually constitutes only a small part of the sum of the water-soluble and exchangeable forms. Anderson *et al.* (1942) found that the exchangeable potassium content of surface samples of six soils from the humid and subhumid region of the United States ranged from 66 to 341 $\mu g/g$ of dry soil (0.17 to 0.87 m.e./100g). If calculated on the same basis, the potassium content of the displaced soil solutions of the same soils ranged from 1 to 8 $\mu g/g$ (0.003 to 0.02 m.e./100 g).

In most experimental work, soil-solution potassium is not measured independently but is included with exchangeable potassium. The two fractions are extracted simultaneously by the salt solutions used to remove exchangeable potassium for analysis, and this is one reason for the common practice. Another reason is that the soil-solution potassium is usually so low in relation to exchangeable potassium and in relation to uptake of potassium by plants that the portion in the soil solution is quantitatively of little significance. In humid regions, the content of exchangeable-plus-soluble potassium in soils usually is less than 200 $\mu g/g$ and rarely is more than 600 $\mu g/g$. In arid regions the amount may be higher.

Although exchangeable potassium may be defined in the same terms used to define exchangeable bases, as in the chapter on that subject, research workers at present are coming to observe two special conventions in determining exchangeable potassium in soils. Both are of significance with regard to the relationship between exchangeable potassium and uptake of potassium from soils by plants.

The first convention is that an ammonium salt is used to displace the potassium. The reasons for this convention are that ammonium salts (particularly ammonium acetate) are commonly used to displace exchangeable bases and that the distinction between exchangeable and nonexchangeable forms of potassium is much more definite with ammonium salts than with other common salts and is less dependent on the technique. The results obtained by use of different displacing agents may be rather different, as shown in Fig. 12.2. One may note from the figure that release of potassium from the soil soon stopped where ammonium acetate was used but that it continued for a longer time and never ceased, within the limits of the experiment, where other cations were used. Ammonium acetate displaced less potassium than any of the other salts except calcium acetate. The low results obtained with calcium acetate and the similarity of the extraction pattern obtained

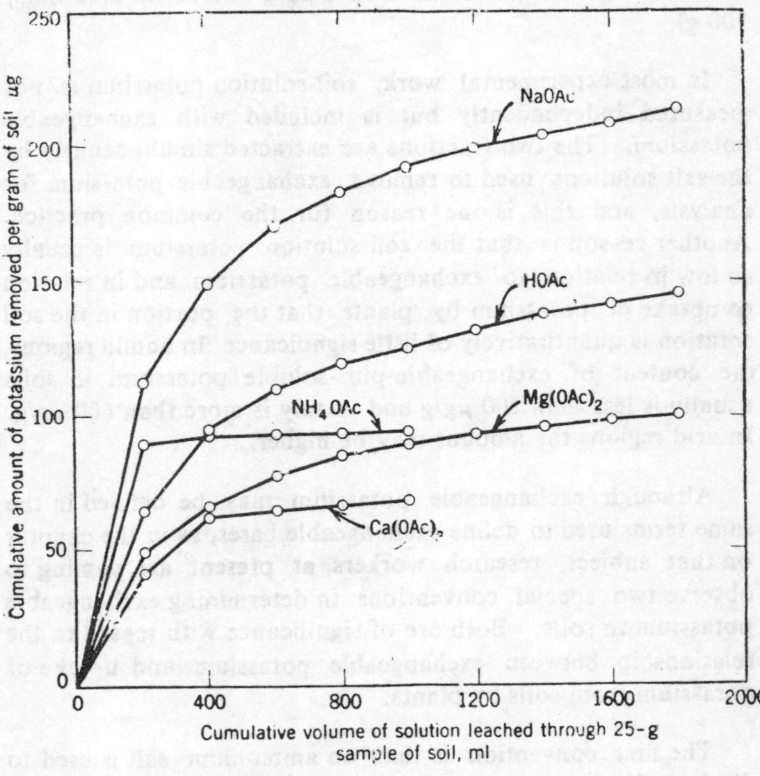


**Fig. 12.2. Cumulative amounts of potassium removed from a soil from New York by repeated leaching with 0.2-normal acetic acid and 0.5 normal solutions of various acetates.**

with calcium and ammonium acetates may be attributed to the potassium present as an impurity in the calcium acetate solution. The authors found that all the soils they tested except the one shown in Fig. 12.2 took up potassium from the calcium acetate solution. Because of the potassium impurity, they found more release of potassium from the soils with a dilute calcium acetate solution (0.01 to 0.1 normal) than with a more concentrated solution (0.5 to 1 normal). In explanation of these results, it may be mentioned here that ammonium and potassium behave similarly and that both are thought to prevent for the most part the release of potassium from interlayer positions in micaceous minerals. The singular behaviour of potassium and ammonium in cation-exchange reactions involving micaceous minerals will be considered further in subsequent sections.

The second convention is that exchangeable potassium is being determined on undried soils. Use of air-dried samples is technically convenient, and earlier analyses were made on samples that were air-dried or occasionally on samples that had been oven-dried.

Many soils yield a different value for exchangeable potassium if the displacing solution is added to a field-moist sample than to an air-dried sample. The tendency is for higher values on air-dried samples than on field-moist samples, although the reverse may occur. Decreases due to drying before analysis are usually encountered in samples high in exchangeable potassium. Subsoils usually are low in exchangeable potassium if analyses are made on field-moist samples. Analyses of subsoils made on dried samples are almost always higher than those on undried samples. Figure 12.3 shows results of analyses made on field-moist and variously dried samples of a silt loam soil from Indiana in which the effect of drying was pronounced.

Luebs *et al.* (1956) obtained a higher correlation between uptake of soil potassium by plants and exchangeable potassium in a group of undried soils in the greenhouse where the latter measurements were made on samples that had not been

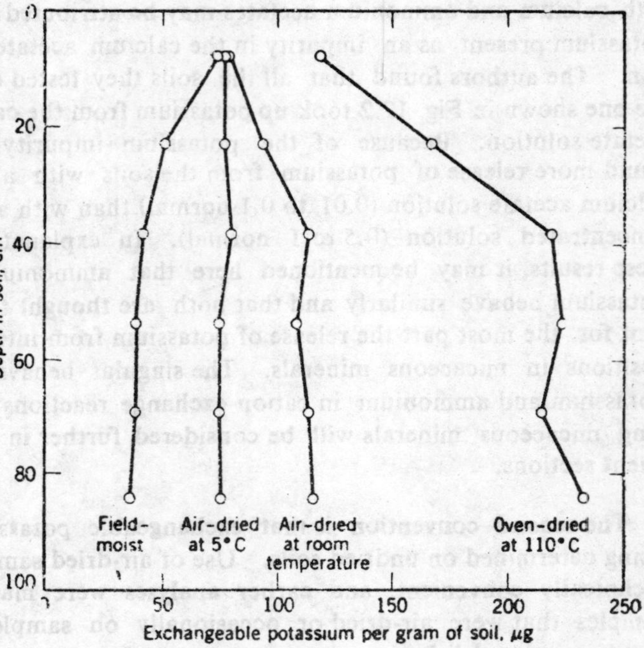


Fig. 12.3. Exchangeable potassium at different depths in a silt loam soil in Indiana as determined on samples that were field-moist or dried in different ways before analysis.

dried than on samples that had been air-dried before analysis. Their finding was subsequently confirmed by Hanway *et al.* (1961, 1962) in a large number of field experiments in north-central United States. Avoidance of drying samples of soil before analysis therefore seems advisable if analyses for exchangeable potassium are to represent the exchangeable potassium existing in soil in the field or if they are to be used as an index of availability to plants of the potassium in soils that have not been dried.

## AVAILABILITY OF POTASSIUM TO PLANTS

### Uptake from Solution

Epstein *et al.* (1963) determined the rate of potassium removal by barley roots from solutions having different concentrations of potassium and explained their observations on the basis that two kinds of carrier sites govern binding and uptake of potassium. One kind of site operates at one-half the maximum velocity at a potassium concentration of about 0.7 μg/ml, and the other operates at one-half the maximum velocity at a potassium concentration of about 625 μg/ml.

Williams (1961) grew barley plants in culture solutions with low concentrations of potassium and found that the plants rapidly removed all the detectable potassium from solution (a concentration of 0.001 μg/ml was said to be detectable) if the potassium supply was not renewed. Where special arrangements were made to replace the potassium continuously as it was removed by plants, however, Williams found that the plants grew well with a concentration of only 0.01 μg/ml. Comparison of Williams' results with those obtained by Epstein *et al.* (1963) indicates that the carrier sites specific for potassium may be operating far below their maximum velocity and still provide enough potassium for plants.

The concentration of potassium in the soil solution varies, but most analyses give values considerably higher than Williams' 0.01 μg/ml. Figure 12.4 gives a frequency distribution of concentrations of potassium in water extracts obtained from samples of many soils after they had been equilibrated with water in an amount to saturate them. The concentrations fell more frequently in the range from 2 to 6 μg/ml than in any other similar range, but the concentrations in almost half the soils exceeded 6 μg/ml. These analytical values no doubt underestimate the concentrations of potassium that would have been present in solution at lower contents of water in the range at which most plants grow. At the same time, the concentrations are probably somewhat high in comparison with those that would have been found at saturation in the field because the process of drying soil in air usually increases the salt content and potassium content of the soil solution and because the process of saturation in the field is usually accompanied by downward movement of salts originally present in the surface layer. In reviews of literature on concentrations of potassium in displaced soil solutions, the range was given as 8 to 391 μg/ml by Fried and Shapiro (1961) and as 3 to 156 μg/ml by Barber (1962).

Although the potassium concentrations mentioned in the preceding paragraph are relatively high in comparison with Williams' value of 0.01 μg/ml, these concentrations are not necessarily maintained when plants are present. In Table 12.2 is an example giving measurements on soil solutions displaced at different times during the growing season of corn. The values are expressed in terms of milligram equivalents per liter to show the stoichiometric relations among the various ions. Because of the relatively large amount of potassium required by corn plants and the low concentration of potassium in solution in comparison with the other ions, potassium was removed selectively. For comparison with the numerical values in the preceding paragraphs, the concentrations of potassium may be changed to the same units used previously, giving values of 3.7, 2.2, and 0.2 μg/ml of soil solution on June 26, July 12, and August 5, respectively. The value 0.2 is the average figure obtained from one field with a concentration of 0.79 μg of potassium per milliliter and four fields with a concentration of 0.00 μg/ml. Reduction of the concentration of potassium in the soil solution by roots provides an explanation for the fact that potassium deficiency occurs more frequently

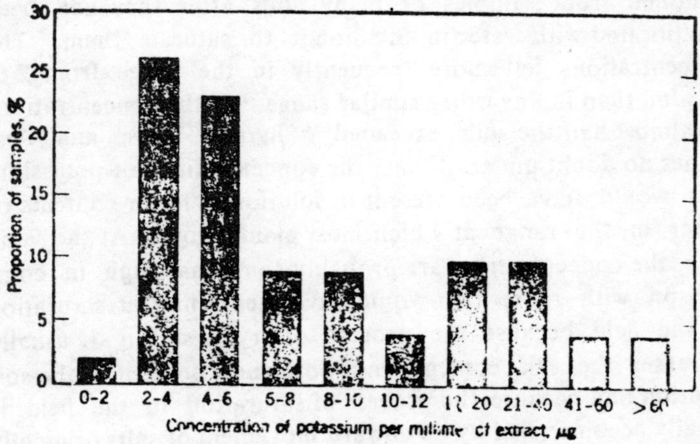


Fig. 12.4. Frequency distribution of concentrations of potassium in aqueous extracts of samples of 142 cultivated soils.

**Table 12.2. Average Concentrations of Potassium and Other Ions in the Solutions.**

| *Date of Sampling* | *Concentration of Indicated Ion per Liter of Displaced Soil Solution, m.e.* | | | |
|---|---|---|---|---|
| | *Potassium* | *Calcium* | *Magnesium* | *Nitrate* |
| June 26 | 0.095 | 13.6 | 4.4 | 14.2 |
| July 12 | 0.056 | 20.3 | 5.3 | 18.9 |
| August 5 | 0.004 | 8.4 | 2.3 | 8.1 |

than would be expected from a comparison of Williams' value of 0.01 μg/liter with the concentrations of potassium found in displaced solutions and other water extracts of soils.

**Loss from Exchangeable and Nonexchangeable Forms**

In this section, emphasis will be placed on relationships between exchangeable and nonexchangeable forms in potassium nutrition of plants. Reference should be made to for information on mechanisms by which plants may take up potassium ions present originally in exchangeable form.

The potassium content of soil solutions is usually small compared with the amounts of potassium absorbed by crops. Hence renewal during the season is needed if plants are to derive their potassium from the soil solution. Because cations in the soil solution remain continuously almost at equilibrium with those in exchangeable form, selective removal of potassium by plants from the soil solution should produce selective removal of potassium from exchangeable form. If, in fact, plants derive potassium from the exchangeable form, one should expect to find that removal of potassium from soil by plants would reduce the exchangeable potassium. Moreover, if exchangeable potassium is the only source other than the soil solution from which plants derive potassium, the reduction in exchangeable-plus-soluble potassium due to plant growth should equal the potassium removed by the plants.

Figure 12.5 shows the results of a field experiment in Hawaii in which eight crops of grass were grown in continuous cropping over a period of somewhat more than 2 years, with

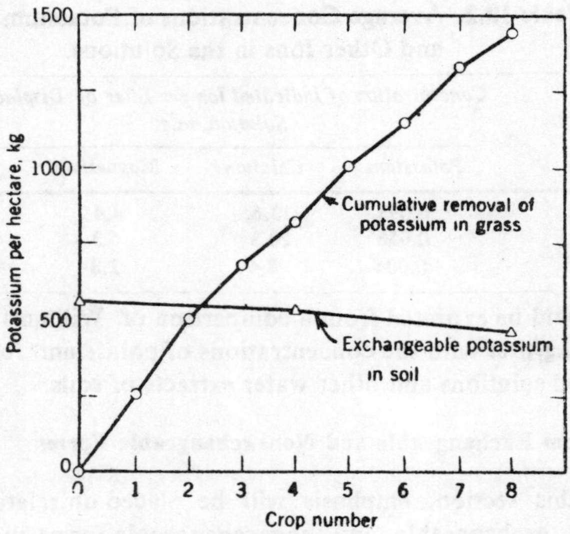


**Fig. 12.5. Cumulative removal of potassium to eight crops of napier-grass and exchangeable potassium in the surface 92 cm of a clay soil**

measurements of potassium removed in the grass and of exchangeable-plus-soluble potassium in the surface 92 cm of soil. The existence of a decrease in exchangeable-plus-soluble potassium with cropping is in agreement with the theory that plants derive potassium from exchangeable form ; however, the fact that the decrease was relatively small in comparison with the removal of potassium by plants provides evidence that most of the potassium harvested in the eight crops was not present in exchangeable form in the surface 92 cm of soil at the beginning of the experiment. Presumably, the potassium in the eight crops in excess of the decrease in exchangeable-plus-soluble potassium in the soil came mostly from the non-exchangeable form. Although there is little information on exchangeable soil potassium and on potassium removed by crops grown in the field over a period of years, the situation shown in Fig. 12.5 must exist rather generally if the potassium balance is determined over a long enough span of years. The exchangeable potassium in the soil within the root zone of ordinary crop plants is seldom sufficient for more than 10 or 15 years if the produce is removed; yet, plants continue to grow, and exchangeable potassium is not exhausted over much longer periods even if no appreciable addition of potassium is made. (Over a short time, however, the potassium removed by crops is sometimes balanced by an almost equivalent reduction in exchangeable potassium in the soil. An experiment with such results will be discussed subsequently, and a theoretical explanation for the differing findings will be developed.)

If, indeed, plants may take up potassium that was present in nonexchangeable form before their growth, an appreciation of the nature of the process is evidently important to an understanding of the potassium nutrition of plants growing in soils. Because weathering of potassium minerals in soils no doubt proceeds continuously, one may suppose that release of potassium by chemical breakdown of minerals such as potassium feldspars might provide the potassium that plants derive from nonexchangeable form. If this supposition is correct, one should expect the content of exchangeable-plus-soluble potassium to increase substantially if soils are kept moist and warm so that weathering will proceed, and if no plants are grown to remove the potassium.

Bray and De Turk (1939) conducted an experiment that provides information on this question. They divided samples of each of ten Illinois soils into two parts, dried one part, and sealed the other part in moist condition. After incubation for 5 years, the moist samples were dried. Then exchangeable potassium in both sets of samples was determined. The results in Fig. 12.6 show that exchangeable potassium did increase in most of the samples during the 5-year period but that the increases were relatively small and by no means sufficient to account for the marked use of potassium from nonexchangeable form observed by Ayres *et al.* (1947) in the experiment mentioned previously. Addition of soluble potassium in the form of fertilizer, however, does increase the content of exchangeable-plus-solution potassium. For example, Reitemeier *et al.* (1951) found that samples of a loam soil from a long-term field experiment in Maine had contents of

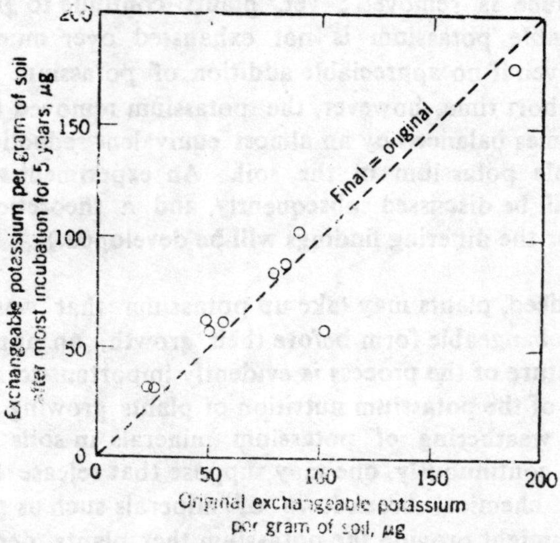


**Fig. 12.6.** **Exchangeable potassium in ten Illinois soils upon drying the samples after incubation in a moist condition for 5 years versus "original" exchangeable potassium determined on comparable samples that had been dried initially and stored in a dry condition for 5 years.**

exchangeable-plus-soluble potassium ranging from 35 μg/g in the control receiving no potassium fertilizer to 456 μg/g in a sample receiving both potassium fertilizer and manure. The indications are therefore that exchangeable potassium would accumulate in quantity if soluble potassium were actually being added by weathering of minerals that contain non-exchangeable potassium. The smallness of the accumulation of exchangeable potassium in the soils in the experiment given in Fig. 12.6 thus is evidence that soluble potassium is not being added by weathering in quantities that would account for the uptake by plants from nonexchangeable form.

Additional information on the behaviour of exchangeable and nonexchangeable potassium during cropping is provided in Fig.12.7, which shows the cumulative loss of these two forms of potassium from a soil and the cumulative uptake of potassium by corn plants in intensive culture in the green-

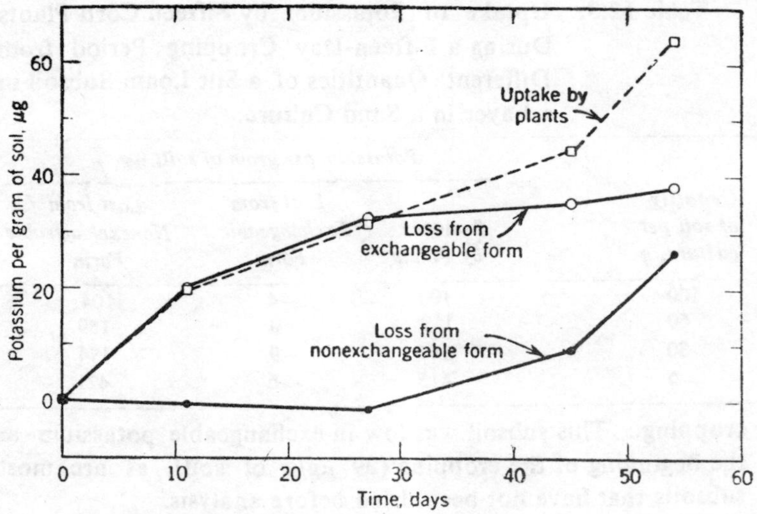


Fig. 12.7. Uptake of potassium from 1 kg of a loam soilby six corn plants in the greenhouse and loss of potassium from exchangeable and nonexchangeable forms in the soil.

house. The figure shows that initially the potassium taken up by the plants could be accounted for almost entirely by a loss of exchangeable potassium. Toward the end of the cropping period, however, most of the potassium taken up by the plants could be accounted for by loss of initially nonexchangeable potassium from the soil. Although the tendency for greater loss of exchangeable potassium initially and greater loss of nonexchangeable potassium later is commonly observed, there is much variation from differences among soils and differences in intensity of cropping. In general, the more intensive the cropping and the lower the exchangeable potassium, the sooner loss of nonexchangeable potassium occurs and the greater is loss from this form relative to that from the exchangeable form. Results of an experiment on intensity of cropping are shown in Table 12.3. Evidently the smaller the quantity of soil supplied per plant, the greater was the loss of potassium from nonexchangeable form. An additional important point to note about this experiment is that an undried subsoil was used, and it lost no exchangeable potassium during

**Table 12.3.** Uptake of Potassium by Fifteen Corn Plants During a Fifteen-Day Cropping Period from Different Quantities of a Silt Loam Subsoil in a Layer in a Sand Culture.

| *Quantity of soil per culture, g* | *Potassium per gram of soil, μg* | | |
|---|---|---|---|
| | *Removed by Plants* | *Lost from Exchangeable Form* | *Lost from Nonexchangeable Form* |
| 100 | 100 | —4 | 104 |
| 60 | 159 | 0 | 159 |
| 30 | 165 | −9 | 174 |
| 5 | 415 | —6 | 421 |

cropping. This subsoil was low in exchangeable potassium at the beginning of the cropping (39 μg/g of soil), as are most subsoils that have not been dried before analysis.

Another significant experiment was conducted by Verma (1963). He subjected three soils to intensive cropping in the greenhouse to reduce the level of exchangeable potassium and to cause loss of much nonexchangeable potassium. He then removed the roots of the plants, allowed the soils to incubate in a moist condition, and analyzed the soils for exchangeable potassium at intervals, with results as shown in Fig. 12.8. The most important point to note in this figure is that the exchangeable potassium remained almost constant in each soil during moist incubation after cropping. Now from the additional fact that from 78 to 91% of the potassium absorbed by the corn plants in this experiment was accounted for by loss of nonexchangeable potassium, one may infer that loss of potassium from nonexchangeable form took place rapidly while the plants were growing but almost ceased when the plants were removed. Presumably, the plants reduced the concentration of potassium in solution below some critical value at which release was inhibited ; and, when they were removed, the concentration rapidly increased to the critical level and then prevented further release from nonexchangeable form. This mechanism will be considered further in the section on release from mineral forms.

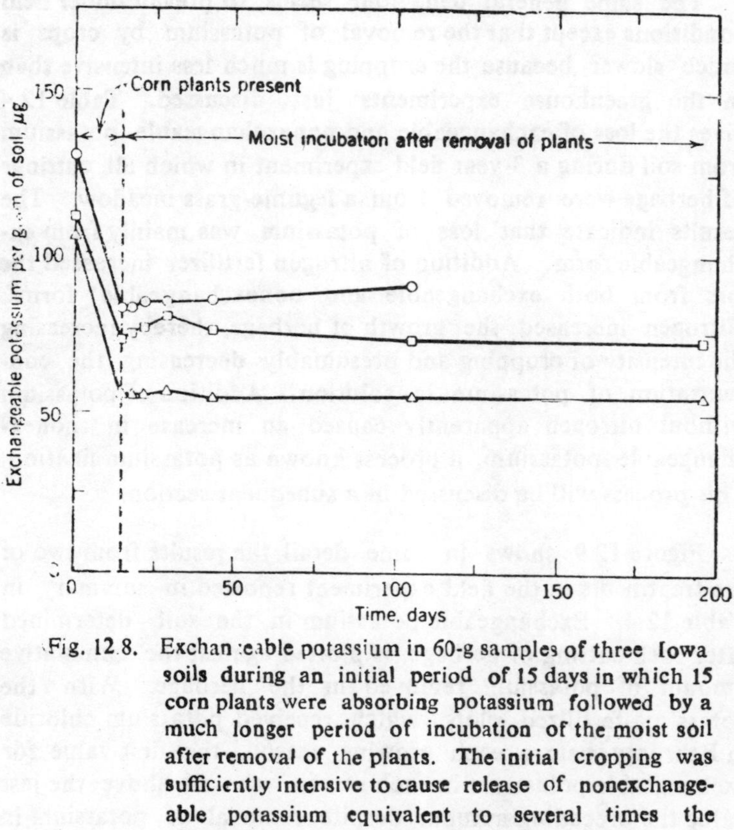


Fig. 12.8. Exchangeable potassium in 60-g samples of three Iowa soils during an initial period of 15 days in which 15 corn plants were absorbing potassium followed by a much longer period of incubation of the moist soil after removal of the plants. The initial cropping was sufficiently intensive to cause release of nonexchangeable potassium equivalent to several times the observed decrease in exchangeable potassium.

**Table 12.4. Loss of Exchangeable and Nonexchangeable Potassium from Soil under Legume-Grass Meadow During a Three-Year Period in Scotland.**

| *Fertilizers Added* | *Loss of Potassium from Soil per Hectare, kg*[1] *Exchangeable* | *Nonexchangeable* |
|---|---|---|
| None | 179 | 36 |
| Nitrogen | 202 | 64 |
| Potassium | 11 | —31 |
| Potassium+nitrogen | 191 | 1 |

[1]Potassium added in fertilizer was assumed to be added to the exchangeable form.

The same general behaviour seems to prevail under field conditions except that the removal of potassium by crops is much slower because the cropping is much less intensive than in the greenhouse experiments just discussed. Table 12.4 gives the loss of exchangeable and nonexchangeable potassium from soil during a 3-year field experiment in which all cuttings of herbage were removed from a legume-grass meadow. The results indicate that loss of potassium was mainly from exchangeable form. Addition of nitrogen fertilizer increased the loss from both exchangeable and nonexchangeable forms. Nitrogen increased the growth of herbage, thereby increasing the intensity of cropping and presumably decreasing the concentration of potassium in solution. Addition of potassium without nitrogen apparently caused an increase in nonexchangeable potassium, a process known as potassium fixation. This process will be discussed in a subequent section.

Figure 12.9 shows in some detail the results from two of the treatments in the field experiment reported in summary in Table 12 4. Exchangeable potassium in the soil, determined after each cutting of herbage, is plotted against the cumulative amount of potassium removed in the herbage. With the potassium-fertilized plots, which received potassium chloride in February before each growing season, the first value for exchangeable potassium in each season was well above the last value the preceding autumn despite removal of potassium in the first cutting of herbage. Transfer of potassium from nonexchangeable to exchangeable form during the winter would have produced the same sort of discontinuity in results with the plots receiving no potassium fertilizer if sufficient transfer had occurred. The results provide evidence of a small release of potassium between the first year and the second but not between the second and the third.

Further information on potassium behaviour under field conditions was obtained by Grimes (1966). He found, like Hemingway (1963), that the exchangeable potassium in soils may decrease considerably during a single growing season in the persence of a crop. His data from three field experiments

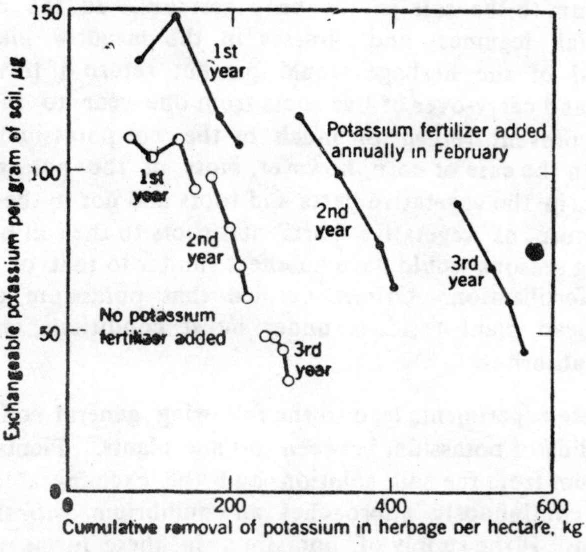


Fig. 12.9. Exchangeable potassium in samples of soil taken after each cutting of legume-grass meadow versus cumulative removal of potassium in herbage, with and without potassium fertilization, in three successive years in a field experiment in Scotland.

on silt loam soils indicated that the decrease in exchangeable soil potassium per gram of soil from spring to autumn in the presence of a crop of corn was as much as 50 μg if the exchangeable potassium was relatively high (125 μg) but was zero if the exchangeable potassium was relatively low (20 μg or less). In his experiments, he found higher values for exchangeable potassium in the spring than in the preceding autumn in soil under corn but not in soil under perennial legume-grass meadows. Moreover, the exchangeable potassium in soil under corn in the spring decreased with an increases in number of cuttings of hay removed the preceding year from the legume-grass meadow. For example, values of 65, 44, and 31 μg of exchangeable potassuim per gram of the surface 15 cm of soil were recorded following a meadow of red clover and timothy in which 0, 1, and 2 cuttings of hay were removed. He interpreted his findings in terms of removal of potassium from the soil by the growing crops and return of potassium to the soil in the crop residues. In the case of perennial legumes and grasses in the meadow plantings, removal of the herbage would prevent return of this potassium ; and carry-over of live roots from one year to the next would prevent return of much of the root potassium to the soil. In the case of corn, however, most of the potassium is present in the vegetative parts and roots and not in the grain ; and return of vegetative parts and roots to the soil between growing seasons would have an effect similar to that of potassium fertilization. Grimes verified that potassium diffuses from dead plant residues under moist conditions and that plants absorb it.

These experiments lead to the following general concept of the cycling of potassium between soil and plants. Plants draw potassium from the soil solution and the exchangeable form, which continnously approaches an equilibrium with the soil solution. If the supply of potassium in these forms is high, they may sustain the entire loss occasioned by growth of plants. If the soil solution and cation-exchange sites are depleted sufficiently of potassium, potassium is released rapidly from nonexchangeable form ; but the release almost stops as soon as the plants stop removing potassium. Once the potassium has been depleted to a level at which release from nonexchangeable forms occurs throughout the soil, the content of exchangeable potassium remains approximately constant despite much additional removal of potassium. As a result of recurrent additions of potassium in the form of plant residues, manures, or fertilizers, soils usually contain potassium in excess of the level needed to inhibit release from nonexchangeable form. Most of the potassium of plants is in the vegetative parts and underground parts ; and if these are removed, soil potassium is depleted rapidly. Removal of only selected parts (such as grain) that contain little potassium causes little depletion.

Subsoils are almost always low in exchangeable potassium. For example, in analyses of profiles of 48 mineral soils in the north-central region of the United States, Hanway *et al.* (1962) found an average content of exchangeable potassium amounting to 102, 52, 42, 38, 35, and 35 $\mu g/g$ of soil in successive 15-cm layers from the surface downward. Subsoils probably contain enough exchangeable and soluble potassium to stop further transfer from nonexchangeable to exchangeable form but possibly little more in most instances. The excess of exchangeable potassium over the level that stops release from nonexchangeable form probably occurs almost entirely in the upper part of the soil to which is added the potassium supplied by plant residues, manures, and fertilizers.

This concept is a generalization based on limited experimental evidence derived from work on soils containing micaceous minerals, the behaviour of which will be considered in more detail in the section on release from mineral forms. Soils that do not contain micaceous minerals will probably behave differently with regard to release of potassium from nonexchangeable form.

**Release from Particle-Size Fractions**

Determination of the source of the potassium released from nonexchangeable form in soils has been the subject of a number of investigations. Several kinds of evidence indicate that the finer mineral fractions are of major importance.

McEwen and Matthews (1958) used the technique of correlating the content of individual particle-size fractions of soils of Ontario with the release of potassium from nonexchangeable form during continuous percolation of water saturated with carbon dioxide through the samples. They found that release of potassium increased with the was closely related to the clay content of the soils.

Merwin and Peech (1951) investigated the release from different particle-size fractions of four soils of New York in another way. They separated the soils into fractions of sand, silt, and clay size and then recombined the fractions in various ways. The recombined particle-size fractions were initially calcium saturated and were then incubated in a moist condition to allow potassium release to proceed. At the end of the incubation, the samples were leached to remove the exchangeable potassium that had accumulated ; and the potassium removed was then determined. The results, summarized in Table 12.5, again indicate the importance of the finer fractions. Although the greatest release was from the clay fraction, release from particles of silt size appears important where the relative release from the different particle-size fractions is calculated on the basis of the proportion in which the size-fractions occurred in the soils.

**Table 12.5.** Relative Quantities of Nonexchangeable Potassium Released to Exchangeable Form from Different Particle-Size Fractions of Four Soils of New York During Incubation Following Removal of Exchangeable Potassium and Saturation with Calcium

| | *Relative Quantities of Nonexchangeable Potassium Released to Exchangeable Form from Indicated Fraction* | | |
|---|---|---|---|
| *Basis of Calculation* | *Sand* | *Silt* | *Clay* |
| Equal weights of all fractions | 3 | 19 | 100 |
| Relative weights of fractions found in the soils | 13 | 62 | 100 |

Still another technique was used by Arnold (1960a). He grew ryegrass on a soil of Great Britain in a greenhouse for 18 months and determined the uptake of potassium by the ryegrass from nonexchangeable form. He separated the soil into different particle-size fractions before and after cropping and analyzed the fractions for total potassium. He found that about 40% of the potassium removed by the crop could be accounted for by loss of potassium from the fraction less than 0.1$\mu$ in diameter, and another 40% could be accounted for by loss of potassium from the 0.1 to 0.3-$\mu$ fraction. These two fractions together contained 25% of the total potassium in the soil. The remaining 20% of the potassium removed by the crop could be accounted for by loss of potassium from the fractions coarser than 0.3$\mu$ in diameter, which contained about 75% of the total soil potassium.

**Release from Mineral Forms**

Although the behaviour of potassium in soils can be determined only by studying soils, much has been learned from studying potassium-bearing minerals that contributes understanding to the behaviour of potassium in soils. A special advantage in work with minerals is that the findings may be related to a particular structure and composition if pure mineral species are used. The two principal mineralogical sources of nonexchangeable potassium in soils are feldspars and micas, and consideration will be limited here to these two classes of minerals.

Potassium in feldspars is imbedded in a crystalline aluminosilicate structure (the chemical formula for orthoclase, for example, may be written as $KAlSi_3O_8$) in which each potassium atom is surrounded by and bonded to oxygen atoms. The bonds in the aluminosilicate part of the crystal structure extend completely around each potassium atom. Hence each potassium atom is isolated from each other potassium atom ; and there is no access of the outside medium to the potassium, and *vice versa* except by disruption of the aluminosilicate network.

Several observations indicate that release of potassium from feldspars is inhibited by development of a protective residual layer on the surface of the particles in response to mild chemical weathering of the type occurring in soils.

1. If potassium feldspars are broken into small pieces and then subjected to chemical weathering in the laboratory, a small portion of the mineral dissolves. But the constituents do not appear in the solution in the same proportion as they occur in the mineral. Less aluminum and silicon are dissolved than would be expected from the potassium released. For example, Correns and von Engelhardt (1938) leached a quantity of finely ground adularia feldspar with distilled water in a reflux apparatus and then measured the constituents removed. The particles of feldspar were less than $2\mu$ in diameter ; and the action of water was continued for 43 days, although the refluxing action (passage of water through the sample) took place for only 218 hours. This treatment removed from the feldspar

12.2% of the potassium, sodium, and calcium but only 1.4% of the silicon and 1.2% of the aluminum. The feldspar residue therefore contained relatively more aluminum and silicon and less potassium, sodium, and calcium than the original adularia.

2. The residue from treatment of finely ground feldspar with water, dilute acids, or salt solutions will retain in exchangeable form cations other than the ones originally present. For example, Nash and Marshall (1957) treated finely ground samples of seven different varieties of feldspar (0.5 to 1$\mu$ in diameter) with normal ammonium chloride solution and then removed the excess ammonium chloride by washing the samples with alcohol. Subsequent treatment of the samples with normal potassium, magnesium, and strontium chlorides released from 0.4 to 3.2 m.e. of ammonium per 100 g of feldspar.

3. Verma (1963) found that release of potassium from nonexchangeable form in orthoclase feldspar under the influence of growing plants was rapid at first and then occurred more slowly (see Fig 12.10), as would be expected from the theory of resistant surface layers. Release continued at a relatively high rate from certain micaceous minerals.

4. Feldspars occur mostly in the sand and silt fractions of soil and are absent from the clay fraction or occur there only in traces. Hence stability is associated with relatively large particle diameter, as would be expected from the theory of resistant residual surface layers.

Potassium in micas is imbedded in a crystalline aluminosilicate structure, but the potassium atoms lie in planes between molecular aluminosilicate networks. Although each potassium atom is surrounded by and bonded to oxygen atoms, two separate sets of bonds are involved, one set in the molecular aluminosilicate layer on one side of the potassium and the other in the aluminosilicate layer on the other side of the potassium. The molecular aluminosilicate layers are bonded together through the potassium ions. No other electrostatic bonds cross the plane containing the potassium atoms.

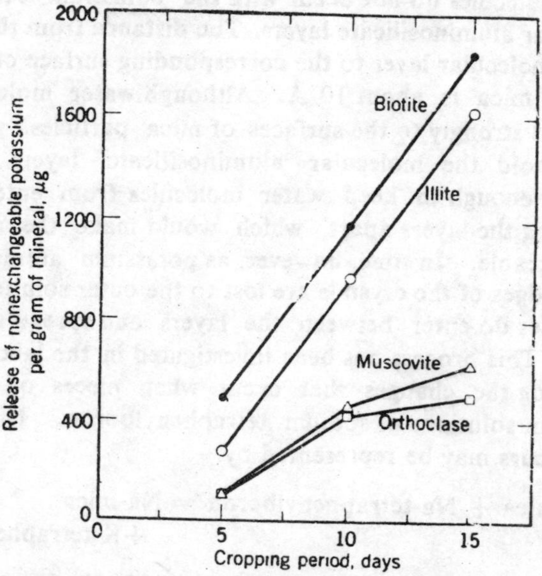


**Fig 12.10.** **Release of nonexchangeable potassium from different minerals during intensive cropping of mixtures of the minerals with quartz sand. Particles of biotite, muscovite, and orthoclase were less than 50 μ in diameter, and those of illite were less than 20 μ in diameter. All minerals were added in quantities to supply the same total amount of potassium per culture.**

**The potassium atoms are isolated from each other and from the external medium as long as the aluminosilicate layers remain drawn together by the intervening potassium atoms. The potassium plane is the weakest part of the crystal structure, however, as evidenced by the ease with which sheets of mica may be split physically into thinner sheets. This splitting occurs along the potassium planes and causes exposure of potassium atoms on the cleaved surfaces. Potassium atoms thus exposed are exchangeable.**

**Under chemical weathering in the soil, as in physical disintegration of mica crystals, the weakest part of the structure is the potassium plane. Access of all the potassium ions in a given plane to the external solution is provided if the enclosing silicate layers are pried apart. In the original mica structure,**

water molecules do not occur with the potassium between the molecular aluminosilicate layers. The distance from the surface of one molecular layer to the corresponding surface of the next layer in mica is about 10 Å. Although water molecules are adsorbed strongly to the surfaces of mica particles, potassium atoms hold the molecular aluminosilicate layers together strongly enough to keep water molecules from entering and spreading the layers apart, which would make the potassium exchangeable. In time, however, as potassium atoms adjacent to the edges of the crystals are lost to the outer solution, water molecules do enter between the layers and spread the layers apart. This process has been investigated in the laboratory by observing the changes that occur when pieces of mica are placed in solutions of sodium tetraphenylboron. The reaction that occurs may be represented by

$$\text{K-mica} + \text{Na-tetraphenylboron} = \text{Na-mica} + \text{K-tetraphenylboron.}$$

Potassium tetraphenylboron is practically insoluble in an excess of sodium tetraphenylboron, and so the reaction proceeds to the right. Provides graphic illustration of the rapidity with which certain micas may be degraded by use of sodium tetraphenylboron, which acts only on the potassium and does not attack the aluminosilicate structure. All that is needed to cause rapid release of potassium is a way to remove the potassium from solution and another cation to maintain electrical neutrality in the depleted residue.

To put the results of these laboratory observations in perspective in terms of the behaviour of micas in soils, it seems reasonable that soil micas should release potassium rapidly provided the external concentration of potassium is low enough. Now, potassium-bearing micas are known to be present in many soils, and evidence indicates that they have been there, subject to chemical weathering, for thousands of years. From the laboratory findings just mentioned, the potassium should have been lost long ago if conditions were sufficiently favourable for potassium removal. Although some potassium has been lost, the conditions for potassium removal have evidently been far less effective than those employed in the laboratory.

Release of potassium from nonexchangeable, interlayer form in micas is primarily a cation-exchange reaction. Accordingly, if one represents the process in terms of interaction with soil solutions, an equation such as

$$\text{K-mica} + (\text{Ca, Mg, Na, Al, H, NH}_4)\text{-anion} = (\text{Ca, Mg, Na, Al, H, NH}_4)\text{-mica} + \text{K-anion}$$

may be written, where "anion" refers collectively to the freely diffusible anions in the soil solution. At equilibrium, the mica will have taken up some of each cation originally in solution and will have released some potassium to the solution. Application of the law of mass action in a qualitative way indicates that the total quantity of potassium released from a given quantity of mica at equilibrium should increase with the concentration of cations other than potassium in solution, with an increase in volume of solution, and with a decrease in concentration of potassium in solution before the reaction. Exchangeable cations enter the picture in that the potassium and other cations in the soil solution are substantially at equilibrium with those in exchangeable form. Ammonium is of special significance in that it behaves much like potassium in this particular reaction, as will be explained further in the next section. High concentrations of ammonium, like those of potassium, tend to keep the reaction from proceeding to the right.

Scott and Smith (1966) conducted an experiment that provides evidence on the concentrations of potassium needed to inhibit the reaction. In this experiment, finely divided micaceous minerals were equilibrated with a 1-normal sodium chloride solution. With each mineral species, the ratio of total potassium added in the mineral to total sodium in solution was varied ; but in all instances the excess of sodium ion was great enough to cause practically quantitative release of all the potassium subject to exchange. Hence the potassium in solution represented practically all the potassium released from nonexchangeable form. Interpolation in Fig. 12.11 shows that the

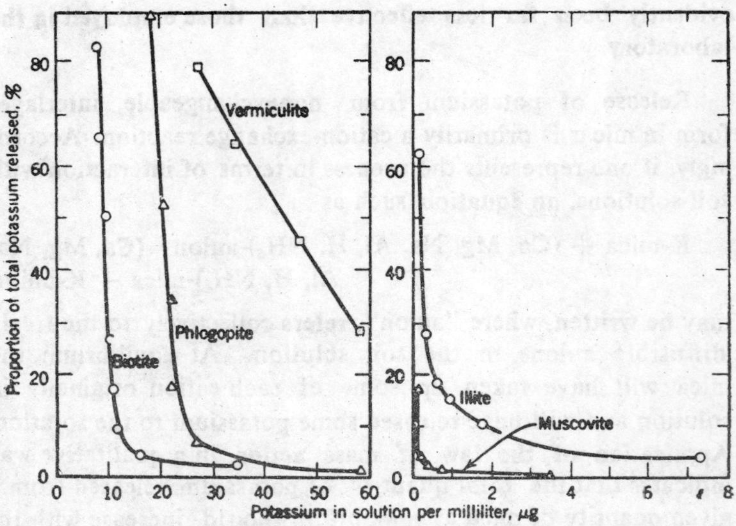


Fig. 12.11. Release of potassium from different micaceous minerals suspended in 1-normal sodium chloride solution versus concentration of potassium in solution at equilibrium. In addition to sodium chloride, the solution in which the illite and muscovite were suspended contained 0.01-normal ethylene diamine tetraacetate. Illite particles were less than 2 μ in effective diameter ; particles of all other minerals were 10 to 20 μ in effective diameter. Note the difference in horizontal scale between the two graphs.

concentration of potassium in solution in micrograms per milliliter associated with a replacement of 5% of the potassium by sodium was about 30 with phlogopite, 15 with biotite, 7 with illite, and 0.3 with muscovite. These concentrations are in the same range as the concentrations of potassium in saturation extracts of soils in Fig. 12.4, which suggests that the concentrations of potassium in many soil solutions are high enough to keep most of the interlayer potassium in micaceous minerals present as such. Further information is needed, however, because saturation extracts are not soil solutions and because the quantitative significance of values measured in 1-normal sodium chloride solution in terms of soil solutions is not known.

Other evidence indicates that, at least qualitatively, the principles discussed in preceding paragraphs apply to the release of potassium from micaceous minerals in the presence of plants. One may note, for example, that the order of decreasing release of potassium from micaceous minerals in sodium chloride solutions (biotite ⊳ illite > muscovite) in Fig. 12.12 is the same as the order of decreasing uptake of potassium from the minerals by plants in Fig. 12.10.

Mortland *et al.* (1956) made a detailed examination of the changes that took place in biotite when four successive crops of wheat were grown on mixtures of quartz sand and finely ground biotite with nutrients other than potassium added artificially. The biotite was added in different quantities to permit different degrees of depletion, and some of the material residual from the biotite was recovered from the sand for analysis after growth of the crops. Table 12.6 shows that cropping reduced the potassium percentage and increased the

**Table 12.6.** Potassium Content and Cation-Exchange Capacity of Finely Ground Biotite and of the Material Residual from the Biotite after Growth of Four Crops of Wheat in Biotite-Quartz Cultures.

| *Biotite Added per Culture, g* | *Total Potassium Content of Residue from Biotite,* % | *Cation-Exchange Capacity per 100 g of Residue from Biotite, m.e.* |
|---|---|---|
| 25 | 2.4 | 54 |
| 50 | 3.6 | 38 |
| 100 | 4.5 | 30 |
| Original biotite | 5.8 | 14 |

cation-exchange capacity of the residue from the biotite. The increase in cation-exchange capacity may be attributed to removal of potassium by plants, with consequent entry of cations other than potassium to satisfy the negative charges on the mineral layers and with spreading of the layers by water.

X-ray analysis showed that the diffraction pattern of biotite given by the original material had largely disappeared during cropping and was replaced by the pattern for vermiculite. The distance from the surface of one molecular layer to the corresponding surface of the next layer is about 10 Å in biotite but 14 Å in vermiculite. Figure 12.12 shows clearly a gradual change from a strong diffraction peak at 10Å with no diffraction peak at 14Å in the original biotite to a weak diffraction peak at 10 Å with a diffraction peak of moderate intensity at 14 Å in the material residual from the application of 25g of biotite per culture. The results of the X-ray analysis thus provide additional understanding of the nature of the alteration that took place in the presence of plants. Vermiculite has a high cation-exchange capacity, the major part of which is attributable to interlayer sites (the ones that held the potassium in the original biotite). Confirmation that the product was vermiculite and not chlorite, another mineral with a 14-Å diffraction peak, was obtained by subjecting to X-ray analysis the material residual from the biotite after growth of plants and after saturating the exchange sites with potassium and heating the material to 110°C. When this was done, the 14-Å diffraction peak was replaced by one at 10 Å, indicating contraction of the layers and trapping of added potassium in the sites between the layers. Chlorite does not contract upon treatment in this way.

The foregoing experimental work was done under artificial conditions suitable for testing the principle of plant-induced release of potassium from micaceous minerals. Detection of analogous changes in potassium-bearing minerals in the field on a short-term basis is difficult because the quantities of potassium removed are so small, comparatively speaking, and the methods of mineralogical analysis for detecting the changes are not highly sensitive. A partial verification of operation of the process under field conditions was obtained by Scheffer *et al.* (1960) in investigating changes in soil of eight plots in a field experiment in Germany. Four of these plots had received an average application of 12.7 kg of potassium per hectare annually over a period of 77 years, and the other four had received no potassium during this time. The cation-exchange

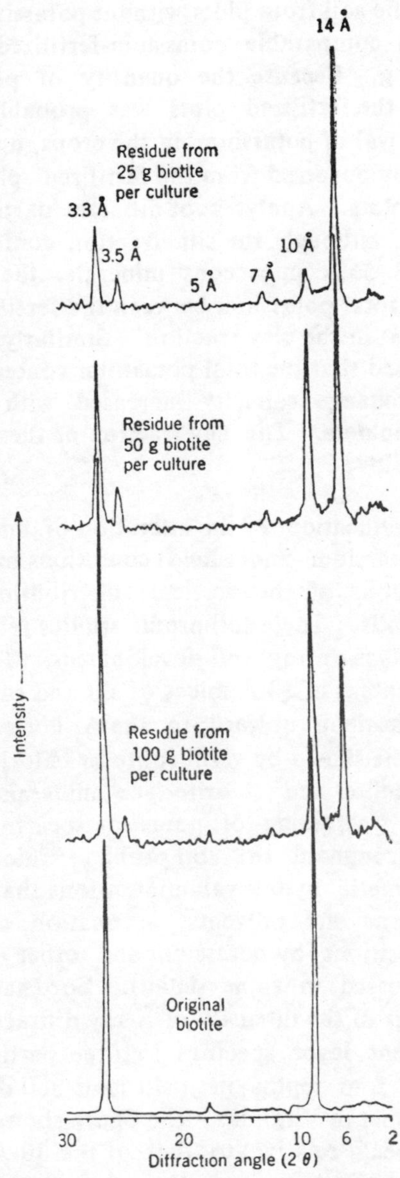


Fig. 12.12. Intensity of X-ray diffraction at different angles from finely ground biotite and from the residues of finely ground biotite after growth of four consecutive crops of wheat on mixtures of quartz and biotite with nutrients other than potassium supplied artificially. The intensities of diffraction from the original biotite should be doubled to make them comparable with intensities from the other samples. Note particularly the gradual change in relative intensities of diffraction corresponding to 10 and 14A° in proceeding from the original biotite to the residue from biotite supplied at 25 g/culture.

capacity of the soil from plots without potassium exceeded that of soil from comparable potassium-fertilized plots by 0.9 to 2.4 m.e./100 g. Because the quantity of potassium applied annually to the fertilized plots was probably well below the average removal of potassium in the crops, a net loss of potassium probably occurred from the fertilized plots as well as the unfertilized plots. Analyses of different particle-size fractions showed that, although the silt fraction contained about 25% feldspar and 35% micaceous minerals, the only significant difference in total potassium between the fertilized and unfertilized plots was in the clay fraction. Similarly, Van Ruymbeke (1963) reported that the total potassium content decreased and the cation-exchange capacity increased with age in soils of Belgian sea polders. The clay fraction of these soils contained about 50% illite.

Clearer verification of the alteration of micaceous minerals by loss of potassium under field conditions has been obtained in investigations of the vertical distribution of micaceous minerals in soils. These soil profile studies reflect changes that have taken place during soil development. The usual finding is that the content of 10-Å micas of silt and sand size decreases from the C horizon upward to the A horizon and that the reverse trend is shown by vermiculite-or chlorite-type minerals. These vermiculite- and chlorite-type minerals presumably are products of weathering of potassium-bearing micas present originally throughout the soil profile. Chlorite has an interlayer of polymeric hydroxyaluminum ions that develops under acid conditions and prevents occupation of the interlayer exchange positions by potassium and other cations (this process was discussed in some detail). For example, Fig. 12.13 shows tracings of the intensity of X-ray diffraction corresponding to different layer spacings in three particle-size fractions from each of four depths in a silt loam soil developed from a muscovite schist in Virginia. The figure shows that the height of the 14-Å peak relative to that of the 10-Å peak increased not only in proceeding from the greatest depth toward the surface but also in proceeding from particle-size fractions of relatively large size of those of small size. Similar trends were observed in samples from two other profiles of the same soil type, but the tracings of X-ray diffraction intensities from each profile showed some characteristic features that differed from the tracings from the other profiles.

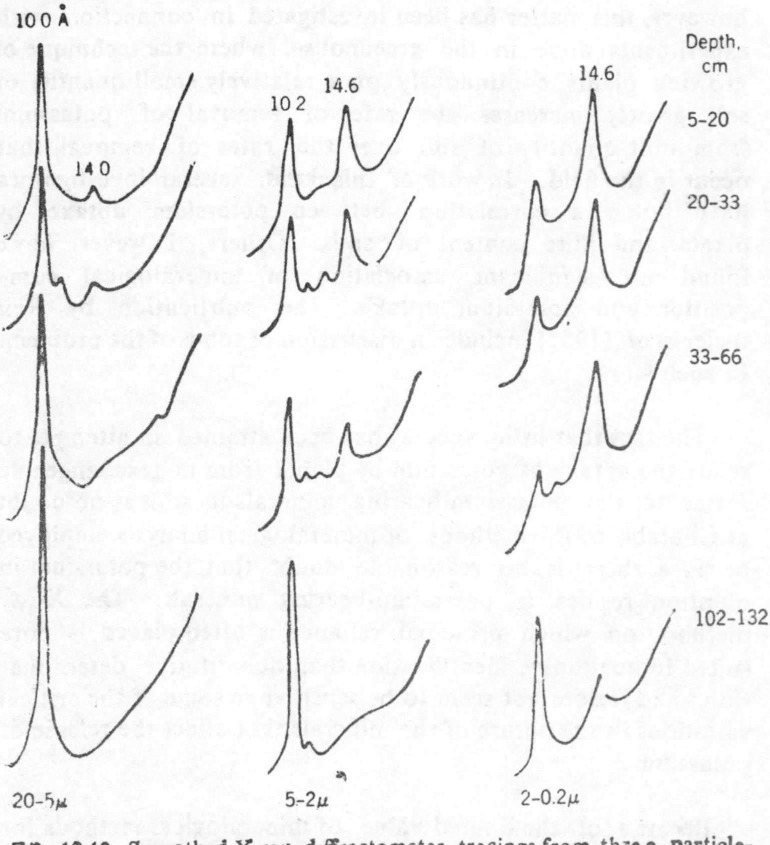


Fig. 12.13. Smoothed X-ray diffractometer tracings from three particle-size fractions from four depths in a silt loam soil developed from muscovite schist in Virginia. The numbers associated with peak diffraction intensities indicate the distance from the surface of one molecular layer to the corresponding surface in the next layer.

On the basis of the foregoing considerations relative to release of potassium from feldspars and micas in soils, one might expect the supply of potassium for plants to be related to the content of these minerals in soils. Some rather broad associations between potassium-bearing minerals in soils and the supply of potassium to plants in the field were reported in literature summarized by Arnold (1960). For the most part, however, this matter has been investigated in connection with experiments done in the greenhouse, where the technique of growing plants continuously on a relatively small quantity of soil greatly increases the rate of removal of potassium from unit quantity of soil over the rates of removal that occur in the field. In work of this kind, several investigators have noted a correlation between potassium uptake by plants and illite content of soils. Others, however, have found no significant association of mineralogical composition and potassium uptake. The publication by Reit meier *et al.* (1951) includes a discussion of some of the problems in such work.

The fact that little success has been attained in attempts to relate the uptake of potassium by plants from nonexchangeable forms to the potassium-bearing minerals in soils is no doubt attributable to the methods of mineralogical analysis employed because there is no reasonable doubt that the potassium in question resides in potassium-bearing minerals. The X-ray method, on which principal reliance is often placed, is more suited to qualitative identification than quantitative determination ; and it does not seem to be sensitive to some of the critical variations in the nature of the minerals that affect the release of potassium.

Because of the limited value of mineralogical methods for estimating the quantities of potassium-bearing minerals in the clay fraction of soils, chemical analysis is sometimes used to estimate the illite content, the usual assumptions being that all the potassium in the clay fraction is present as illite and that illite contains 6% potassium. The values for potassium are ordinarily used as such in investigations of potassium supply for plants. Arnold (1960a) used the total-potassium method in work on a group of 19 soils in Great Britain. He grew ryegrass on the soils for 18 months in the greenhouse, determined the uptake of potassium by the ryegrass and the changes in potassium fractions in the soil, and plotted the results in a form similar to that shown in Fig. 12.14. This figure gives a plot of potassium percentage in the fine clay fraction against the percentage content of fine clay in the soil. The data points representing individual soils are coded into three groups according to the amount of originally nonexchangeable potassium taken up by the ryegrass in the 18 months of intensive cropping. With the exception of results for two soils, the data may be seen to fall into three groups. A given uptake of potassium could be obtained from a soil in which the potassium percentage in the fine clay was high and the percentage content of fine clay was low or in soils in which the potassium percentage in the fine clay was low and the percentage content of fine clay was high.

One of the reasons for reproducing Fig. 12.14 here is that the author provided certain additional information to explain the

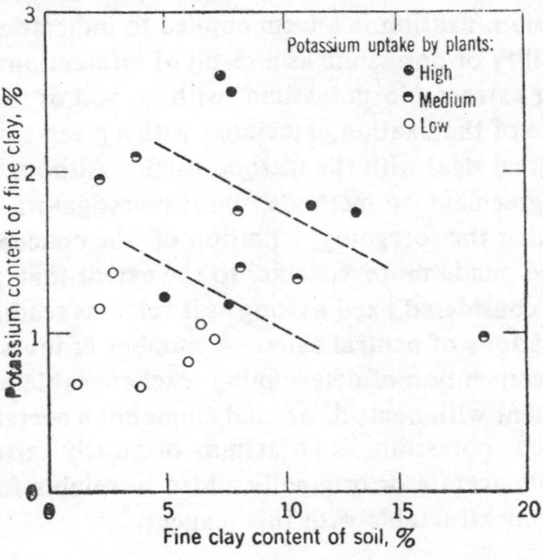


Fig. 12.14. Potassium percentage in fine clay ($<0.1\mu$ in effective diameter) versus fine clay content of 19 soils of Britain, with values classified according to relative uptake of potassium from nonexchangeable forms in 18 months of intensive cropping by perennial ryegrass grown on the soils in the greenhouse. Potassium uptakes in micrograms per gram of soil are classified as follows : $>500$=high, 100—500=medium, and $<100$=low.

behaviour of the two exceptional soils in which the potassium uptake was high but in which analyses of the fine clay caused the soils to be classified with others that supplied little potassium to the ryegrass. The silt fraction of one of the exceptional soils was found to release considerable potassium when incubated with a hydrogen-saturated cation-exchange resin. Much of the potassium taken up from this soil by the ryegrass thus may have come from the silt fraction. The other exceptional soil was from an experimental field that had been heavily fertilized with potassium Presumably, (1) the nonexchangeable potassium that developed in the fine clay of this soil from fertilization was released to plants more readily than an equivalent amount of nonexchangeable potassium in the other soils, or (2) the potassium was present mostly in particle-size fractions other than fine clay.

**Fixation**

Potassium fixation is a term applied to indicate a decrease in extractability of potassium as a result of interaction of originally soluble or extractable potassium with a soil or mineral. The magnitude of the fixation associated with a given soil or mineral varies a great deal with the method used. Although there is no general agreement on methods, most investigators would seem to agree that the foregoing definition of the concept of fixation should be made more specific to the extent that potassium is not to be considered fixed as long as it remains readily exchangeable by cations of neutral salts. A number of investigators employ the convention of determining exchangeable potassium by displacement with neutral, normal ammonium acetate and hence define fixed potassium as potassium originally extractable with ammonium acetate or originally added in soluble form that has become nonextractable with this reagent.

Potassium fixation is of interest in mineralogy in connection with the structure and behaviour of minerals, and it is of interest in soil-plant relationships from the standpoint of availability of potassium to plants. In this section, attention will be directed first to the nature of the process and then to the significance of the process in the potassium nutrition of plants.

A great deal of research has been done on potassium fixation ; and the status of knowledge cn the subject is set forth in papers by Schuffelen and van der Marel (1955), Agarwal (1960), and Weaver (1958). The first two papers are reviews, and the third is primarily interpretive.

*Nature of Potassium Fixation*. Potassium is fixed in various minerals, which may be classified into three groups : (1) those that fix potassium by ion-exchange in sites that hold potassium in the unweathered mineral (examples are vermiculites and surface-weathered potassium feldspar) ; (2) those that fix potassium by ion-exchange in sites that do not necessarily hold potassium in the unweathered mineral (examples are certain zeolites) ; and (3) those that are newly synthesized (examples are complex phosphates of iron, aluminum, or both, containing also potassium, ammonium, or both).

The potassium-fixing properties of various minerals in these three classes have been discovered by special experiments with fairly pure minerals or in other ways. As yet, however, it is not possible to determine potassium fixation on a soil and allocate the fixation quantitatively among various minerals.

Schuffelen and van der Marel (1955) summarized the information in the literature on potassium fixation by various minerals and included original data on fixation by many minerals collected from different parts of the world. Of all the minerals they tested, the ones that fixed the most potassium per 100 g of mineral were vermiculites and zeolites. Zeolites are hydrous aluminosilicates with small channels in which the potassium presumably may be trapped. Zeolites do not seem to be common in soils, and little attention is accorded these minerals in research on potassium fixation. Vermiculite, as mentioned previously, is a layer-type mineral similar to micas except that (1) interlayer potassium has been replaced by other cations that remain exchangeable and (2) the molecular aluminosilicate layers are spread farther apart than they are in micas. Various other layer-type clay minerals with expanded layers fix smaller quantities of potassium, and they may fix substantial quantities under special conditions. Most of the research on potassium fixation has been done with vermiculite and other expanded clay minerals.

As was mentioned in the preceding section, potassium ions lie in a plane between the molecular aluminosilicate layers of micas. They are imbedded in openings in the surface layers of oxygen atoms in adjacent layers and are bonded electrostatically to the oxygens in both layers. The fit of the potassium ions is such that they spread the layers slightly as compared to the spacings found by X-ray diffraction if the micaceous minerals contain no interlayer potassium and are fully dehydrated. If micaceous minerals are depleted of potassium without loss of integrity of the aluminosilicate layers, other cations enter and balance the electrostatic charges that were originally satisfied by potassium ; and water molecules enter and spread the molecular layers apart, producing an expanded mineral known as vermiculite. The water molecules are bonded strongly to the inner surfaces, as evidenced by the fact that they are not driven off even when the mineral is heated at 110°C. If the mineral is heated at 750°C, the water is driven off ; the molecular layers then collapse to produce a spacing slightly less than that observed when potassium ions are present. The cations that permit such collapse (calcium, magnesium, sodium) are smaller than potassium and presumably fit readily into the openings in the oxygen layers with space to spare. (Incidentally, the gross expansion of vermiculite produced by heat treatment is thought to result from mechanical breakage of some of the layers by the steam produced when the interlayer water is volatilized and not from uniform spreading of the molecular layers. X-ray diffraction shows the molecular layers to be collapsed.)

If the cations that have replaced potassium in formation of vermiculite from mica are subsequently replaced by potassium, the molecular layers are drawn together ; and the potassium is again entrapped in nonexchangeable form. This is the general theory of potassium fixation by expanded minerals. The process is simple enough in general outline, but the fixation that occurs seems to depend a great deal on the precise nature of the mineral and on the conditions.

With regard to the nature of the mineral, several factors are thought to be important. First, it should be mentioned that vermiculite and the other expanded clays that fix potassium all have the same basic molecular structure. All have the surface layers of oxygen atoms with the openings into which potassium can fit, and all have a negative charge on the molecular aluminosilicate layers that is balanced by cations located outside the layers ; however, the minerals differ in charge per unit area of the molecular layers. The electrostatic charge per unit area is commonly called the "charge-density" and is expressed in units of milligram equivalents per 100 g, like cation-exchange capacity. In fact, if the balancing cations are exchangeable, the charge-density is equivalent to the cation-exchange capacity. According to the theory that electrostatic charges on the aluminosilicate layers cause bonding of adjacent layers through intervening potassium ions, the strength of bonding between adjacent layers should increase with the number of charges per unit area. In agreement with this concept, micas and vermiculites, which have a relatively high charge-density, hold potassium in nonexchangeable form to a much greater extent than does montmorillonite, which has a lower charge density.

Also involved in the bonding between adjacent layers and hence in the fixation of potassium is thought to be a charge-strength factor associated with the origin of the negative charges. Three general concepts are involved in this idea. First, the negative charges on the aluminosilicate layers result from a deficiency of cations in these layers (the principal cations are silicon, aluminum, and iron) in relation to oxygens and hydroxyls. Second, an individual negative charge may result from a cation deficiency in the outer, tetrahedral sheet of the aluminosilicate layer or from a cation deficiency in the inner, octahedral sheet. Third, the strength of the bond at the surface of the layer, where it is balanced by potassium, will be greater if the charge originates in the outer, tetrahedral sheet than in the inner, octahedral sheet.

Three accessory conditions have been found to be of importance with regard to fixation of potassium by layer-silicate minerals : the nature of the balancing cations between layers, the particle size, and the degree of drying and heating. These will be discussed in turn.

If the balancing cations are polymeric hydroxyaluminum ions fixation of potassium will be reduced or eliminated because the hydroxyaluminum ions keep the molecular aluminosilicate layers permanently expanded. On the other hand, if the balancing cations have the same charge and about the same size as potassium, they will tend to be nonexchangeable themselves and for this reason will inhibit fixation of potassium. Effects of ionic size and charge are illustrated in Fig. 12.15, which shows the fixation of various monovalent and divalent cations by clay separated from a soil in Ohio. The monovalent ions seem to

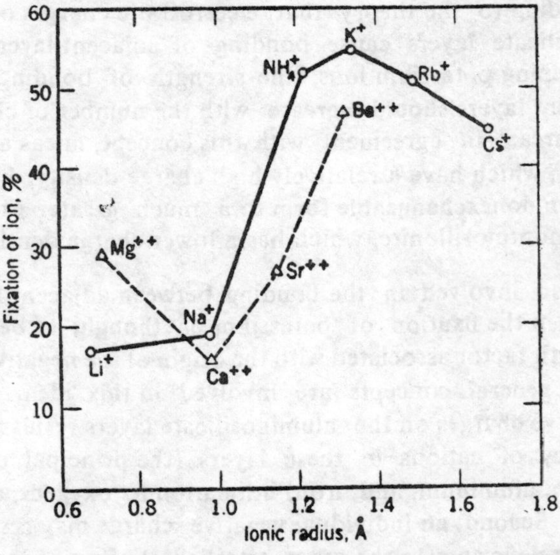


Fig. 12 15. Percentage fixation of different cations by a soil clay versus ionic radius. Percentage fixation is defined as the decrease in the quantity of the cation released from the clay to soluble form caused by drying the clay at 100°C before treatment with hydrochloric acid equivalent to the cation-exchange capacity expressed as a percentage of the quantity of the cation released in the same way from the undried clay saturated with the cation in question. This method of expressing the results tends to separate the fixation effect from differences in replaceability of the individual cations by the hydrogen ion.

form a regular series according to ionic size with peak fixation at the potassium size and decreasing fixation of smaller or larger cations, but the divalent cations do not follow this pattern. As indicated in the figure, the size of the ammonium ion is similar to that of the potassium ion, and the fixation of ammonium by the soil clay used was similar to the fixation of potassium.

Considerable research has been done on ammonium fixation not only because experiments involving ammonium alone or ammonium with potassium may provide information about potassium behaviour but also because of the possible significance of ammonium fixation in connection with nitrogen behaviour in soils. Ammonium fixation in the latter context was discussed at some length.

At this point, consideration of ammonium fixation will be limited to introduction of Fig. 12.16, which shows the results of two experiments that indicate the similarity of behaviour of ammonium and potassium. In Fig. 12.16*a*, fixation of potassium by samples of eight soils is similar to but, on the average, somewhat greater than the fixation of ammonium determined in the same manner except by addition of an ammonium salt instead of a potassium salt. In Fig. 12.16*b*, the amounts of ammonium and potassium fixed by samples of a single soil are shown where the ammonium was added first in quantities of 0 to 15 m.e,/100 g of soil and the potassium was added subsequently in the quantity of 2.5 m.e./100 g of soil. Fixation of ammonium increased with the quantity of ammonium added, and fixation of potassium decreased with an increase in the quantity of ammonium fixed previously ; but the sum of ammonium fixed and potassium fixed remained almost constant. The results of this experiment indicate that ammonium and potassium were fixed by the same sites and that if ammonium was fixed first, potassium merely occupied the remaining potassium-fixing sites that were not already occupied by ammonium. In an additional experiment, not shown in the figure, potassium was added first, in quantities of 0 to 15 m.e./100 g ; and ammonium was added second, in quantities

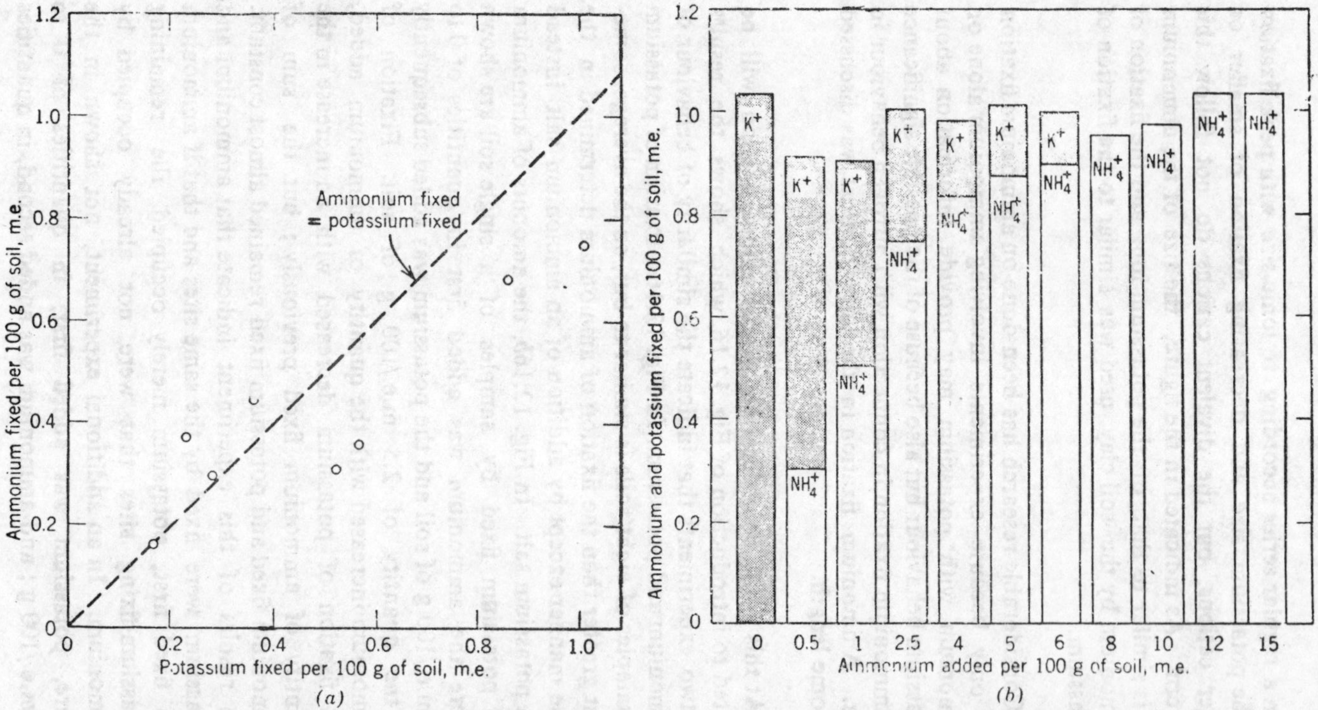


**Fig. 9.16.** Fixation of potassium and ammonium by moist soils. (*a*) Ammonium fixation versus potassium fixation by eight different potassium-fixing soils of Iowa where ammonium and potassium were added singly as the chlorides in quantities of 2.5 m.e./100 g of soil. (*b*) Ammonium fixed from different additions of ammonium chloride to samples of a single, potassium-fixing, silty clay loan soil of Iowa : and potassium fixed from a uniform addition of 2.5 m.e. of potassium chloride per 100 g of soil after the ammonium chloride had been added.

of 10 m.e./100 g. Prior potassium fixation then reduced ammonium fixation, but the sum of the two increased somewhat with the quantity of potasium added. At a given concentration of fixable cations, there was greater fixation of potassium than of ammonium, in accordance with the observations shown in Fig. 12.16*a* and Fig. 12.15.

A second accessory condition that affects the fixation of potassium by expanded layer-silicate minerals is the particle size. This factor has been investigated with vermiculite, which may be obtained in particles of the order of a centimeter or more in diameter. Barshad (1954) found that sodium was far more effective than potassium or ammonium in replacing other exchangeable bases from vermiculite particles of large (gravel) size but that in particles small enough to pass a sieve with openings 0.1 mm in diameter the effect was practically absent. He explained these results by the theory that addition of potassium or ammonium promoted contraction of the molecular layers of the vermiculite at the edges of the particles where the exchange began and that the cations in the more remote inner locations were then trapped and nonexchangeable because of the restricted movement of cations in and out through the peripheral contracted zones. He noted the same trapping effect in an experiment on a subsoil that contained coarse, weathered biotite and had a high capacity to fix potassium and ammonium. He saturated the soil with barium, removed the excess barium salt with methyl alcohol, and then leached separate portions with neutral, 1-normal ammonium acetate, with neutral, 1-normal sodium chloride, and with 1-normal hydrochloric acid until the filtrate was free of barium. The latter two reagents replaced 31 m.e. of barium per 100 g of soil, but the ammonium acetate replaced only 25 m.e. The difference of 6 m.e. represents the trapped barium.

A third accessory condition that influences the fixation of potassium by expanded layer-silicate minerals is the degree of drying and heating to which the material is subjected before the fixation is measured. In general, drying and heating markedly increase the fixation. The effects of drying and heating of layer silicates on potassium fixation are explained theoretically on the basis of loss of absorbed water from interlayer positions and closure of the lattice layers about the potassium ions. Vermiculite, with a spacing of about 14 Å from the surface of one aluminosilicate layer to the corresponding surface of adjacent layers, contains absorbed water between the molecular layers. If potassium ions are exchanged for the other balancing cations without drying the sample in the process, the spacing is decreased, often to about 12 Å, which must mean that the increase in strength of bonding between adjacent layers from introduction of potassium has caused some of the adsorbed water to be squeezed out. The potassium is not completely fixed when the mineral remains in this hydrated condition. If the material is air-dried at room temperature or at higher temperature, however, the spacing between the layers decreases; and more of the potassium becomes nonexchangeable. Existence of adsorbed water between the layers of micaceous minerals of high charge-density after saturation with potassium seems to be a metastable condition because the illites that presumably develop when potassium-depleted micaceous minerals take up potassium from sea water have the 10-Å spacing characteristic of muscovite and biotite. Montmorillonite, which has a relatively low charge-density, can be caused to fix potassium by saturating the exchange positions with potassium followed by oven-drying the mineral. Layer spacings of samples treated in this way are greater than those found in vermiculites after similar treatment, and subsequent release of potassium and expansion of the layers occur more readily than in vermiculites.

In research on potassium fixation in soils, considerable importance is attached to the distinction between fixation on drying and fixation occurring under moist conditions because soils remain for the most part in a moist condition as regards potassium fixation even though at times they may not contain enough water to support higher plants. Hence the fixation that occurs under moist conditions in the laboratory is inferred to be a phenomenon that would occur also in the field if an excess of potassium were added. If one disregards fixation of potassium in trace quantities, experiments indicate that some soils fix potassium under moist conditions and others do not. Perhaps most soils would fix some potassium under moist conditions with long exposure to a great enough excess of potassium. Many more soils show potassium fixation if they are dried and heated after treatment with excess potassium than if they are maintained in a moist condition.

One curious finding that has intrigued a number of investigators is that, although soils usually fix added potassium if they are dried or dried and heated, these same treatments often cause release of potassium from nonexchangeable form if the soil is dried without addition of potassium. A widely accepted theory to account for this behaviour is that the content of exchangeable and soluble potassium in the potassium-treated soil is above an equilibrium condition in relation to nonexchangeable potassium, whereas the untreated soil is below it, and that drying and heating promote attainment of equilibrium.

The theory just mentioned seems now to be inadequate to account for experimental observations that have been made on potassium fixation and release. Luebs *et al.* (1956) found that the release of potassium from soils on drying took place when the soils they investigated approached the air-dry state. In laboratory experiments with samples of soil maintained in contact with air containing water vapour at different relative humidities, they found that the exchangeable potassium would remain relatively high at low humidities and low at high humidities. If the relative humidity was increased, the exchangeable potassium would decrease and *vice versa*. These observations verify the concept of an equilibrium between exchangeable and nonexchangeable potassium because the exchangeable potassium first approached a certain value at a given relative humidity and then remained approximately constant for a long time. But because the value for exchangeable potassium depended on the relative humidity, there was evidently no single equilibrium level.

The same authors made further analyses for exchangeable potassium in soils in the field to see if the values they obtained would vary from time to time with the water content of the soil. As shown in Fig. 12.17, the values obtained for exchangeable potassium were by no means constant and tended to vary

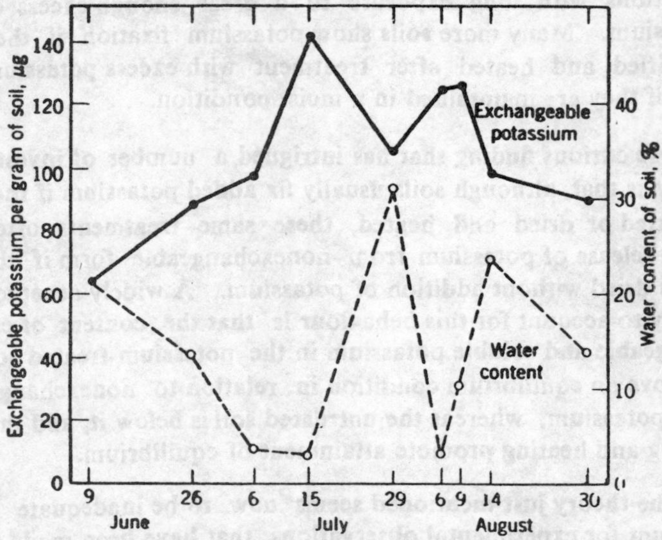


Fig. 12.17. Content of exchangeable potassium and water in the surface 1.3 cm of a loam soil sampled in the field at different dates duriug a summer season in Iowa.

inversely with the water content of the soil, as had been found in the laboratory work. This effect was largely confined to the upper part of the soil ; the range of values for exchangeable potassium in micrograms per gram of soil was 64 to 141 in the surface 1.3 cm, 57 to 87 in the second 1.3 cm, and 52 to 65 in the next 2.5 cm. The lower layers did not become as dry as the surface layer.

The effect of drying on exchangeable potassium may be of some significance to plants under field conditions if plant roots intermittently absorb potassium from the surface layer. As indicated by Fig. 12.18, the increase in exchangeable potassium produced when a soil was dried before growth of corn plants was associated with a substantial increase in uptake of potassium by the plants as compared with soil that was not dried. In the instance shown in the figure, the effect of drying was

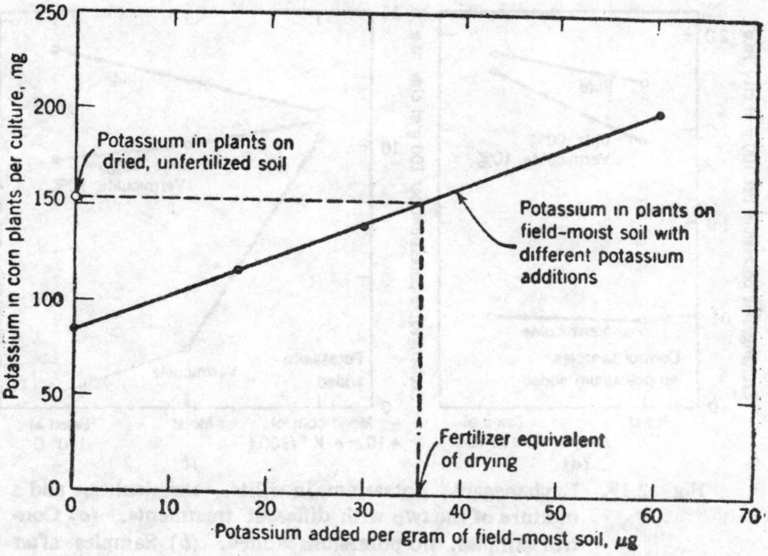


**Fig. 12.18. Biological assay of effect of drying on supply of native potassium for corn plants in a loam soil in greenhouse cultures. Drying the soil before cropping increased the exchangeable potassium, as determined by the neutral, normal, ammonium acetate method, from 47 to 85 μg/g.**

equivalent to the increase in potassium uptake obtained from addition of 36.5 μg of potassium as potassium chloride per gram of undried soil. In analogous trials with ten other soils of Iowa, the increases in uptake of potassium resulting from drying the soil before growth of plants were found to be equivalent to 8 to 60 μg/g. The potassium status of soil thus seems to be affected long enough by drying to influence the uptake of potassium by plants grown after the soil has been remoistened.

Development of knowledge of potassium behaviour in relation to mineralogy has resulted in a new theory to the effect that potassium fixation is due to potassium-fixing minerals and release is due to potassium-bearing minerals and that both processes may occur in a given soil under appropriate conditions. Figure 12.19 shows the results of an experiment that

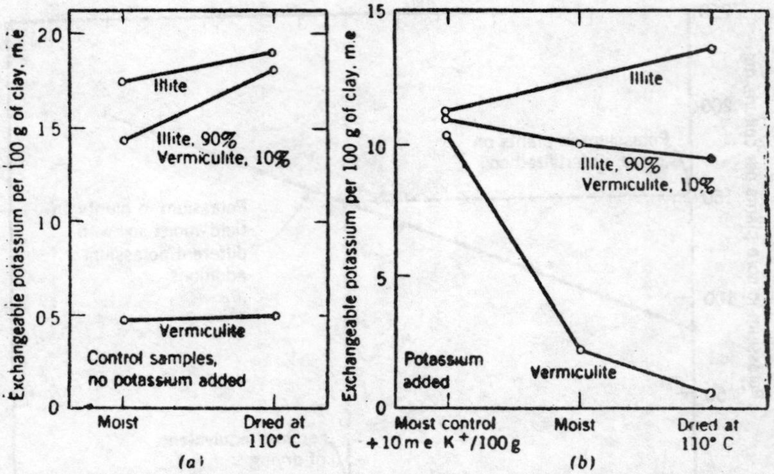


**Fig. 12.19.** Exchangeable potassium in illite, vermiculite, and a mixture of the two with different treatments. (*a*) Control samples, no potassium added. (*b*) Samples after addition of 10 m.e. of potassium per 100 g. Values for "moist control + 10 m.e. $K^+$ per 100 g" are calculated.

verify this theory. In this experiment, 0.5-normal sodium acetate was used to displace exchangeable potassium from illite alone, vermiculite alone, and a mixture of the two minerals with and without addition of potassium. Illite (a potassium-bearing mineral) released potassium from nonexchangeable form when dried at 110°C in either the presence or absence of added potassium. Vermiculite (a potassium-fixing mineral), on the other hand, maintained a low, approximately constant value of native exchangeable potassium despite the drying and heating treatment but fixed most of the added potassium. A mixture of 90% illite with 10% vermiculite behaved like soils in that it released potassium on drying of the control samples but fixed potassium if potassium was added. As part of the same investigation, DeMumbrum and Hoover (1958) found that drying two soil clays and heating them at 110°C after addition of potassium caused release of potassium from nonexchangeable form from one clay containing illite and montmorillonite and fixation of potassium by the other clay containing illite, montmorillonite, and vermiculite.

As yet, the cause of release of potassium from nonexchangeable form that occurs when soils are dried is uncertain. A.D. Scott (unpublished) developed the theory that, although the X-ray diffraction pattern of micaceous minerals indicates a decrease of interlayer spacing when the expanded, weathered minerals are dried, certain of the interlayer spacings are increased by curling of the mineral layers when the mineral dries enough to cause loss of some of the interlayer water. The originally nonexchangeable potassium exposed on these more open layers then becomes exchangeable. The X-rays presumably do not show these widened interlayer spaces because there are only a few of them and their spacing varies. Diffraction of X-rays with enough intensity to be registered requires repetition of a given spacing. Represents evidence on a macro basis for the theory. The figure shows that drying actually increased the over-all thickness of a particle of mica that had been partially degraded at the edges by immersion in a solution containing sodium tetraphenylboron. This theory is consonant with the observation that heating vermiculite at a high temperature causes loss or water and a decrease of the interlayer distance, as indicated by X-ray diffraction, at the same time as it produces gross expansion of the vermiculite particles.

The theory that the potassium fixation in soils under moist conditions occurs mostly in vermiculite-type minerals implies that this capacity to fix potassium is a consequence of prior removal of potassium (because vermiculite is thought to be derived from potassium-bearing micas). Thus, with greater prior removal of potassium, the fixation capacity should be increased. Welte and Niederbudde (1963) verified this inference in an experiment on samples of soil from a field plot in Germany that fixed no potassium under moist conditions, according to the procedure they used, in consequence of prior heavy fertilization with potassium. The soil was mixed with quartz sand in different proportions, with the total weight remaining constant at 11.5 kg per culture. The cultures were planted to ryegrass, and four cuttings were taken. Potassium analyses on the crop and soil and potassium fixation determinations on the soil then yielded the results shown in Fig, 12.20.

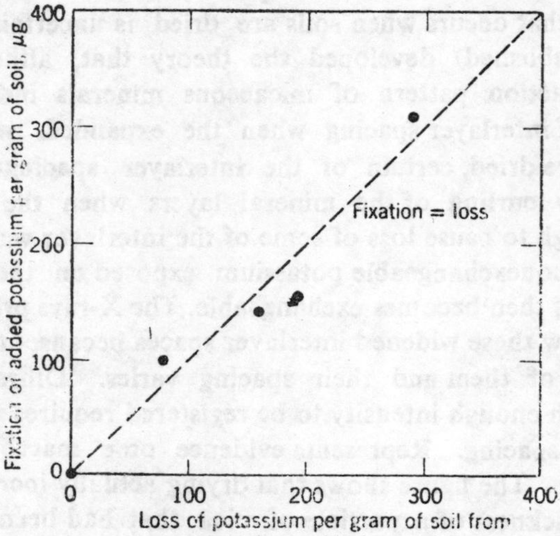


**Fig. 12.20. Fixation of added potassium under moist conditions, determined after growth of ryegrass (four cuttings), versus loss of potassium from nonexchangeable form in the soil during growth of the crop. The different experimental values were obtained with a single soil at different cropping intensities produced by varying the proportions of soil and sand in the culture vessels. The values of zero fixation and zero loss of potassium from nonexchangeable a form were obtained on uncropped soil.**

Evidently cropping did generate capacity for potassium fixation in this soil, and the potassium-fixing capacity produced was about equal to the loss of potassium from nonexchangeable form during growth of the crop.

In consequence of the increase in fixing capacity of soil around the roots, subsequent equilibration may be expected in which potassium released from other locations is reintroduced into some of the newly formed exchange sites from which originally nonexchangeable potassium was removed. Detection of such adjustments in soils under natural conditions is not readily accomplished ; but existence of the process in principle was verified by Mortland (1961), who suspended equal quantities of biotite mica and sodium-saturated soil clays in a large volume of 0.02-normal sodium chloride solution for 21 days and then determined the total potassium content of the clays after the mica had been removed by sieving. He found (Fig. 12.21) that the potassium percentage increased in all clays in the presence of biotite and that the lower the original potassium percentage in the clay the greater was the increase. The increases in potassium content of the clays were so great in relation to the concentration of potassium in the sodium chloride solution (7 to 9 $\mu$g/ml) that most of the increase must have been non-exchangeable potassium. X-ray diffraction analysis of the residue from the mica at the end of the equilibration period indicated the presence of vermiculite.

With regard to association of potassium fixation with other soil characteristics, three properties have been found significant : clay content, *p*H, and percentage potassium saturation. These will be discussed in turn.

Kaila (1965) found that potassium fixation by samples of many soils of Finland increased with the clay content. Schachtschabel and Koster (1960) found that potassium fixation under moist conditions by marsh soils of Germany increased with the clay content but that, as shown in Fig, 12.22*a*, the fixation at a given clay content tended to be relatively low in the salt-water-marsh samples, intermediate in the lake-marsh samples, and high in the river-marsh samples. The clay content of soil may be considered a capacity or quantity factor in potassium fixation. Most of the fixation takes place in this fraction. Nevertheless, as indicated in Fig. 12.22 and as would be expected from the theory of potassium fixation, potassium fixation per gram of clay may differ enough from one soil to another to make clay content as such of limited value as an index of potassium fixation.

The effect of soil *p*H on potassium fixation has been investigated principally in connection with experiments on application of calcium carbonate. In some instances liming has increased potassium fixation, and in others it has had little effect. Kaila (1965) examined the *p*H effect in measurements of potassium fixation by a large number of soil samples from Finland and found a significant increase in fixation with an increase in soil *p*H.

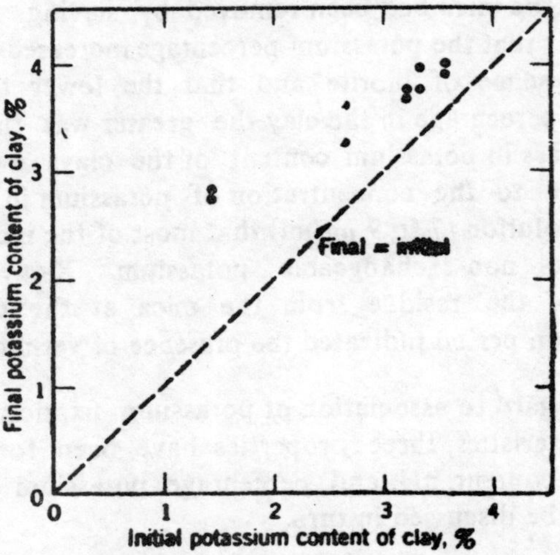


Fig. 12.21. Potassium content of the clay from five soils of Michigan after contact for 21 days with an equal weight of biotite in 0.02-normal sodium chloride solution versus original potassium content of the clays. At the end of the reaction period, the biotite was removed by sieving before analysis of the clay for potassium. The two data points for each soil represent results obtained with different particle-size fractions of biotite. The greater increase in potassium content of four of the clays during equilibration occurred with the finer particle-size fraction.

Page and Ganje (1964) found that fixation of potassium under moist conditions by soils containing vermiculite-type minerals was little influenced by *p*H in the range from 3 to 7 as long as the concentration of aluminum remained low. High concentrations of nonexchangeable interlayer aluminum resulting from leaching the soils with dilute acid prevented fixation of potassium at *p*H 3. Fixation occurred again when the acid-treated soils were adjusted to *p*H 7. Existence of a blocking effect of aluminum may be inferred also from earlier work by Barshad (1954) on fixation of ammonium by vermiculite. He found that fixation of

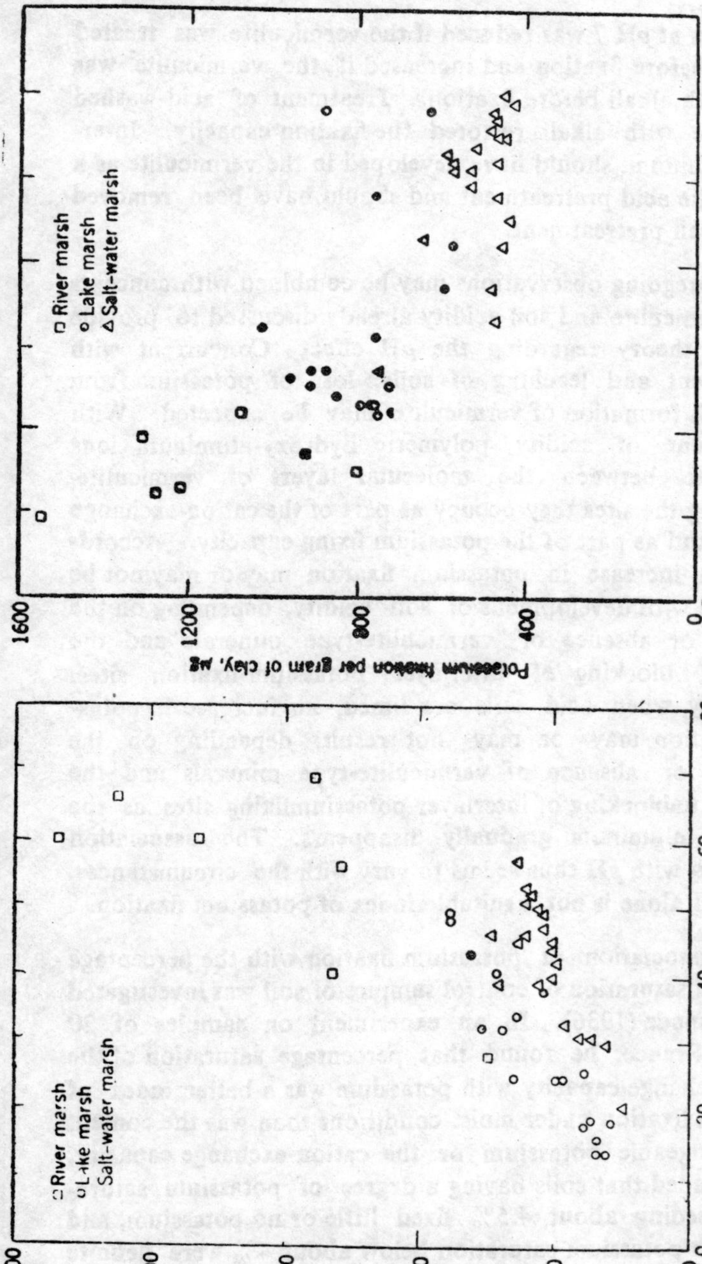


**Fig. 12.22.** **Potassium fixation under moist conditions by soils from three kinds of marshes in Germany. (*a*) Potassium fixation versus clay content of the soils. (*b*) Potassium fixation versus percentage saturation of the cation-exchange capacity of the soils with potassium.**

ammonium at *p*H 7 was reduced if the vermiculite was treated with acid before fixation and increased if the vermiculite was treated with alkali before fixation. Treatment of acid-washed vermiculite with alkali restored the fixation capacity. Interlayer aluminum should have developed in the vermiculite as a result of the acid pretreatment and should have been removed by the alkali pretreatment.

The foregoing observations may be combined with concepts about vermiculite and soil acidity already discussed to provide a general theory regarding the *p*H effect : Concurrent with development and leaching of soils, loss of potassium from micas with formation of vermiculite may be expected. With development of acidity, polymeric hydroxy-aluminum ions accumulate between the molecular layers of vermiculite, eliminating the sites they occupy as part of the cation-exchange capacity and as part of the potassium fixing capacity. Accordingly, an increase in potassium fixation may or may not be associated with development of soil acidity, depending on the presence or absence of vermiculite-type minerals and the degree of blocking of interlayer potassium-fixation sites. Conversely, when acid soils are limed, an increase in potassium fixation may or may not result, depending on the presence or absence of vermiculite-type minerals and the degree of unblocking of interlayer potassiumfixing sites as the interlayer aluminum gradually disappears. The association of fixation with *p*H thus seems to vary with the circumstances, and so *p*H alone is not a suitable index of potassium fixation.

The association of potassium fixation with the percentage potassium saturation of control samples of soil was investigated by Chaminade (1936). In an experiment on samples of 20 soils of France, he found that percentage saturation of the cation-exchange capacity with potassium was a better index of potassium fixation under moist conditions than was the content of exchangeable potassium or the cation-exchange capacity. He concluded that soils having a degree of potassium saturation exceeding about 4.5% fixed little or no potassium, and those with potassium saturation below about 4% were definite potassium fixers.

Schachtschabel and Koster (1960) likewise noted that fixation of potassium under moist conditions by marsh soils of Germany decreased with an increase in percentage saturation of the cation-exchange capacity with potassium. Their data in Fig. 12.22*b* show that values for the three classes of marsh soils tend to fall in different ranges but that all the results may be expressed approximately by a single curve. The seeming existence of a single relationship in Fig. 12.22*b* for the same soils that gave somewhat different relationships in part *a* of the figure suggests that percentage potassium saturation is a more valuable index of the potassium-fixing tendency of soils than is clay content. Percentage potassium saturation is related to the proportion of potassium in the soluble cations, and the latter has application in the mass-law concept of potassium fixation and release discussed briefly in the preceding section. A similar plot of data by van der Marel and Venekamp (1955) on samples of 14 soils derived from river sediments and 13 soils derived from marine sediments in the Netherlands shows that potassium fixation decreased with an increase in percentage saturation of the cation-exchange capacity with potassium, with no indication of existence of separate relationships for the two groups of soils.

An additional experiment of significance in connection with the effect of percentage potassium saturation was conducted by Chaminade (1936). A field plot on low-potassium soil was fertilized each year with potassium sulfate equivalent to 208 kg of potassium per hectare, and a sample of soil was taken in late autumn of each year for measurement of exchangeable potassium and percentage potassium saturation. The results in Fig. 12.23 show that the annual increase in exchangeable potassium from the treatment was relatively small until the percentage potassium saturation reached a value of about 4%, and then it increased rapidly. Chaminade attributed the small increases in exchangeable potassium during the earlier years to fixation of the added potassium in nonexchangeable form and the larger increase at the end to lesser fixation at the higher percentage potassium saturation.

In summary, fixation of potassium in nonexchangeable form in soils is now thought to be a property of many minerals, of which expanded layer-silicates in the clay fraction are by far the most important. In these minerals, the potassium is thought to be held electrostatically on the outside of the molecular aluminosilicate layers as a balancing cation and to remain exchangeable if present on outer surfaces but to be subject to confinement in nonexchangeable form in closely fitting voids between the layers. If the expanded layer-silicate has a sufficiently high cationexchange capacity (vermiculite), partial closure of the layers with loss of part of the interlayer water and fixation of much of the interlayer potassium occurs in an aqueous environment ; and full closure to the spacing

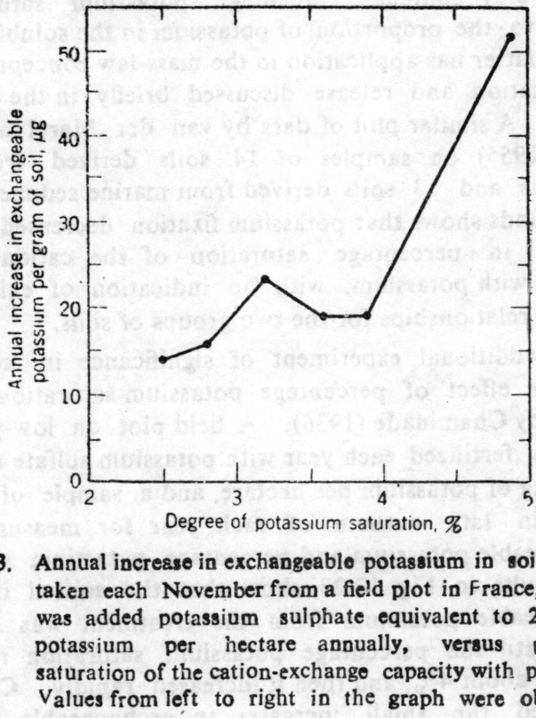


**Fig. 12.23.** **Annual increase in exchangeable potassium in soil samples taken each November from a field plot in France, to which was added potassium sulphate equivalent to 208 kg of potassium per hectare annually, versus percentage saturation of the cation-exchange capacity with potassium. Values from left to right in the graph were obtained in successive years.**

of micas, with loss of most or all of the remaining interlayer water and fixation of more potassium, occur if the mineral is dried and heated. If the expanded layer-silicate has a lower cation-exchange capacity (montmorillonite or related minerals), fixation of potassium occurs mostly when the mineral is dried and heated ; and re-expansion and release of potassium occur comparatively readily.

The potentials to release potassium and to fix potassium coexist in soil. Release of potassium from nonexchangeable, interlayer form occurs rapidly if the external concentration of potassium is low enough and if the concentration of other cations is high enough. The cation-exchange positions thus produced are potential potassium-fixing sites as long as they are not blocked. Blockage may occur from the presence of ammonium or polymeric hydroxyaluminum ions. Uptake of potassium in nonexchangeable form by vermiculite-type minerals occurs even at potassium concentrations of the order of a few micrograms per milliliter, and vermiculite can probably be largely resaturated with nonexchangeable potassium at such concentrations. Higher concentrations may be expected to cause somewhat more fixation by increasing the proportion of interlayer exchange sites occupied by potassium, with consequent contraction of more molecular layers. All layers do not behave the same even in a single particle, perhaps because of differences in density of charge on the molecular aluminosilicate layers. Ocean water contains about 380 $\mu g$ of potassium per milliliter, which accounts for regeneration of micas in ocean sediments.

The capacity of some soils to maintain for months or years, without loss by fixation, a level of exchangeable potassium in excess of that at which potassium is released from nonexchangeable form by micaceous minerals may be explained on the basis that the strongly fixing, vermiculite-type minerals are largely saturated with nonexchangeable potassium, and a much higher concentration of potassium and a much higher proportion of potassium in the soluble cations would be needed to cause much fixation by other less strongly fixing minerals that may be present. Another possibility is that uptake of potassium by vermiculite-type minerals at high concentrations of potassium in the external solution is inhibited by closure of molecular layers around the periphery of the particles before complete exchange has occurred.

***Soil-Plant Relationships in Potassium Fixation.*** To conclude the subject of potassium fixation, an attempt will be made to place in perspective the biological significance of the process. First, some experiments described in a paper by Chaminade (1936) on potassium fixation will be considered. These provide different forms of evidence and at the same time illustrate techniques by which the biological significance of potassium fixation may be investigated.

In the first experiment, Chaminade added 830 $\mu$g of potassium as potassium chloride per gram of soil to a sample of potassium-fixing soil and incubated this sample along with an untreated control sample for 2 months. During this time, fixation of potassium in nonexchangeable form amounted to 294 $\mu$g/g of soil. At the end of the incubation, portions of both samples were leached to different degrees with calcium chloride to obtain three different levels of exchangeable potassium, with one series containing 294 $\mu$g of fixed potassium per gram of soil and the other series containing no fixed potassium. All samples were then subjected to intensive cropping by rye seedlings, and the seedlings were analyzed for potassium. The results in Fig. 12.24 show not only that the

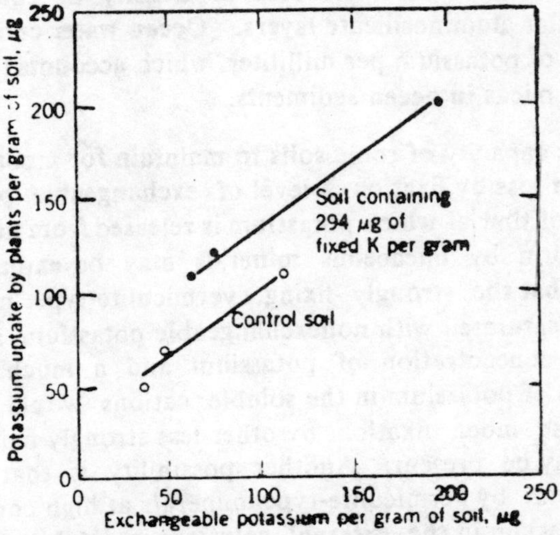


Fig. 12.24. Uptake of potassium by rye seedlings from equal quantities of control soil and soil containing fixed potassium versus exchangeable potassium in the two samples.

potassium-treated soil contained more exchangeable potassium than the control soil but also that the quantity of exchangeable potassium needed to obtain a given uptake of potassium was less where plant were grown on the potassium-treated soil than on the control. Presumably, the plants were making use of the potassium that had been fixed from the potassium added during the 2-month incubation period. If the difference in quantity of exchangeable potassium required to obtain equal uptake of potassium from the control and potassium-treated samples did in fact result from use of some of the fixed potassium, one may estimate the relative value of the exchangeable and fixed potassium in the following way : The horizontal distance between the two lines in the figure is equivalent to about 50 $\mu$g of exchangeable potassium per gram of soil. Dividing 50 by 294, the number of micrograms of fixed potassium per gram of soil, gives 0.17. The value 0.17 is sometimes called the availability-coefficient ratio because it gives the ratio of the biological effectiveness of one unit of the nutrient in one from to that of one unit of the nutrient in another form taken as the standard. In this instance, exchangeable potassium is taken as the standard, and one unit of fixed potassium had a biological value 0.17 as great as that of one unit of exchangeable potassium.

In the second experiment, Chaminade employed a potassium-fixing soil and added different quantities of potassium as potassium chloride to one series of samples in the autumn and maintained the samples in a moist condition during the winter. To another series of samples, which had been kept moist during the winter but with no potassium treatment, he added similar quantities of potassium in the spring. All containers were then planted immediately to barley, and the barley was grown to maturity. The results of this experiment are shown in Fig. 12.25. In this experiment, application of the equivalent of 52.4 $\mu$g of potassium per gram of soil in the autumn produced the same yield of barley as about 16 $\mu$g applied in the spring, which means that the availability coefficient of the added potassium as a whole was a little less than one-third as great if the addition was made in the autumn as if it was made in the spring. The decrease in availability due to autumn application may be said

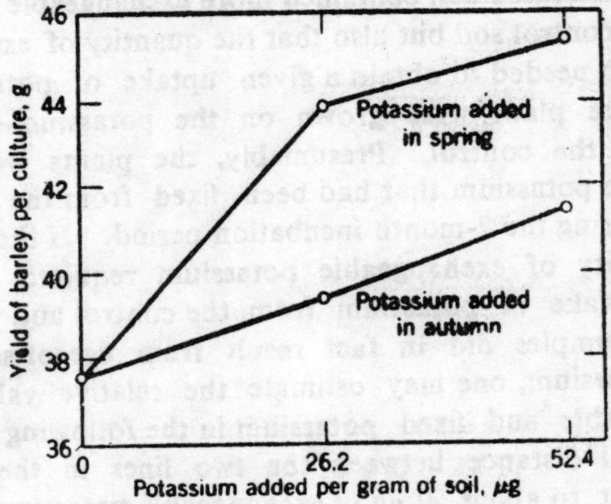


Fig. 12.25. Yield of barley on soil with different additions of potassium as potassium chloride in the spring just before planting or in the preceding autumn.

to result from fixation because the soil samples were not subjected to cropping or leaching during the incubation period.

The experiment just described was done with a soil that was known to fix added potassium. In another experiment on eight soils, Chaminade found that incubation for a year before cropping did not decrease the value of added potassium to plants unless the soil fixed potassium in nonexchangeable form.

Potassium fixation may be evidenced in several ways in the behaviour of plants in the field. Many crop plants are deficient in potassium on potassium-fixing soils and do not respond in the usual way to potassium fertilization. If potassium fertilizer is applied to soils that produce potassium-deficient crops but fix little or no potassium, a plot of crop yield against potassium applied gives a curve that rises rapidly with small applications of potassium (of the order of 20 to 60 kg/hectare) and slowly with larger applications. If, on the other hand, potassium fertilizer is applied to potassium-fixing soils that produce crops similarly deficient in potassium, the yield does not increase so rapidly per unit of potassium added. For example, Page *et al.* (1963) reported that correction of potassium deficiency for cotton required quantities of fertilizer potassium in excess of 170 kg/hectare on potassium-fixing soils in the San Joaquin Valley in California in contrast to quantities of the order of one-third as much under usual circumstances of potassium deficiency. If potassium fixation is moderate, the response curve may have the same general shape as it does if there is little or no fixation except that the quantities required to reach the maximum yield are greater. With strong potassium fixation, the response curve may have a different shape, rising slowly with small applications and more rapidly with larger applications and finally reaching a maximum with still larger applications. Fig. 12.26 shows the lower part of such a response curve. The potassium content of the leaf blades of the cotton plants 90 days after planting was only 0.9% of the dry matter with the heaviest application of potassium, which verifies that the plants were not high in potassium. This experiment was done in the greenhouse with one of the soils of the San Joaquin Valley that produced potassium-deficient cotton in the field and required heavy applications of potassium fertilizer. The soil in question contained vermiculite, and laboratory work on samples of soil from the various cultures in the greenhouse test showed that the cation-exchange capacity was reduced from 3 2 to 2.5 m.e./100g upon application of the largest quantity of potassium employed. This reduction in cation-exchange capacity, calculated as potassium fixation, is equivalent to transfer of 65% of the added

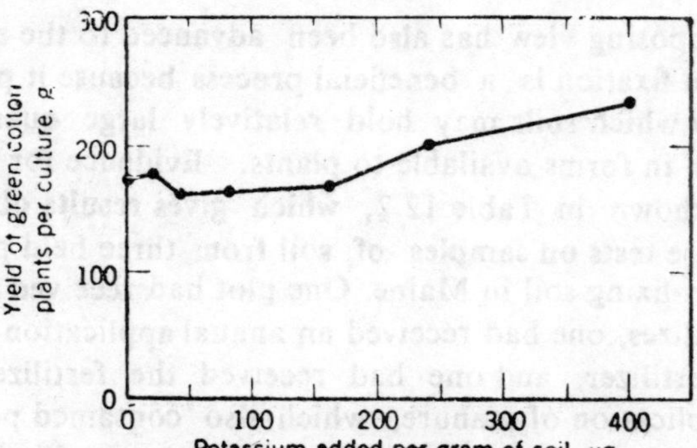


Fig. 12.26. Yield of cotton plants with different applications of potassium as potassium sulphate to a potassium-fixing, fine sandy loam soil of California.

potassium to nonexchangeable form. Probably a response curve similar to the one shown in Fig. 12.26 would have been obtained in the field by Van Dijk (quoted by Middelburg, 1955) if a wide enough range of quantities of potassium had been applied. Van Dijk found that cassava grown on a clay soil in Java did not respond to potassium until potassium fertilization had been practiced for four consecutive years.

In addition to the gross effects evidenced by plant yields, there may be internal effects that give additional information on the effect of potassium fixation. Perhaps the most sensitive of these is the potassium content of the plant, to which may be attributed the yield effects more commonly determined. Breemhaar (1956) found that the potassium percentage in leaves of potato plants increased much more with a given application of potassium to a field in which the soil had little tendency to fix potassium than it did with an equal application of potassium to another field in which the soil fixed potassium strongly.

The experiments just discussed provide evidence for the view that potassium fixation has the detrimental effect of decreasing the availability coefficient of potassium added to soil. That is to say, if equal quantities of added potassium are present initially in nonexchangeable and exchangeable form, the former will be less effective than the latter because plants use potassium preferentially from the exchangeable form.

An opposing view has also been advanced to the effect that potassium fixation is a beneficial process because it provides a means by which soils may hold relatively large quantities of potassium in forms available to plants. Evidence for this position is shown in Table 12.7, which gives results obtained in greenhouse tests on samples of soil from three field plots on a potassium-fixing soil in Maine. One plot had received no potassium fertilizes, one had received an annual application of potassium in fertilizer, and one had received the fertilizer plus a heavy application of manure, which also contained potassium. The amount of potassium lost from nonexchangeable form during cropping evidently increased with the amount of potassium added in previous field treatments and with the accumulation of nonexchangeable potassium released from the soil upon treatment with 1-normal nitric acid.

These opposing views may be advanced from the same factual background, although some experiments would be more favourable to one and some to the other. The views represent, in effect, judgments about what is and is not beneficial ; and these judgments involve considerations not inherent in the experimental data.

The view that potassium fixation is detrimental may be supported on two bases : (1) Over a period of years following potassium application, fixation will decrease the uptake of potassium by plants per kilogram added. Accordingly, more

**Table 12.7.** Loss of Potassium from Exchangeable and Nonexchangeable Forms During Greenhouse Cropping of Soil from Field Plots in Maine That Had Accumulated Different Amounts of Potassium as a Result of Previous Field Treatments.

| *Potassium Applied Per Hectare Annually in the Field, kg* | *Potassium per Gram of Soil, µg* — *Present in Soil at Planting*: *Exchangeable* | *Present in Soil at Planting*: *Nonexchangeable Released by 1-N $HNO_3$* | *Loss from Soil During Growth of Ladino Clover (15 Cuttings)*: *Exchangeable* | *Loss from Soil During Growth of Ladino Clover (15 Cuttings)*: *Nonexchangeable* |
|---|---|---|---|---|
| 0 | 36 | 146 | 5 | 49 |
| 74 | 92 | 201 | 59 | 66 |
| 74[1] | 457 | 327 | 420 | 234 |

1. Plus potassium added in 2.2 metric tons of manure per hectare.

potassium will be required, at greater total cost, to eliminate the deficiency of potassium. (2) Although the potassium that has been fixed in nonexchangeable form may eventually be used, its principal use will occur after the exchangeable potassium has been reduced to a level low enough that failure to add more fertilizer potassium may be uneconomic. Moreover, return in crop value from money expended for the fixed potassium will be delayed ; this also has the effect of decreasing the total return because the earning power of money invested in fixed potassium is zero as long as the potassium remains fixed.

The alternative view that potassium fixation is beneficial may be supported by arguments such as the following : (1) Potassium fixation retards loss of potassium by leaching and hence increases the efficiency of use of fertilizer. (2) Plants take up potassium much in excess of their needs if enough is present in exchangeable form. Potassium fixation reduces this excessive uptake and hence increases the efficiency of use of the fertilizer.

There are several ways in which the efficiency of potassium use on potassium-fixing soils may be improved : (1) changing the method of application of the fertilizer from the usual one of mixing the fertilizer with the soil to (*a*) localized application, (*b*) spray application to above-ground plant parts, or (*c*) injection into the plants ; and (2) changing the cropping pattern from crops that require a high level of exchangeable potassium to those that grow well with a low level of exchangeable potassium. With regard to method of application, (*a*) seems to be employed most commonly, (*b*) to a small extent, and (*c*) rarely if ever. Adjustments in cropping pattern are widely employed as farmers observe how different crops behave on their fields. In general, however, the decision to use any of these practices is probably rarely based on explicit knowledge of potassium fixation by particular soils because too little information is available.

The significance of potassium fixation to plants growing in the field has received little attention except by Dutch scientists. Schuffelen and van der Marel (1955), who reviewed the work on this subject, were not convinced that the phenomenon has much effect on plant yields. In discussing Dutch field experiments, they noted that soils that fix potassium strongly produce normal yields of sugar beets, mangolds, oats, barley, and wheat if applications of 250 to 330 kg of potassium are made per hectare. Potatoes are more sensitive to potassium deficiency. In this connection, they cited experimental findings on two fields, one on soil that fixed much added potassium in non-exchangeable form and the other on soil that fixed little potassium. Annual applications of 241 kg of potassium per hectare were needed for 8 to 10 years to increase potato yields on the former soil to the level of those on the latter.

In interpreting the views of Schuffelen and van der Marel (1955) in terms of conditions elsewhere, it may be worth bearing in mind that applications of fertilizer per hectare in the Netherlands are among the highest in the world. Quantities of 100 to 150 kg of potassium per hectare, which they classed as normal annual applications for clay soils developed on marine and river sediments, would be considered high applications in most parts of the world. Accordingly, in an area where common annual applications are of the order of one-fourth of this range, potassium fixation might be considered more important.

**Water Content and Aeration of Soil**

Of the accessory factors that may affect the availability of soil potassium to plants, the water content of the soil has probably received most attention. Figure 12.27 shows the results of a field experiment in the Netherlands in which potatoes were grown with and without an application of potassium fertilizer in each of a number of consecutive years. The aspect of the results that seems to characterize the relation between water supply and availability of soil potassium is the gradually increasing response to potassium fertilization as the number of rainless days increased. Because the heavy annual applications of potassium should have eliminated potassium deficiency as a limiting factor in growth of potatoes on the fertilized plots in all years, the trend of the results signifies that the deficiency of soil potassium in the control plots became more pronounced as the soil became drier. Similarly, Bruns (1935) in Germany found that the ratio of yield of field beans on control plots to the yield on potassium-fertilized plots was 0.37 in 5 moist years and 0.22 in 6 dry years. In England, Russell and Watson (1938) found that the increase in yield from an application of the sulphates of potassium, sodium, and magnesium increased

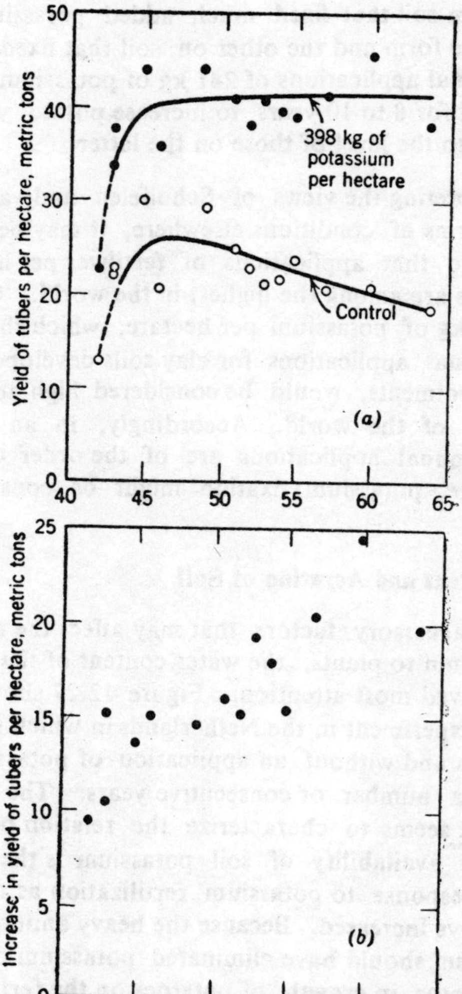


**Fig. 12.27. Response of potatoes to potassium fertilization versus number of rainless days in May, June, and July in 14 consecutive years in a field experiment in the Netherlands. (*a*) Yield of tubers with and without potassium fertilization. (*b*) Increase in yield from potassium fertilization.**

as the spring and summer rainfall decreased. In Ireland, Walsh and Cullinan (1945) observed that crops which developed severe symptoms of potassium deficiency during a drought would recover, sometimes completely, during a subsequent wet period. Richards and Wadleigh (1952) reviewed results of field experiments by several investigators showing an increase in potassium percentage in plants with an increase in water content of the soil, and van der Paauw (1958) made a similar observation in connection with the experiment on potatoes in Fig. 12.27.

Although the effects just mentioned may be obvious, the manner in which they arise is not so obvious. Various theories may be proposed to account for the field behaviour. In the field experiments, comparisons were usually made between years or between different periods within years ; thus one cannot be certain that the effects are attributable strictly to the water content of the soil because such factors as temperature and sunshine were uncontrolled.

Experiments by Danielson and Russell (1957), Mederski and Stackhouse (1961), and Place and Barber (1964), in which water content was altered without varying the other conditions mentioned, have verified that potassium does move more readily through soil to plant roots if the water content approaches the one-third-bar percentage (field capacity) than if it approaches the fifteen-bar percentage (permanent wilting percentage). Place and Barber (1964) found that uptake of rubidium (sometimes used as a tracer for potassium) by plants from soil at different water contents increased linearly with the rate of diffusion of rubidium through the soil. From these experiments is derived the theory that potassium availability increases with water content of soil because both the proportion of the soil volume effective for diffusion of potassium and the volume of water that can carry potassium to roots by mass movement increase with the proportion of the soil volume occupied by water.

Another theory may be proposed on the basis of observations reported by Schuffelen (1954). He found that the molar ratio of potassium to magnesium in solution in equilibrium with a sandy soil increased from 0.5 at 15% water to 1.7 at 60% water. In a clay soil, the molar ratio increased from 0.2 at 11% water to 0.5 at 28% water. From these observations is derived the theory that an ample supply of water in soil increases the proportion of potassium in the soluble cations and hence increases the availability of potassium relative to calcium and magnesium. Schuffelen advanced the data on potassium and magnesium in explanation of the observation that magnesium deficiencies in plants in the Netherlands are more pronounced in wet springs than in dry springs. This observation is connected with potassium behaviour in that magnesium deficiencies are frequently induced when soils low in magnesium are fertilized with potassium.

Finally, a third theory, based on the vertical distribution of water and exchangeable potassium, may be mentioned. If the water supply is ample throughout the soil, the availability of potassium to plants will generally be greatest in the surface layer because the highest concentration of exchangeable potassium and the highest concentration of roots are usually found there. During rainless periods, the soil dries from the surface downward ; hence the availability of potassium in the most valuable portion of the soil is reduced, and plants become largely dependent on potassium in the lower horizons, where the availability is lower.

Considerations thus far have been limited to what seems to be the middle of the range of water content. It is unreasonable to expect, for example, that the absolute increase in yield from potassium fertilization observed by van der Paauw (1958) in Fig. 12.27 should continue to become greater as the number of rainless days increased. When conditions are dry enough, the absolute increase in yield from potassium fertilization should decrease. Bruns (1935) in Germany found that the absolute increase in yield from fertilization of field beans with potassium was less in 6 dry years than in 5 moist years ; but the ratio of yield on the control plots to the yield on the fertilized plots was nevertheless lower in the dry years than in the moist years, *signifying that availability of the potassium* was reduced by soil dryness. Whether or not the ratio of yield on control plots to yield on potassium-fertilized plots would continue to decrease as soil becomes increasingly dry seems doubtful. Data from work by Barber (1961) in Indiana, shown in Fig. 12.28, suggest that it does not; however, the number of observations is insufficient to provide clear evidence one way or the other.

The middle part of the range in Fig. 12.28, where the points are connected by solid lines, may be interpreted as the portion of the range of water supply in which potassium availability increases with an increase in water content of the soil. The opposite trend with high rainfall, denoted by a broken line, may be interpreted as a decrease in availability of soil potassium with an increase in water content of the soil. Barber accounted for this trend on the basis that a high content of water in the soil resulted in poor aeration, and the poor aera-

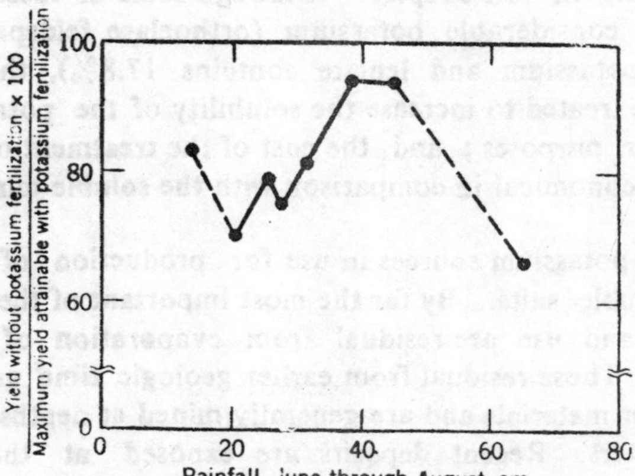


Fig. 12.28. Yield of corn without potassium fertilization as percentage of maximum yield attainable with potassium fertilization in 8 years differing in summer rainfall in Indiana.

tion in turn decreased the availability of soil potassium to plants. Although the trend in question is a consequence of results of only one year that do not agree with the others, and hence cannot be considered significant, the results and Barber's interpretation are in agreement with other work done under controlled conditions that indicates the importance of aeration in uptake and retention of potassium by plants.

For a general review on this subject, a paper by Wallace (1958) may be consulted. A brief review of the aeration effect was published by Barber and Humbert (1963).

## FERTILIZERS AND FERTILIZATION

### Potassium Sources

Potassium is widely distributed, and large deposits of potential value for fertilizer production are found in many countries. For present purposes, potassium-bearing minerals may be considered to consist of one group with moderate to high solubility in water and another group with extremely low solubility in water. The low-solubility group includes the feldspars, micas, and certain other silicate minerals not mentioned previously in this chapter. Although some of these minerals contain considerable potassium (orthoclase feldspar contains 13.9% potassium and leucite contains 17.8%), the minerals must be treated to increase the solubility of the potassium for fertilizer purposes ; and the cost of the treatment makes their use uneconomical in comparison with the soluble minerals.

The potassium sources in use for production of fertilizers are soluble salts. By far the most important of these in both extent and use are residual from evaporation of bodies of water. Those residual from earlier geologic time are covered by other materials and are generally mined at depths exceeding 200 meters. Recent deposits are exposed at the surface. According to estimates in a paper by MacDonald (1960), the countries with the largest known reserves of soluble potassium salts may be listed in order of decreasing reserves as follows : Germany, Canada, Israel-Jordan, Soviet Union, United States, France, and Spain. A number of other countries also are known to have deposits. In North America, large buried deposits of ancient origin are found in New Mexico, Texas, eastern Utah, and Saskatchewan. Surface deposits of more recent geologic time are being worked at Searles Lake in California and at the Bonneville salt flats in western Utah.

Most of the production is from the New Mexico deposits. An account of these and other potassium sources was published by Reed (1953), and papers including information on the recently developed Saskatchewan deposits were published by MacDonald (1960) and Kapusta and Wendt (1963).

**Fertilizer Processing**

The potassium-bearing salt deposits contain a number of different forms of potassium. A table listing the minerals and their composition was published by Reed (1953) and MacDonald (1960). The principal ore in commercial use is sylvinite, which is a mixture of crystals of sylvite (potassium chloride) and halite (sodium chloride). Although the underground salt deposits are often thick, and sylvinite occurs in thin layers. These are mined mechanically, and the solid material is brought to the surface for processing. The potassium chloride is separated from the sodium chloride by flotation (a physical separation of the solids) or by fractional crystallization of the potassium chloride from solution. These processes were described by Harley (1953), MacDonald (1960), and Kapusta and Wendt (1963). The product usually contains about 50% potassium (pure potassium chloride contains 52.4% potassium).

Potassium sulphate is produced commercially in the United States from langbeinite ($K_2SO_4 \cdot 2MgSO_4$) by treatment with potassium chloride followed by crystallization, from burkeite ($Na_2CO_3 \cdot 2Na_2SO_4$) by treatment with potassium chloride followed by crystallization, from potassium chloride and sulphur, and from potassium chloride and sulphuric acid. The processes were described by Harley (1953), MacDonald (1960), and Kapusta and Wendt (1963). Fertilizer-grade potassium sulphate usually contains 41 to 44% potassium (pure potassium sulphate contains 44.9% potassium). Potassium sulphate is marketed also as the mineral langbeinite which, in the New Mexico deposits, is mixed with halite and sylvite. The latter two minerals dissolve more rapidly than the langbeinite and are largely removed by fractional dissolution. The product contains about 18% potassium.

**Fertilizer Statistics**

According to statistics collected by the Food and Agriculture Organization of the United Nations (1965), 88% of the potassium in fertilizers manufactured in 1962-63 in France, Germany, Spain, and the United States, was in the form of potassium chloride, 8% was in the form of potassium sulphate, and 4% was in other forms. In the United States, 93% was in the form of potassium chloride, 6% was in the form of potassium sulphate, and 1% was in other forms.

The same FAO source shows that, in 1962-63, 53% of the potassium used in fertilizer was consumed in Europe, 27% in North and Central America, 9% in the Soviet Union, 8% in Asia, and 1% each in Africa, Oceania, and South America. In the world as a whole (excluding mainland China and North Korea), an average of 2 kg of potassium was applied in fertilizer per hectare to agricultural land. The applications were greatest in Europe (17 kg) and least in Africa (0.1 kg). The figure for North and Central America was 3 kg. Individual countries with heavy use of fertilizer potassium included Belgium (94 kg), Barbados (8 kg), Germany (66 kg), and Japan (64 kg).

**Fertilizers and Their Effects**

A wide variety of potassium salts can be made and could be applied as fertilizer if desired From the standpoint of benefit to plants and efficiency in transportation, the anion of the potassium salt should preferably contain a plant nutrient. Nutrients that occur in anion form include borate, molybdate, chloride, sulphate, nitrate, and phosphate. The first two are immediately ruled out on the basis of toxicity if the salt is to be used in quantities needed as a source of potassium. Some of the first potassium fertilizer produced from deposits in the United States had disastrous effects on crops because of toxicity of borate present as an impurity.

The need for chlorine as a nutrient is extremely small, and occurrence of chlorine deficiency in plants under practical conditions seems unlikely. Hence, from the standpoint of supplying something useful, the chloride salt appears to be a poor choice. The reason most of the potassium fertilizer is marketed as the chloride salt is, of course, that this form is generally the most economical to produce and transport. Although sulphur is a plant nutrient, the quantities of sulphur supplied as calcium sulphate in ordinary superphosphate meet the needs in many areas that would otherwise be deficient. The gradual replacement of ordinary superphosphate by concentrated superphosphate may be expected to increase the need for another source of sulphur such as potassium sulphate.

The chloride ion is not absorbed to a significant degree by soils and is absorbed readily by plants, in which it is thought to occur mostly as an ion in solution, like potassium. A high content of chloride in plants increases the succulence, increases the thickness of the cuticle, and decreases the size of the epidermal cells. Sulphate is absorbed to some extent by soils, is absorbed by plants less readily than is chloride, and in plants occurs in part in organic combination in proteins and amino acids and in part as sulphate ion in solution.

The comparative value of potassium chloride and potassium sulphate as fertilizers has been investigated in many experiments. Occasionally the chloride salt seems preferable to the sulphate, as in work by Kramer (1963) on strawberries. He noted that the yield of a variety that ripened during a dry period was nearly 30% higher where potassium was applied as the chloride than where it was applied as the sulphate. With two other varieties that ripened following rains, the yields where potassium was applied as the chloride differed by less than 10% from the yields obtained with the sulphate, one being above and the other below the yield with potassium sulphate. With cereals, there is usually no significant difference in results obtained with the two forms of potassium. For the most part, where recognizable differences do occur, crop yields and quality are better with the sulphate salt than with the chloride. Nevertheless, the fact that almost all the potassium used as fertilizer is applied as the chloride indicates that the benefits from the sulphate are not considered important enough to justify the higher cost of potassium sulphate. With two crops, tobacco and potatoes, however, the sulphate is rather commonly used. Tobacco leaf has poor fire-holding quality if it contains much chloride, and the yield and starch content of potato tubers are greater with the sulphate than with the chloride. A review of effects of chloride and sulphate was published by Burghardt (1962).

Potassium nitrate and potassium phosphate are superior for general use to all the other salts mentioned because both these substances supply two macronutrients that are frequently applied in large quantities in fertilizers. The disadvantage to use of potassium nitrate and potassium phosphate is economics. At present, their preparation is based on processing as a raw material the potassium chloride that is salable in its own right as a finished product in fertilizers.

**Fertilizer Application**

According to Ignatieff and Page (1958), potassium fertilizers should usually be applied at or just before the time of planting of annual crops, and the fertilizer should be in localized bands near the plants if the quantities applied are small (less than, say, 40 kg of potassium per hectare). This generalization is in agreement with de Wit's (1953) theory of fertilizer application, which predicts the greatest benefit from localized placement where small applications of fertilizer are made. The greatest benefit from localized application near the plants should be obtained on soils that will fix much potassium in none-exchangeable form. In experiments on potassium fertilization of cereals on potassium-fixing soils in the Netherlands, Prummel (1957) found that band applications had an efficiency averaging 3.65 times as great as broadcast applications. These are among the highest efficiencies that have been recorded for localized applications of potassium fertilizer. On similar soils, however, efficiencies of localized and broadcast applications were about equal in experiments on potatoes and beets, which indicates that the nature of the crop is a significant factor. As yet, insufficient research has been done to correlate the relative efficiencies of localized and broadcast applications of potassium fertilizers with the potassium-fixing properties of soils.

Localized placement of potassium chloride or potassium sulphate produces a high local concentration of soluble salts. Hence, if the fertilizer is placed too near the seed, severe damage to the seedlings and loss of stand may result because of insufficient time for reduction of the salt concentration to a safe level before the seed starts to germinate. Applications with, directly above, ordirectly below the seed are especially unsuitable. Most of the movement of water is in a vertical direction, and so potassium salts placed above or below the seed may be moved into the seed zone with movement of water. Placement 5 cm or so to the side of the seed is preferable because then the seed zone will not be affected, and root development can occur in soil that does not contain excess salt. Within a short time, movement of the potassium salt into the surrounding soil will have reduced the salt concentration to a safe level. The marked local increase in potassium supply that has occurred may then increase the growth of roots in the fertilizer-affected soil. An illustration of this effect may be found in a paper by Cooke (1954).

For crops with a long growing period and for sands with low cation-exchange capacity, application of part of the potassium as a delayed side dressing may prove beneficial. For perennial crops, small annual applications are preferable to large, infrequent applications. In tropical regions, perennial crops of high value may receive potassium fertilizer several times annually.

The recovery of applied potassium in the first annual crop following application is not uncommonly 50% or more. Potassium thus ranks along with nitrogen and considerably above phosphorus with regard to efficiency of use. This difference in efficiency is probably one reason why the ratio of potassium applied to potassium in the crop is characteristically lower than the ratio of phosphorus applied to phosphorus in the crop. The most suitable applications of potassium vary greatly with the conditions, and much information on practices in different parts of the world was summarized by Ignatieff and Page (1958). Generally, in relation to world practice, annual applications up to 40 kg of potassium per hectare might be considered low, applications of 40 to 100 kg moderate, and applications exceeding 100 kg high.

Among crop plants, considerable difference exists in sensitivity to potassium deficiency. Crops may be compared in this respect in different ways. One way is to determine how much the yield of the various species or the content of potassium increases upon application of a given quantity of potassium fertilizer under comparable conditions. Results are preferably expressed on a relative basis. Figure 12.29 gives an example of differences among certain crop plants based on results of field experiments in England and Wales. In this example, the data are averages obtained from numerous experiments, but the crops were grown individually and hence not

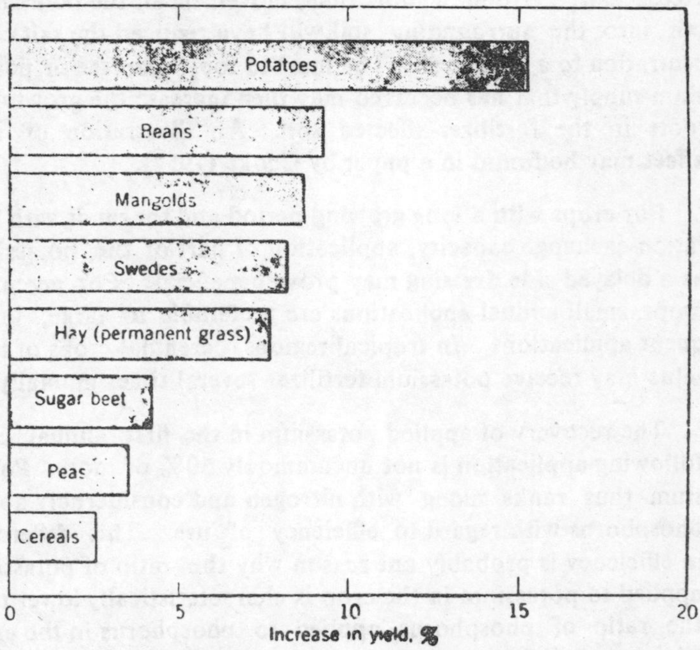


Fig. 12.29. **Percentage increase of yield of different crops from fertilization with 52 kg of potassium per hectare in experiments in England and Wales.**

necessarily under entirely comparable conditions. For this reason, the ratings should be qualified by representing thems a the responses of the crops for the conditions under which the crops were grown.

Another rating may be obtained by growing the desired crops under comparable conditions and determining the yield

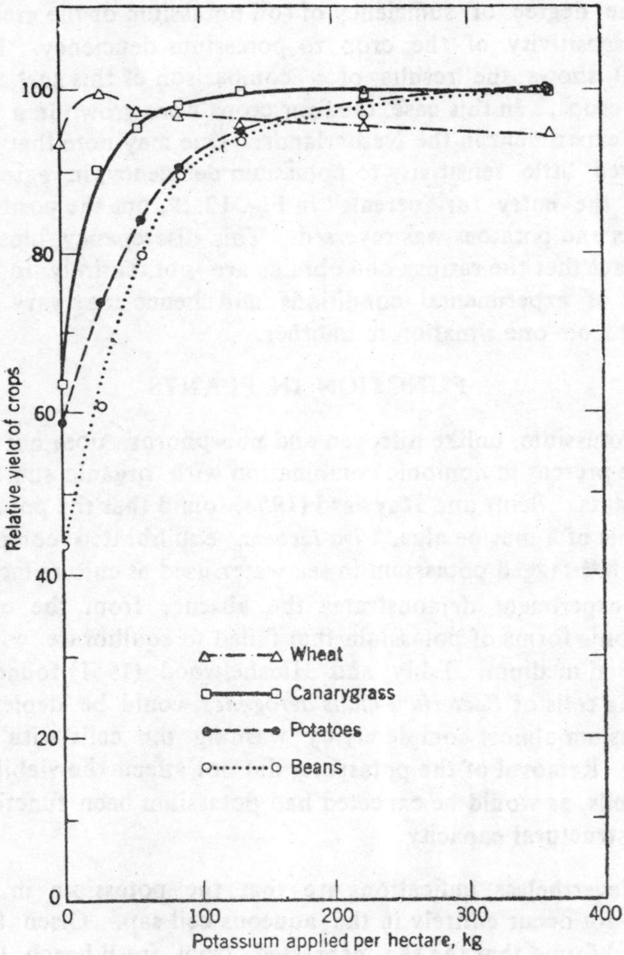


Fig. 12.30. Relative yields of different crops with different annual applications of potassium in fertilizer in field experiments in the Netherlands. The maximum yield obtained from each crop on application of potassium fertilizer is given a value of 100.

of the controls without added potassium and the maximum yields attainable by application of potassium. The yields are then recalculated on a relative basis with the maximum yield of each crop taken as 100 or some other arbitrary figure. The lower the yield of the control on this relative basis, the lower is the degree of sufficiency of soil potassium or the greater is the sensitivity of the crop to potassium deficiency. Figure 12.30 shows the results of a comparison of this sort among four crops. In this case, the four crops were grown in a single field experiment in the Netherlands. One may note that wheat showed little sensitivity to potassium deficiency, in agreement with the entry for "cereals" in Fig. 12.29, but the position of beans and potatoes was reversed. This discrepancy illustrates the fact that the ratings one obtains are not entirely independent of experimental conditions and hence may vary somewhat from one situation to another.

## FUNCTION IN PLANTS

Potassium, unlike nitrogen and phosphorus, does not seem to be present in nonionic combination with organic substances in plants. Scott and Hayward (1953) found that the potassium in cells of a marine alga, *Ulva lactuca*, equilibrated completely with $K^{42}$-tagged potassium in sea water used as culture medium. This experiment demonstrates the absence from the cells of nonionic forms of potassium that failed to equilibrate with the external medium. Eddy and Hinshelwood (1951) found that resting cells of *Bacterium lactis aerogenes* could be depleted of potassium almost completely by washing the cells with dilute acid. Removal of the potassium did not affect the viability of the cells, as would be expected had potassium been functioning in a structural capacity.

Nevertheless, indications are that the potassium in plants does not occur entirely in the aqueous cell sap. Olsen (1948, 1948*a*) found that the sap expressed from fresh beech leaves, after the leaves had been frozen to kill the tissue and destroy the differential permeability of the cells, contained only 78% of the total potassium, due allowance being made for the portion of the sap not extracted. More of the potassium was extracted as he added increasing amounts of water as an extractant, but 2% of the potassium was still unextracted where 400 ml of water were added to 10 g of previously frozen leaf tissue. Upon addition of 50 ml of 0.1-normal calcium chloride solution, however, he obtained quantitative extraction of the potassium. These experiments were performed on leaves sampled early in September. Tests on samples taken late in October just before leaf fall, when much of the nitrogen had migrated from the leaves into the wood, showed that all the potassium was present in the cell sap. Accordingly, he proposed the theory that the potassium not present in the cell sap during the summer is adsorbed by proteins and that this binding disappears with hydrolysis of the proteins before leaf fall. Olsen's theory that the potassium in plants is adsorbed in part by proteins is consonant with the current view that potassium functions in enzymatic reactions.

Two other lines of evidence substantiate the view that potassium plays a part in metabolic processes. First, the potassium content of plants increases with the rate of metabolism. Work of Hoagland and Broyer (1936) and Luttkus and Botticher (1939) may be cited in this connection. Second, potassium is known to be required in three types of enzymatic reactions. Potassium is essential in certain enzymatically catalyzed reactions involving adenosine phosphates. Adenosine phosphates are the energy carriers in metabolic processes in both plants and animals, as mentioned on phosphorus. In carbohydrate metabolism, the reactions requiring potassium are steps in the process by which energy is obtained from sugar. Potassium is necessary also in some of the syntheses. Webster and Varner (1954) found that potassium is required in the coupling of certain amino acids to form peptides, thus suggesting that potassium is essential in protein synthesis. Further work has verified the importance of potassium in protein synthesis. Finally, work by Latzko (1959) indicates that potassium activates one of the first reactions in photosynthesis in which part of the light energy is captured in a reaction between adenosine diphosphate and inorganic phosphate to form adenosine triphosphate.

The action of potassium in these enzymatic processes has been verified in simple systems involving substrate, enzyme, and different concentrations of salts of potassium or other metals. The catalysis is thought to require the presence of potassium in ionic combination with the enzymes, the potassium being subject to exchange with other potassium ions in the medium or with cations other than potassium. Hence a high internal concentration of potassium and a high proportion of potassium in the total internal cations should increase the rate of transformation of the substrate under otherwise equal conditions.

Much biochemical and general physiological research has been done in relation to potassium supply to plants, and there is evidence for definite malfunction of metabolism under conditions of potassium deficiency. The complexity of metabolic processes and the inability to isolate potassium-bearing organic compounds, however, makes it difficult to interpret such experiments in terms of the manner in which potassium functions. Some of the observations made in this work will be mentioned in succeeding sections.

## SODIUM-POTASSIUM RELATIONS

Sodium is of importance in plant nutrition not only because sodium is required by at least a few plants and is detrimental in excess, but also because of its relation to potassium. Sodium and potassium are the two principal monovalent metallic cations in plants, and an increase in one generally brings about a decrease in the other. Sometimes the change in proportions of sodium and potassium in plants is brought about with relatively little alteration of the total equivalent concentration of metallic cations in the tissue. Practical interest in sodium-potassium relations stems principally from a potassium-sparing action sometimes observed, which suggests the possibility of using sodium salts to replace a part or all of the more costly potassium salts used as fertilizer. In Great Britain, for example, Tinker (1965) found in 42 field experiments that either potassium chloride or sodium chloride, applied singly, produced a profitable increase in yield of sugar from sugar beets. In the presence of sodium chloride, however, the increase in yield from potassium chloride was comparatively small and no longer profitable.

Plant species behave rather differently with regard to uptake of sodium. Figure 12.31 shows the relative content of sodium and potassium in 21 species of plants grown together in the same nutrient solution. Initially, the molar concentrations of

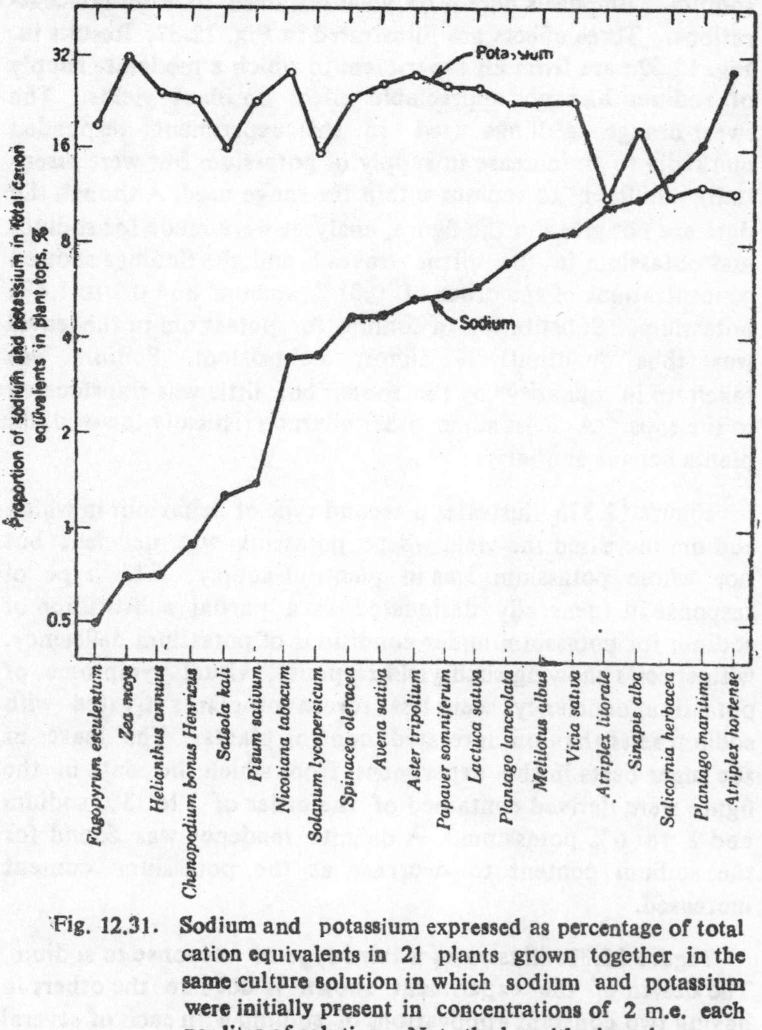


Fig. 12.31. Sodium and potassium expressed as percentage of total cation equivalents in 21 plants grown together in the same culture solution in which sodium and potassium were initially present in concentrations of 2 m.e. each per liter of solution. Note the logarithmic scale.

sodium and potassium in the solution were equal; yet 18 of the plants contained less sodium than potassium, and 6 contained less than 10% as many equivalents of sodium as potassium. Only 3 contained more equivalents of sodium than of potassium.

Plant species differ also with regard to yield response to sodium. Emphasis here is on yield response to lesser concentrations. Three effeots are illustrated in Fig. 12.32. Results in. Fig. 12.32*a* are from an experiment in which a moderate supply of sodium had no appreciable effect on plant yields. The sweet-orange seedlings used in this experiment responded markedly to an increase in supply of potassium but were essentially indifferent to sodium within the range used. Although the data are not given in the figure, analyses were made for sodium and potassium in the citrus leaves; and the findings showed concentrations of the order of 0.01% sodium and 0.2 to 1.4% potassium. Substitution of sodium for potassium in the leaves was thus quantitatively almost nonexistent. Sodium was taken up in quantity by the roots, but little was translocated to the tops. A least some other characteristically low-sodium plants behave similarly.

Figure 12.32*b* illustrates a second type of behaviour in which sodium increased the yield where potassium was deficient but not whese potassium was in plentiful supply. This type of response in generally designated as a partial substitution of sodium for potassium under conditions of potassium deficiency. with species showing such yield response, visual symptoms of potassium deficiency seem less severe on plants treated with sodium salts than on untreated control plants. The leaves of the sugar beets in the experiment from which the data in the figure were derived contained of the order of 1 to 3% sodium and 2 to 6% potassium. A definite tendency was found for the sodium content to decrease as the potassium content increased.

Figure 12.32*c* illustrates a third type of response to sodium. The design of the experiment shown is not like the others in having two constant applications of sodium with each of several

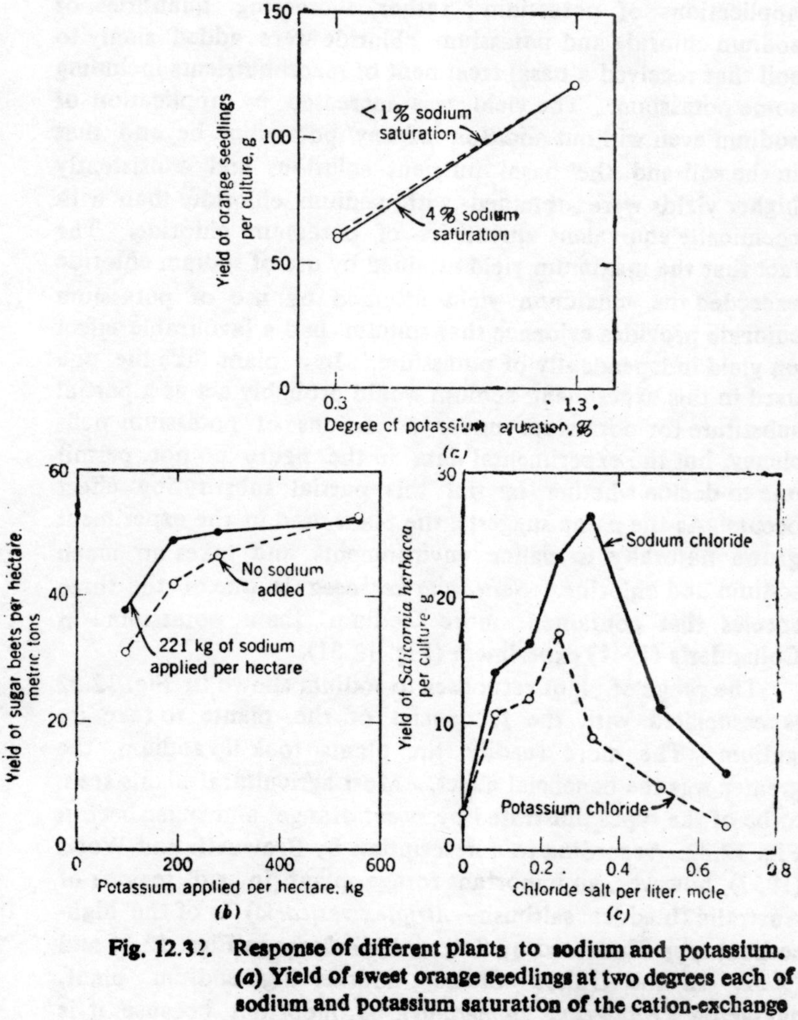


**Fig. 12.32. Response of different plants to sodium and potassium. (*a*) Yield of sweet orange seedlings at two degrees each of sodium and potassium saturation of the cation-exchange capacity of a loam soil of California. (*b*) Yield of sugar beets in Michigan with different applications of potassium to a much soil with and without a supplemental application of sodium as sodium chloride. (*c*) Yield of *Salicornia herbacea* in the Netherlands on 5 kg of soil treated with 2 liters of a nutrient solution containing different concentrations of sodium chloride or potassium chloride. The basal solution applied to all caltures contained, in addition, 0.0086 gram mole of potassium per liter and nutrients other than potassium.**

applications of potassium ; rather, increasing quantities of sodium chloride and potassium chloride were added singly to soil that received a basal treatment of macronutrients including some potassium. The yield was increased by application of sodium even without addition of any potassium beyond that in the soil and the basal nutrient solution, and consistently higher yields were obtained with sodium chloride than with chemically equivalent quantities of potassium chloride. The fact that the maximum yield attained by use of sodium chloride exceeded the maximum yield attained by use of potassium chloride provides evidence that sodium had a favourable effect on yield independently of potassium. In a plant like the one used in this experiment, sodium would probably act as a partial substitute for potassium under conditions of potassium deficiency, but the experimental data in the figure do not permit one to decide whether or not this partial substitution effect occurs. As the name suggests, the plant used in the experiment grows naturally in saline environments and takes up much sodium and chlorine. *Salicornia herbacea* is one of the three species that contained more sodium than potassium in Collander's (1941) experiment (Fig. 12.31).

The range of plant responses to sodium shown in Fig. 12.32 is associated with the tendencies of the plants to take up sodium. The more readily the plants took up sodium, the greater was the beneficial effect. Most agricultural plants seem to be of the types illustrated by sweet orange and sugar beet in Fig. 12.32. According to a description by Brownell and Wood (1957), however, an important forage plant in arid regions of Australia (bladder saltbush—*Atriplex vesicaria*) is of the high-sodium type illustrated by *Salicornia herbacea* in Figs. 12.31 and 12.32. In the United States, another high-sodium plant, halogeton (*Halogeton glomeratus*), is important because it is widely distributed in western grazing areas and is toxic to livestock on account of the high concentrations of oxalate it contains Classifications of crop plants with regard to response to sodium were published by Harmer and Benne (1945) and Lehr (1953).

The general subject of sodium-potassium relations under conditions of potassium deficiency was reviewed briefly by

Baumeister (1958). An analysis of sodium-potassium relations where sodium salts are present in the salinity range was published by Heimann (1958).

Sodium-potassium relations have received far more attention than lithium-, cesium-, and rubidium-potassium relations because of the greater importance of sodium under natural conditions and the low cost of sodium salts. None of these alkali cations have been found to replace potassium completely for growth of higher plants. Nevertheless, as far as absorption is concerned, plants do not make much distinction between rubidium and potassium ; and rubidium is sometimes used as a tracer for potassium because of its more convenient radioactive isotope. With regard to lower plants, Kellner (1955) described a mutant of the alga *Ankistrodesmus braunii* that would grow without potassium if rubidium was present.

## POTASSIUM SUPPLY AND PLANT BEHAVIOUR

### Deficiency Symptoms

The two most commonly observed symptoms of potassium deficiency are (1) shortening of stems, producing a squat or bushy appearance in plants such as cereals that normally produce an extended stem or a rosette appearance in plants such as beet that normally grow from a crown without an extended stem and (2) browning and death of leaf tissue, commonly occurring at leaf margins (popularly known as "leaf scorch") or tips (known as "tip burn") but in some plants as spots in interveinal areas. Leaves may be narrow and crinkled with a tendency for the margin to curl toward the upper or lower surface.

Browning and death of the leaf tissue is usually preceded by other less marked symptoms, including chlorosis and tinting with bluish-green, purple, or orange colours, visible macroscopically, and by the shrinking, collapse, and browning of individual cells, visible microscopically. The necrotic or dead spots that characterize potassium deficiency are groups of cells that have collapsed. Bussler (1964) made an extensive investigation of the comparative microscopic morphology of tissues of potassium-deficient and normal plants.

Individual species have distinctive symptoms that are helpful in visual diagnosis. Descriptions of potassium-deficiency symptoms for a number of crops and reproductions of numerous colored photographs of deficiency symptoms may be found in a book by Wallace (1961) and in a book edited by Sprague (1964).

The occurrence of potassium deficiency symptoms is not a sign of incipient deficiency but rather an indication that potassium is a seriously limiting factor. Potassium fertilization thus may be of distinct benefit even though no foliar symptoms of deficiency appear. On the other hand, severe potassium deficiency symptoms early in the season (which indicate a real deficiency for growth at that time) may later disappear, with the consequence that the response of the harvested crop to potassium fertilization is relatively small. Interpretation of the symptoms in terms of remedial treatment thus may not be entirely straightforward. Walsh (1954) described some of the difficulties related to use of deficiency symptoms as a guide to fertilization of crops with potassium in Ireland.

Analysis of the plant or some index tissue may be helpful in refining the diagnosis of potassium deficiency beyond that permitted by visual symptoms of deficiency. In the case of sugarcane, for example, Schmehl and Humbert (1964) gave 1.5% potassium in leaf sheaths three to six (measured from the top down) as the upper limit at which deficiency symptoms occurred but noted that increases in yield from potassium fertilization could be obtained until the potassium content of this tissue reached 2.25%.

Development of symptoms of potassium deficiency is related to the internal behaviour of potassium in plants. Retention of potassium in plant cells seems to require continuous expenditure of metabolic energy. For example, Luttkus and Botticher (1939) observed that plant roots alternately lost potassium to a nutrient solution when maintained for a day or more in darkness and gained potassium from the solution when maintained a day or more in light. By use of radioactive potassium, Mengel (1964) found that potassium moved out of roots as well as into roots during net uptake and that net loss of potassium to the surrounding solution occurred upon addition of sodium azide, an inhibitor of metabolic reactions. Rathje (1961) found that the potassium concentration in the leaf sap in the afternoon was relatively high in the interveinal tissue and low in the veins of leaves of sugar beet and potato grown in the field in Germany. At night the concentration decreased in the interveinal tissue and increased in the veins. (Sodium incidentally, manifested precisely the opposite behaviour.) Kidson (1943) ashed an intact apple leaf showing marginal browning due to potassium deficiency and then sprayed the ash skeleton with a solution of sodium cobaltinitrite, which produces a bright yellow-orange precipitate with potassium. The result was a sharp differentiation between the original marginal brown area and the inner green area. The inner area yielded a much more dense precipitate than the outer area. The ash skeleton of a green control leaf from a tree that had been fertilized with potassium developed a comparatively uniform precipitate of potassium sodium cobaltinitrite over the entire surface. Potassium thus is not inactivated in cells but, under some circumstances, may move from one place to another.

If analyses are made of leaves at different position on a single plant, definite differences in potassium content are sometimes found. As might be expected from the mobility of potassium and its function in metabolism, there is a tendency for potassium to occur in higher concentration in younger leaves than in older leaves. An illustration may be found in Table 12.8, which contains values for the potassium content in lower, middle, and upper leaves of field-grown tobacco and includes observations on the severity of symptoms of potassium deficiency on the plants as a whole. The highest potassium percentage evidently occurred in the upper leaves where deficiency symptoms were present and in the lower leaves where deficiency symptoms were ~~absent. Liebhardt and~~

Table 12.8. Deficiency Symptoms and Potassium Content of Leaves of Tobacco with Different Applications of Potassium as Potassium Sulfate.

| *Potassium Applied in Fertilizer per Hectare, kg* | *Symptoms of Potassium Deficiency* | *Potassium Content of Dry Matter, %* | | |
|---|---|---|---|---|
| | | *Lower Leaves* | *Middle Leaves* | *Upper Leaves* |
| 0 | Very severe | 0.6 | 0.6 | 1.3 |
| 22 | Severe | 1.0 | 1.6 | 2.2 |
| 45 | Moderate | 2.4 | 2.5 | 2.8 |
| 112 | None | 5.3 | 4.4 | 3.1 |
| 157 | None | 6.2 | 5.4 | 4 7 |

Murdock (1965) found potassium to behave similarly in corn. Appling and Giddens (1954) observed that, where soil potassium was deficient for cotton, potassium accumulation was greatest in the upper, immature leaves ; and sodium accumulation was greatest in the lower leaves.

The higher concentration of potassium in the younger vegetative tissue may be produced in part at the expense of the older tissue. Greenway and Pitman (1965) found, for example, that 51 of the 129 $\mu$g of potassium entering per day into the youngest leaf of barley in the three-leaf stage came from sources within the plant. At the same time the oldest leaf was losing 63 $\mu$g of potassium per day. The selective efflux of potassium from the older leaves provides a partial explanation for the tendency of symptoms of potassium deficiency to appear soonest and to be most severe on the older leaves. Withdrawal of potassium may hasten senescence of the older leaves, as suggested by the finding that the rate of photosynthesis in older leaves declined more rapidly where the supply of potassium was low than where it was high.

Although translocation of potassium from old to new growth is no doubt a link in the chain of events that leads to development of visual symptoms of potassium deficiency in the old growth, the loss of potassium does not appear to be the immediate cause of the symptoms. There is evidence for action of specific toxic organic substances in this regard. When potassium is deficient, protein synthesis is limited and breakdown products accumulate. The latter include various amino acids that are normal components of proteins as well as other nitrogenous substances that are presumably derived from the primary components. Richards and Coleman (1952) discovered that barley plants deficient in potassium contained putrescine ($NH_2—CH_2—CH_2—CH_2—CH_2—NH_2$), a substance present only in traces in plants well supplied with potassium. The content increased with the severity of the visual symptoms of potassium deficiency, reaching maximum concentrations estimated at 0.15 to 0.2%. Coleman and Richards (1956) then added putrescine to plants in various ways and found that symptoms characteristic of potassium deficiency were thus produced even in plants high in potassium. Addition of potassium to deficient plants resulted in a reduction of their putrescine content. According to Richards (1954), low-potassium barley that received sodium, rubidium, or lithium developed lower concentrations of putrescine and had less severe visual symptoms than did barley without these other alkali metals. All this evidence supports the view that leaf tissue of barley is sensitive to putrescine and dies if the concentration is great enough.

Smith (1963) tested 13 plant species, including barley, and found putrescine in all, whether potassium was deficient or not. Also present was agmatine, a chemically related substance. In 11 species, potassium deficiency increased the concentration of putrescine, and in 6 species potassium deficiency increased the concentration of agmatine. Neither substance accumulated in potassium-deficient groundsel (*Senecio vulgaris*) or flax. Coleman and Richards (1956) had previously found that potassium-deficient flax accumulates arginine. The evidence now available indicates that arginine is formed first and that putrescine and agmatine are formed from it. According to work by Smith (1963), the accumulation of putrescine and agmatine results from action of an enzyme that removes a carboxyl group from arginine. His work showed that activity of this enzyme was greater in potassium-deficient barley than in barley well supplied with potassium. Presumably, therefore, the failure of potassium-deficient groundsel and flax to accumulate putrescine and agmatine is related to a low activity of the enzyme that acts on arginine. In tobacco, Takahashi and Yoshida (1960) found a marked accumulation of putrescine in plants deficient in phosphorus as well as in those deficient in potassium.

No doubt much remains to be learned about the biochemical processes leading to development of potassium-deficiency symptoms. For the present objective, however, the details of the processes are perhaps not of as much significance as the general concept that the visual symptoms are, in some plants, and perhaps in all, a consequence of toxic substances produced normally but present in much greater concentrations in plants deficient in potassium than in those well supplied with potassium.

**Water Relations**

The literature records many observations and statements to the effect that potassium fertilization "improves" the water relations of plants. Perhaps the most consistently reported observation is an increase in the water content and turgidity of plants with an increase in supply of potassium. Fig. 12.33 gives an example, containing averages of many measurements on field-grown sugarcane in Hawaii. Humbert (1958) noted that leaves of potassium-deficient sugarcane have relatively low turgidity, and the tips commonly become frayed from blowing in the wind. Similarly, Anderson *et al.* (1929, 1930) in Connecticut noted that tobacco plants grown on field plots without potassium fertilization wilted sooner and more severely on hot days than did plants on comparable plots fertilized with potassium. Although the plants that did not receive potassium fertilizer were reduced in size, they showed no acute symptoms of potassium deficiency. Eckstein *et al.* (1937, pp. 38-39) reviewed similar observations by other investigators.

Turgidity effects such as those just mentioned could conceivably result from a relatively high water content in high-

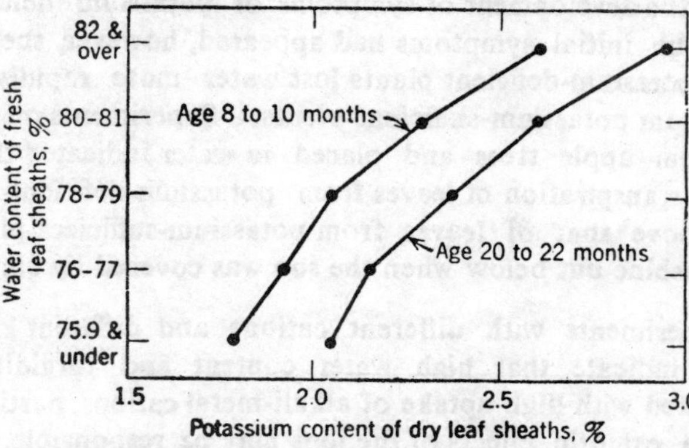


**Fig. 12.33.** **Water content versus potassium content of leaf sheaths of sugarcane in two age groups in Hawaii. Values given are averages of numerous measurements.**

potassium plants independently of loss of water by transpiration or from a relatively low loss of water from plants high in potassium. Evidence exists for both possibilities, but observations are not consistent, particularly those related to potassium supply and loss of water by transpiration. Experiments by Arland and Zwicker (1959) support both possibilities. These investigators found that the yield of expressed sap of potassium-deficient oat plants was much lower from the lower leaves than from the upper leaves even where the plants were well watered and the samples were taken early in the morning before the daytime transpiration could have had much effect. When the plants were detached from the roots and exposed to evaporation under uniform conditions, loss of water as a percentage of initial plant weight decreased with an increase in the prior supply of potassium to the plants. Thus, despite a greater initial content of water in the high-potassium plants, loss of water by transpiration took place more readily from low-potassium plants. Mann (1924) obtained conflicting evidence with regard to the second possibility. He determined the loss of water from leaves detached from gooseberry plants grown in sand cultures supplied with a minus-potassium solution and with a complete nutrient solution. He found that the rate of loss was about the same if he used green leaves before the development of symptoms of potassium deficiency. After the initial symptoms had appeared, however, the leaves from potassium-deficient plants lost water more rapidly than those from potassium-sufficient plants. Experiments on shoots cut from apple trees and placed in water indicated that the rate of transpiration of leaves from potassium-deficient plants was above that of leaves from potassium-sufficient plants in the sunshine but below when the sun was covered by clouds.

Experiments with different cations and different kinds of plants indicate that high water content and turgidity are associated with high uptake of alkali-metal cations, particularly sodium, osmotic effects of the ions may be responsible. With regard to the effect of potassium supply on loss of water by transpiration from individual leaves or plants with equal exposure, the conflicting observations have resulted in various theories. Literature on this subject was reviewed by Biebl (1958).

Some of the observations of beneficial effects of potassium on the water economy of plants seem now to be merely an indirect consequence of the greater growth of the potassium-treated plants and not a specific potassium effect. As noted the amount of water lost by evapotranspiration per unit weight of plant decreases with an increase in production of plant tissue, whether the latter results from an extra supply of potassium or nitrogen or in some other way.

**Time of maturity**

Frank (1931) recorded the observations shown in Fig. 12.34 for oats grown in Norway on soils differing in supply of potassium. The maturity evidently was delayed where potassium was deficient. Alten and Gottwick (1939) found that the length of the growing period of soybeans was 181 days under conditions of potassium deficiency and 157 days where potassium fertilizer was applied. Yields of soybeans were 562 and 967 kg/hectare, respectively. Lagatu and Maume (1932) noted that grapes developed less rapidly and matured less completely where potassium was deficient than where the supply was ample. Unpublished data of L.C. Dumenil at the Iowa Agricultural Experiment Station indicate that potassium deficiency delays the maturity of corn.

In contrast to the foregoing observations, which represent the type reported most frequently, deficiency of potassium sometimes causes plants to mature earlier; and this is true of cotton. The response of cotton is well shown in Fig. 12.35. At the first two pickings, the potassium-fertilized plants yielded less than the controls. At later pickings, the potassium-fertilized

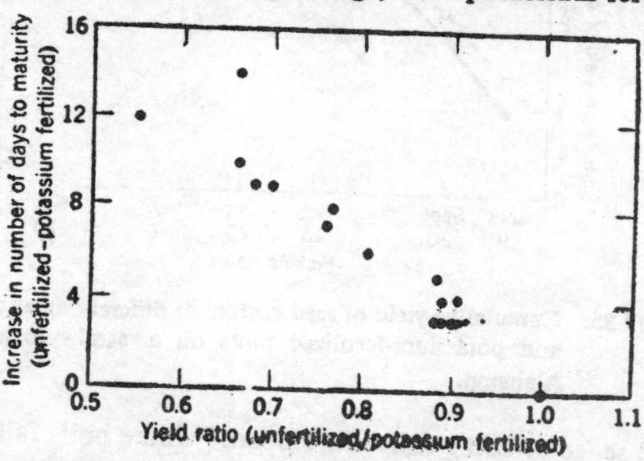


Fig. 12.34. Increase in number of days required for oats to reach maturity without potassium fertilization versus ratio of yield of unfertilized to yield of potassium-fertilized oats grown on different soils.

plants outyielded the controls; and they continued to produce cotton for a month after the controls had ceased production. Two related phenomena contribute to the "early maturity" of potassium-deficient cotton plants. First, the flowering stage of these plants is cut short by potassium deficiency, so that the process of boll maturation begins while the potassium-sufficient plants continue to produce new flowers and bolls. Second, the leaves of the potassium-deficient plants fall early, so that carbohydrate production and translocation come to an early halt. (In the experiment from which Fig. 12.35 was derived, the plants on the control plots had lost all their leaves by September 1.) Although the bolls become dry, they are not necessarily mature

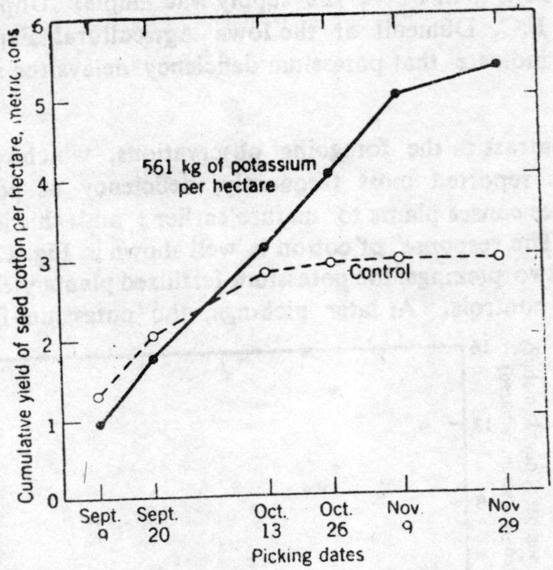


**Fig. 12.35.** **Cumulative yield of seed cotton at different dates on control and potassium-fertilized plots on a sandy loam soil in Alabama.**

**in the sense of being fully developed. Some bolls fail to open. The behaviour of potassium-deficient cotton is thus partly a matter of early maturity and partly a matter of premature death. Fig. 12.35, which shows the cumulative yields obtained at different dates, thus shows maturity in one sense but not in the sense of the degree of completion of development of individual fruits. In the latter sense, potassium fertilization may be observed to increase the proportion of mature bolls and fibers ; however, it is not known to what extent this was the case in the experiment described in Fig. 12.35.**

●●